POWER SYSTEMS RESTRUCTURING:
Engineering and Economics

THE KLUWER INTERNATIONAL SERIES IN ENGINEERING AND COMPUTER SCIENCE

Power Electronics and Power Systems
Consulting Editors
Thomas A. Lipo and M. A. Pai

Other books in the series:

CRYOGENIC OPERATION OF SILICON POWER DEVICES, Ranbir Singh and B. Jayant Baliga, ISBN: 0-7923-8157-2
VOLTAGE STABILITY OF ELECTRIC POWER SYSTEMS, Thierry Van Cutsem and Costas Vournas, ISBN: 0-7923-8139-4
AUTOMATIC LEARNING TECHNIQUES IN POWER SYSTEMS, Louis A. Wehenkel, ISBN: 0-7923-8068-1
ENERGY FUNCTION ANALYSIS FOR POWER SYSTEM STABILITY, M. A. Pai, ISBN: 0-7923-9035-0
ELECTROMAGNETIC MODELLING OF POWER ELECTRONIC CONVERTERS, J. A. Ferreira, ISBN: 0-7923-9034-2
MODERN POWER SYSTEMS CONTROL AND OPERATION, A. S. Debs, ISBN: 0-89838-265-3
RELIABILITY ASSESSMENT OF LARGE ELECTRIC POWER SYSTEMS, R. Billington, R. N. Allan, ISBN: 0-89838-266-1
SPOT PRICING OF ELECTRICITY, F. C. Schweppe, M. C. Caramanis, R. D. Tabors, R. E. Bohn, ISBN: 0-89838-260-2
INDUSTRIAL ENERGY MANAGEMENT: *Principles and Applications*, Giovanni Petrecca, ISBN: 0-7923-9305-8
THE FIELD ORIENTATION PRINCIPLE IN CONTROL OF INDUCTION MOTORS, Andrzej M. Trzynadlowski, ISBN: 0-7923-9420-8
FINITE ELEMENT ANALYSIS OF ELECTRICAL MACHINES, S. J. Salon, ISBN: 0-7923-9594-8

POWER SYSTEMS RESTRUCTURING:
Engineering and Economics

by

MARIJA ILIC
Massachusetts Institute of Technology
Cambridge, MA

FRANCISCO GALIANA
McGill University
Montreal, Quebec, Canada

LESTER FINK
Kema-ECC, Inc., Retired
Fairfax, VA

KLUWER ACADEMIC PUBLISHERS
Boston/Dordrecht/London

Distributors for North, Central and South America:
Kluwer Academic Publishers
101 Philip Drive
Assinippi Park
Norwell, Massachusetts 02061 USA

Distributors for all other countries:
Kluwer Academic Publishers Group
Distribution Centre
Post Office Box 322
3300 AH Dordrecht, THE NETHERLANDS

Library of Congress Cataloging-in-Publication Data

A C.I.P. Catalogue record for this book is available
from the Library of Congress.

Copyright © 1998 by Kluwer Academic Publishers

All rights reserved. No part of this publication may be reproduced, stored in a retrieval system or transmitted in any form or by any means, mechanical, photocopying, recording, or otherwise, without the prior written permission of the publisher, Kluwer Academic Publishers, 101 Philip Drive, Assinippi Park, Norwell, Massachusetts 02061

Printed on acid-free paper.

Printed in the United States of America

Contents

Preface vii

1. Setting the Stage 1
   Leonard Hyman

Part I: Theoretical Challenges in Real-Time Operation ... 13

2. Power Systems Operation: Old vs. New 15
   Marija Ilic and Frank Galiana

3. Framework and Methods for the Analysis of Bilateral Transactions ... 109
   Frank Galiana and Marija Ilic

Part II: Industry Experiences and Challenges ... 129

4. The Political Economy of the Pool 131
   Richard Green

5. Practical Requirements for ISO Systems 167
   Ralph Masiello

Part III: Markets of the Future ... 185

6. Agent Based Economics 187
   Gerald Sheblé

7. One-Part Markets for Electric Power: Ensuring the Benefits of Competition ... 243
   Frank C. Graves, E. Grant Read, Philip Q. Hanser, and Robert L. Earle

Part IV: Planning In the New Industry 281

8. System Planning Under Competition 283
    Ray Coxe and Marija Ilic

9. Transmission Networks and Market Power 335
    Ziad Younes and Marija Ilic

10. Competitive Electric Services and Efficiency 385
    Stephen R. Connors

Part V: Power Systems Control in the New Industry 403

11. New Control Paradigms for Deregulation 405
    Lester H. Fink

12. The Control and Operation of Distributed Generation in a Competitive Electric Market 451
    Judy Cardell and Marija Ilic

13. Application of Dynamic Generation Control for Predatory Competitive Advantage in Electric Power Markets 517
    Thomas Gorski and Christopher DeMarco

Contributors 547

Index 553

Preface

The writing of this book was largely motivated by the ongoing unprecedented world-wide restructuring of the power industry. This move away from the traditional monopolies and toward greater competition, in the form of increased numbers of independent power producers and an unbundling of the main services that were until now provided by the utilities, has been building up for over a decade. This change was driven by the large disparities in electricity tariffs across regions, by technological developments that make it possible for small producers to compete with large ones, and by a widely held belief that competition will be beneficial in a broad sense. All of this together with the political will to push through the necessary legislative reforms has created a climate conducive to restructuring in the electric power industry.

Consequently, since the beginning of this decade dramatic changes have taken place in an ever-increasing list of nations, from the pioneering moves in the United Kingdom, Chile and Scandinavia, to today's highly fluid power industry throughout North and South America, as well as in the European Community. The drive to restructure and take advantage of the potential economic benefits has, in our view, forced the industry to take actions and make choices at a hurried pace, without the usual deliberation and thorough analysis of possible implications.

We must admit that to speak of "the industry" at this juncture is perhaps disingenuous, even misleading. The traditional power industry is fragmenting with the emergence of independent generation, transmission, and distribution entities, and new parties such as "independent system operators," marketers, traders, and large industrial customers (retail customers thus far seem to have no effective representation). The debate, moreover, also includes participation, and ultimately dicta, from regulators and legislators. Among all these disparate parties, there is a wide variety of sometimes conflicting interests, and no unanimity as to the best path to follow in this restructuring exercise, or about the models for and degree of competition, about the regulation of and tariffs for the use of transmission networks, and about the mechanisms for maintaining high levels of reliability, both on-line and in the long term.

The multilateral debate has been extensive, boisterous and inconclusive, often lacking a clear vision. The various parties have, in varying degree, consulted with economists and power systems engineers. The abundant and growing literature, accordingly but unfortunately, is fragmented and spread between the two quite distinct disciplines of engineering and economics. Unfamiliarity, within each of them, with basic concepts, methods, and terminology of the other has made communication and dissemination of knowledge difficult. More importantly, basic assumptions on each side of the debate are just that, and not subject to analytical proof or disproof. The only way such assumptions can be discussed or evaluated is in terms of their implications, which often can be worked out only by experience or, in some cases, by simulation.

What are the major areas of agreement and disagreement? What are the basic reasons for the latter? In what respect are they at the level of axiomatic assumptions? To what extent do they reflect misunderstanding of basic principles - of economics by engineers, of electric systems by economists? This book does not attempt to develop a consensus view, an Hegelian synthesis, but to discuss rather than to assert, to sort out areas (or items) of agreement and disagreement, and to clarify the basic nature of the latter. The main focus is on formal problem posing, including statement of critical assumptions. The main goal is to narrow the gap separating the fields of power systems engineering and economics in the context of the restructuring of the power industry.

The book has brought together an international group of experts to create a broad and incisive coverage of the main issues. The intent has been to provide the reader with a good deal of depth but without excessive specialization, to avoid a purely qualitative treatment by including some analytical and numerical methods, and to offer, whenever possible, discussions of real cases, simple examples and tutorial explanations. The book is intended for power engineers, economists, investment analysts, marketers, system operators and managers, planners and graduate students.

Basic organization of this volume

Chapter 1 by Hyman sets the stage by providing an overall perspective of the ongoing changes. Part I of this book is concerned with the theoretical challenges of power systems operation. In Chapter 2, Ilic and Galiana revisit the main tasks of power systems operation in today's industry, and provide preliminary formulations for, and possible approaches to, performing these tasks in the new industry. In Chapter 3, they introduce a new approach to the modeling and analysis of bilateral markets; a bilateral market structure creates a significant challenge to the present operating paradigms based on coordinated generation scheduling. In Part II, Chapters 4 and 5, by Green and Masiello, respectively, present a comprehensive experience with the UK industry restructuring and the challenges ahead as software is being developed for future ISOs in the United States.

Part III provides views of the future primary electricity markets. Chapter 6 by Sheblé summarizes the new trend of experimental economics for competitive electricity markets and describes agent-based economics for the new industry. Chapter 7 by Graves *et al* describes the role of forward markets in ensuring reliable service at efficient and nonvolatile electricity prices.

In Part IV, planning aspects of the new industry are treated. Coxe *et al* in Chapter 8 provide a semi-tutorial review of the underlying planning principles in today's power industry, and how are these likely to change. In Chapter 9, Younes *et al* define possible market power problems related to weak transmission provision, and the implications of this situation on the price of electricity. In Chapter 10, Connors looks at the impact of competition at the customer level.

Finally, in Part V of this volume, the topic of power systems control is revisited. Fink quite forcefully suggests what should be obvious but is not given much attention at the present stage of the industry debate: power systems control should be designed as a function of market structure, and one should not assume more or less identical operating and control paradigms as the industry structure undergoes a drastic change. Cardell *et al* illustrate in Chapter 12 the role of price feedback for effective use of distributed generation. In Chapter 13, the team from Wisconsin provides us with a warning that strictly technical control design could be used to disable competitors from active participation in power delivery.

The book as a whole suggests that we must develop tools for modeling, designing technical and price feedback and analyzing performance of different designs for the expected system conditions, much in the same way as we strive to do for today's industry.

Although technical and economic processes evolve at different rates, they influence each other. The opinion, too often expressed, that the market would be responsible for efficiency and that the independent system operators, in forming, are responsible for technical performance, leads to a dangerous disconnect. Competitive electricity markets must have carefully designed technical and economic feedback so that they can be adjusted to changes and converge to a near optimal long-term performance.

Acknowledgments

Our plans for this book solidified in 1996, a period during which we had the opportunity to discuss and work on many of the issues included here. The importance of the subject matter, the absence of general textbooks integrating its principal notions and methods, and the great interest on the part of the power and economics community, including graduate students, led us to carry this project through.

We very much appreciate the willingness of all the authors to write their chapters. Without them, this book would not exist.

Ms. Sara Wolfson was in charge of compiling the volume as different parts were sent to her. We very much appreciate her interaction with individual contributors, graduate students and the publisher, and her work in taking on the technical responsibility for this effort.

Particular thanks for helping prepare this volume go to Yong Tae Yoon and Ziad Younes. They have worked very closely with Sara on creating this volume in its present form.

Last but not least, we thank our families for yet another stage of patience and love as we continue our quest to understand the new world of the industry that between the three of us we have been part of for a total of nine decades.

Marija Ilic, Francisco Galiana and Lester Fink
MIT, Cambridge, MA

1 SETTING THE STAGE
Leonard S. Hyman

Senior Industry Advisor
Smith Barney, Inc.

The United States has embarked on the most massive, complicated industrial reorganization since the end of World War II: the deregulation of the electric utility industry (see Table 1). The Eastern Interconnection alone, one part of the North American power grid, has been described as the most complex machine ever devised by man[1], and it is nine times bigger than the largest power system deregulated to date, that of the United Kingdom (see Table 2). The organization of the industry in North America further complicates the process. The North American power grid operates under the jurisdiction of three countries and more than 55 state or provincial governments. Hundreds of firms, investor- and government-owned, of different sizes and cost structures, supply electricity. (The United Kingdom's industry, on the other hand, had one owner, one large fuel supplier, one control area, one regulator, and one government.) The politics of the process could prove as difficult as the technical issues.

---

[1] Attributed to Clark Gellings of the Electric Power Research Institute.

2   POWER SYSTEMS RESTRUCTURING

Table 1: Size of U.S. Industry at Time of Deregulation

Industry	Year of Deregulation	Annual Revenue in Year of Deregulation ($ billions)	Revenues as % of GDP in Year of Deregulation
Motor	1979	$ 41	1.6%
Carrier	1979	27	1.1
Airline	1980	28	1.0
Railroad	1984	74	2.0
Telephone	1985	63	1.6
Natural gas	1996	220	2.9
Electric			

Table 2: Size of Electricity Networks (1995)

Network	Demand at Peak (1,000 MW)	% of World Demand	Control Areas
USA and Canada:			
Eastern	482,587	20	103
Interconnection	128,139	5	32
Western States	45,538	2	9
Texas	30,890	1	1
Hydro-Quebec	270,000	11	N/A
UCPTE*	55,000	3	1
United Kingdom			

*Germany, France, Italy, Spain, Austria, Netherlands, Switzerland, Belgium, Yugoslavia, Greece, Portugal.

Source: Tampa Electric Co.," Interconnected Services: Defining System Requirements Under Open Access", March 1996, pp. 13, 23.

Not every deregulation proceeds smoothly. Rushed, poorly conceived programs lead to unintended consequences. After deregulation, the savings and loan industry of the United States came close to collapsing after a spree of reckless financing and fraud unequaled since the Roaring Twenties.² The Canadian government opened long distance telecommunication in a manner that allowed the incumbents to nearly wipe out the new entrants within a year.³ A decade after beginning the process of reorganization, during which thousands of electrical workers lost their jobs and a large part of the British coal industry closed down, most electricity consumers still cannot choose their suppliers.⁴ And most disquieting, 17 years after the deregulation of the airline industry, the Valujet crash prompted a Federal safety administrator to admit that his agency had trouble keeping tabs on the activities of new carriers.⁵

Reliable electricity supply is vital to a modern civilization. Breakdowns of supply can halt the economy and provoke social turmoil. The existing system seems to work. Why not follow Sam Rayburn's favorite dictum, "If it ain't broke, don't fix it?" What accounts for the push to change the rules that have regulated the industry for almost a century?

Begin at the Beginning

The electric utility business started out in the form of isolated power stations serving limited geographic areas, operating under the jurisdiction of municipal governments. Large consumers of electricity, such as traction companies and industrial firms, produced their own electricity (see Table 3). Electricity suppliers offered a bewildering array of individualized services for different types of load.

---

[2] Edward J. Kane, "The Unending Deposit Insurance Mess", 27 *Science* 451 (October 1989).
[3] John A. Arcate and John B. McCloskey, *Telecommunications in Canada*, (North Tarrytown, NY: Complan Associates, May 1995).
[4] Alex Henney, *A Study of the Privatization of Electricity in England* (London: EEE Ltd., 1994).
[5] Robert Davis, "Chief: FAA took too long on Valujet", *USA Today*, June 25, 1996, p. 1.

4 POWER SYSTEMS RESTRUCTURING

Table 3: Electricity Supply 1902-1932

Generating Capacity (%)	1902	1912	1922	1932
Utility	40	47	69	80
Other	60	53	31	20
	100	100	100	100
Output				
Utility	42	47	71	80
Other	58	53	29	20
	100	100	100	100

Source: Leonard S. Hyman, *America's Electric Utilities: Past Present and Future* (Vienna, VA: Public Utilities Reports, 1997), p. 11.

Eventually, the industry standardized its offerings based on the universal system concept of 1895. The development of alternating current (AC) transmission enabled power stations to serve wider areas. Industry leaders, especially Samuel Insull, then made two discoveries. First, by serving diverse loads (rural, urban, traction, office, residential, industrial), they could sell more electricity from the same machine, because the different customer groups took the electricity at different times of day, week and year. Second, as size of power station increased, production costs decreased. Therefore, one electricity supplier, operating the largest possible generators at maximum output, could produce at a lower cost than many competing, small generators.

Electricity supply, the utility executives said, constituted a natural monopoly. The distribution lines showed network economies. That is, one network serving all customers had lower costs per customer than several duplicating networks that each served only some customers. And, from a practical point of view, city governments did not want to have many companies digging up the streets or covering the sky with wires. The generation function, however, exhibited economies of scale: the larger the facility, the lower the cost per unit of output. If the market were reserved to one generator, that generator could achieve the maximum economies of scale. Only the government, however, could prohibit others from entering the market.

In 1907, after nearly a decade of lobbying by the public utilities, several states set up public utility commissions.[6] The states granted territorial monopolies to the electric utility, so that the utility could reach maximum output and lowest cost. In order to assure that the public rather than the monopolist reaped the benefits from economies of scale, the regulator set a cap on the utility's profit.

Technology

From the advent of regulation to around 1960, the electric utility industry produced a remarkably consistent set of trends. In almost every year, the industry installed larger power plants, generating efficiency increased, production costs fell, and price of electricity declined because regulators required the utilities to pass on cost savings to customers (see Table 4). Around 1960, however, economies of scale petered out. Efficiency had reached the maximum allowed by the Rankine Cycle. Building larger generating stations no longer produced lower costs. That, in itself, did not undermine the utility's claim to a natural monopoly. After all, the utility's large stations still produced at lower costs than any potential competitor.

---

[6] Leonard S. Hyman, *America's Electric Utilities: Past, Present and Future* (Vienna, VA: Public Utilities Reports, 1997), pp. 95-97.

Table 4: Generating Plant Size, Electricity Price and Operating Efficiency Measures 1882-1995

Year	Index of Real Price of Electricity (1902 = 100)	Average Size of Utility Plant (kW)	Cost Equivalent Burned kWh Generated (lbs)	Heat Rate (Btu/kWh)
1882	132	3 E	- -	- -
1902	100	539	6.5	79,900 E
1912	62	1,467	4.0	49,200 E
1922	36	3,813	2.5	30,700 E
1932	46	8,539	1.5	18,450
1945	19	13,002	1.3	15,800
1960	9	48,203	0.9	10,701
1970	6	135,000	0.9	10,508
1980	5	241,300	1.0	10,489
1990	5	237,100	1.0	10,367
1995	4	240,000 E	1.0	10,173

E - Author's estimate

Source: Hyman, *op. cit.*

What undermined the utility's claim to a natural monopoly in generation and set the stage for deregulation, was another technological development. In 1961, an electric utility installed the first stationary gas turbine, essentially a jet engine modified to produce electricity. In the ensuing decades, manufacturers improved the gas turbine, and then developed combined cycle units, which utilize waste heat from the generators to produce additional usable energy. In time, the gas turbine and combined cycle units reached efficiency levels that exceeded those of conventional power stations (see Figure 1). The new generating units had numerous advantages. They were smaller, easier to install, produced less pollution, could be put on line within months of order and had lower capital and operating costs. With the new technology, many small generators could produce electricity for less than the large utility generators.

In effect, the rationale for keeping out competing generators had evaporated. Before, regulation assured that consumers rather than the monopolist received the benefit derived from economies of scale. Now, regulation threatened to protect the monopolists from competition and deprive consumers of the benefits of new technology.

Laws and Ideologies

The Carter Administration favored energy conservation and the introduction of competition into regulated industries. In 1978, Congress passed the Public Utility Regulatory Policies Act (PURPA), which resulted from the perception that existing utilities would not embrace new, energy saving technologies. The law forced utilities to buy power from independent energy producers that either utilized renewable resources to produce electricity, or generated electricity from fossil fuels but made use of the waste heat for industrial processes (co-generation). The bulk of the independent power firms installed gas turbines for their projects. Two decades of PURPA experience validated the worth of the gas turbine, demonstrated that non-utilities were capable of producing electricity reliably and economically, and created an industry that now accounts for roughly one-tenth of America's electricity output.

Meanwhile under the Reagan Administration, the Federal government had begun the process of deregulating parts of the telecommunications and natural gas industries, the Thatcher government in the United Kingdom and the Pinochet dictatorship in Chile moved forward in the process of deregulating utilities, the conservative revival in the USA encouraged skepticism about the accomplishments of regulation, and the fall of the Soviet Union put free marketeers in the driver's seat all over the world.

In the waning days of the Bush Administration, Congress passed the Energy Policy Act of 1992, opening the generation market to all comers and mandating that transmission owners had to provide access over their lines to other producers and sellers of energy. Five years later, the Federal Energy Regulatory Commission had issued detailed rules on transmission access, many firms had entered the business of trading electricity, more than half the states had underway studies of restructuring the electricity supply industry and several utilities had begun the process of selling off generating stations.

8 POWER SYSTEMS RESTRUCTURING

Deregulatory planners begin with the assumption that generation of electricity has become a competitive business that should not require regulatory oversight. Transmission and distribution activities retain monopoly characteristics and should remain under regulatory supervision. The actual supply of electricity to the consumer might fall into the hands of unregulated virtual utilities that, in effect, rent the conduits from the utilities. In order to assure that utilities do not use their regulated monopoly conduits to favor their own regulated affiliates over competitors, regulators have encouraged the separation of functions within the utility, or even sale of generating assets to unrelated parties.

The impetus for deregulation has come from the citizens of states in which the price of electricity is high (notably California), from industrial customers who believed that regulation placed too much of the cost of service on them, from those who believe that consumers are entitled to choose their suppliers, and from those who advocate competition as the means to force electricity suppliers to operate more efficiently.

Issues

The existing electricity supply system was not designed to meet the needs of a transaction-oriented market (in the past 10 years, transactions have increased four fold and wholesale transactions now amount to half of total sales) or of the present patterns of population and economic development. Reserve margins have trended downward. Expansion of the transmission network has not kept up with growth in demand (see Table 4). That increasingly inadequate network, and its stressed owners, will have to deal with the rigors of deregulation.

Table 4: Transmission System Expansion 1975-1995

	1975	1985	1995	Annual % Growth 1975-1985	1985-1995
Transmission circuit miles (thousands)	513.9	607.1	672.2	1.7	1.0
Generating capacity (1,000 MW)	527.6	711.6	817.2	3.0	1.4
Peak load (1,000 MW)	356.8	460.5	620.5	2.6	3.0
kWh generated and imported (billions)	2014.8	2614.2	3450.9	2.6	2.8
Capital spending on transmission (Millions of 1975 dollars)*	1771.0	994.0	954.1	-5.5	-0.5
kWh generated and imported per transmission circuit mile (kWh)	3919.4	4306.0	5133.7	0.9	1.8

* Investor-owned utilities. GNP deflator.

Source: Edison Electric Institute

Electricity consumers, producers and regulators must solve a number of transitional problems before achieving an operationally reliable and financially viable competitive electricity supply sector. The problems are financial, organizational, regulatory and operational.

- **Financial** - As a starter, electric utilities have assets on their books at values greater than their values in a competitive market. They also have obligations to purchase power from others at prices, often set by past regulators, that exceed the value of the electricity in the marketplace. Utilities refer to those potential losses as "stranded costs". One estimate puts those costs at $135 billion.[7] Many utilities have managed to defer the effective date of competition until they can recover stranded costs through surcharges that could cause distortions in the market. In addition, utilities, as a whole, have debt-laden financial structures and dividend pay-out policies inappropriate for competitive industry.[8] The quick advent of a competitive market could produce numerous dividend cuts as well as some defaults on debt.

- **Organizational** - Presently, most utilities are fully integrated entities that produce, transport and distribute electricity to consumers, or buy electricity wholesale and then sell it retail to consumers, at a bundled price that includes all services. The companies must separate regulated from competitive functions. Some will do so by selling generating assets. Others may place regulated and unregulated into different entities under a holding company. To make matters more complicated, those utilities that retain generation assets may be forced to place the operation of their transmission in the hands of an independent operator, in order to remove the possibility that the utility will manipulate the management of its transmission system to disadvantage the competitors of its generating business.

- **Regulatory** - Regulators face several challenges. First, they must assure that a truly competitive market, rather than one dominated by the existing utilities, emerges. They will attempt to devise regulatory schemes that allow utilities to recapture stranded costs. They need to develop methods to regulate the residual monopoly in a manner that encourages efficient operation. And, finally, they have to know when to let go, when regulation no longer serves a productive purpose because the competitive market is functioning.

- **Operational** - In the fully regulated environment, utilities cooperated voluntarily to maintain reliable operations, and, acted as if the cost shifting between interconnected utilities would

---

[7] Moody's Investors Service, *Electric Utility Sourcebook* (NY: Moody's Investors Service, October 1995), p. 35.

[8] Leonard S. Hyman, "Fearless Forecast: Electric Utilities in 2007", in Proceeding of the AIMR Seminar *Deregulation of the Electric Utility Industry* (Charlottesville, VA): Association for Investment Management and Research, 1997), pp. 65-66.

average out over time. In a competitive environment, some electricity suppliers will be capable of providing the services that maintain reliability and others will not. Users of services will have to pay providers of services. Inadequate payments set by regulators might lead to withdrawal of services and consequent deterioration of reliability. Most important, though, system operators have to work out the rules of the road that determine which of many transactions are executed on the constrained facilities available.

Declaring that the goal is a competitive market and reaching that goal in a timely fashion are two different matters. Successfully dealing with transitional issues within the brief time frame set by proposed deregulatory laws will not be an easy task.

Conclusion

Deregulation of other industries has led to substantial cost savings on the part of industry participants, lower prices to consumers, and a plethora of new products. A study made by Mitchell Diamond of Booz-Allen-Hamilton, for instance, asserts that if electric utilities simply adopted best practices within the industry, they could reduce costs $24 billion per year, or 11% off the average electric bill.[9] Those savings, if so applied, could pay down stranded costs in less than six years. With potential savings of that magnitude, we should encourage the fastest possible transition to a competitive market. At the same time, everyone wants deregulation without any degradation in the reliability of the power system. The consequence of power system failure is far greater than from overbooking an airplane or subjecting a telephone caller to busy signals. The organizers of the new system must get the operational and incentive signals right. Wishing won't make it so.

---

[9] Mitchell Diamond, "Prometheus Unbound: Electricity in Era of Competition", presentation to the Smith Barney Energy Conference, Miami, FL, February 7, 1997.

I THEORETICAL CHALLENGES IN REAL-TIME OPERATION

2 POWER SYSTEMS OPERATION: OLD VS. NEW

Marija Ilić* and Francisco Galiana**

*Department of Electrical Engineering and Computer Science
Massachusetts Institute of Technology
Cambridge, MA 02139, USA

ilic@mit.edu

**Department of Electrical and Computer Engineering
McGill University
Montreal, Quebec, Canada

galiana@pele.ee.mcgill.ca

INTRODUCTION

In this chapter, we describe the basic tasks of power system operation by first reviewing these for today's industry, and then describing possible solutions for the evolving industry. We suggest that the new solutions strongly depend on the type of the electricity market (system structure) in place and illustrate possible solutions in the context of specific representative system structures.

We formulate the basic objectives of power systems operation assuming first so-called perfect market conditions. The impact of how well operation tasks are designed and performed is directly measurable with respect to this ideal market equilibrium. The charges for *all* other operation tasks, such as transmission loss compensation and ensuring reserve for reliability, create price deviations from the single ideal market price and are generally different for different system users.

However, an electric power system is never at its equilibrium. Consequently, the primary supply/demand market is not likely to be close to its (economic equilibrium). At later stages of this chapter we define the objectives of performing operation tasks for this more realistic setup of imperfect market conditions. We emphasize how, in this case, the *intertemporal effects* among subprocesses in a real-time electricity market have a significant impact on the characteristics of the electricity price dynamics. We point out that, because of the very unique features of the electricity market being predominantly a *real-time market*, significant tradeoff between the real-time electricity price volatility and efficient electricity pricing could be anticipated. While the question concerning the impact of specific market structure (poolco vs. bilateral, for example) is common to any other industry, the intertemporal effects are potentially much more significant in the power industry than in the others where inventories are a means of dealing with real-time supply/demand imbalances.

In order to provide a theoretical framework for comparing the performance of several market structures under consideration, we must define performance objectives for the various unbundled operation tasks and analyze how they affect the overall system performance. So far, most of the industry debate has been concerned with the organization of the primary electricity market. Much less attention has been paid to the unique challenges of the power industry beyond the basic supply/demand process. In this chapter we describe variations in industry structures as a function of paradigms chosen for transmission loss compensation, frequency regulation, reliability provision, etc. We point out that the electricity price will strongly depend on how efficiently these functions are organized to serve the primary market.

NEW SYSTEM STRUCTURES UNDER COMPETITION

With the opening of the power industry to competition, new organizational structures have evolved (or are still evolving) in order to handle the changing needs with regard to administration, planning and operation. Inevitably, these new structures and terminology are still emerging and may differ from one area of the continent to the other or according to whom one is speaking. Still, such deliberations and evolution are a normal and, arguably, healthy situation.

In this section we briefly summarize various proposed and existing industry structures under competition as well as their components, definitions and functions. We also suggest a general framework of a *transactions network* as a means of representing and analyzing the proposed competitive market structures.[1]

Traditional organizational structures

Generally speaking, the traditional power system is subdivided according to geographical size into utilities, control areas, pools and coordinating councils. Utilities typically own generation, transmission and, possibly, distribution over a wide geographical area.

Each utility contains one or more control centers for the purpose of monitoring and dispatching the equipment under its control. Typically, an energy management system carries out functions like load forecasting, state estimation, generation dispatch, reactive power control, automatic generation control, unit commitment, as well as preventive and emergency security control. The next section in this chapter discusses these functions in greater detail.

Control areas are bodies set up for the purpose of regulating power exchanges across tie-lines interconnecting pairs or groups of utilities. Another important function of control areas is to regulate system frequency.

The pool structure can be an amalgamation of utilities and areas whose resources are pooled to increase reserve margins, thereby cutting down generation capacity while maintaining system security levels.

Finally, coordinating councils are bodies with a broad geographical jurisdiction whose role is to establish common reliability criteria among its control area members as well as to exchange system data for the purpose of carrying out planning studies over the breadth of their territory. Coordinating councils were set up following the major blackouts of the sixties and seventies to prevent a recurrence of such events. In addition, following the most recent blackouts in the Western United States, over twenty security coordinators were formed as bodies ultimately responsible for coordinating decisions in real time to relieve line congestion as many transactions actively take place across large geographical areas [1].

Competitive organizational structures

Unbundling of vertically integrated utilities. One of the principal characteristics of a competitive structure is the identification and separation of the various tasks which are normally carried out within the traditional organization so that these tasks can be open to competition whenever practical and profitable. This process is called "unbundling". An unbundled structure contrasts with the so-called vertically integrated utility of today where all tasks are coordinated jointly under one umbrella with one common goal, that is, to minimize the total costs of operating the utility.

Cost minimization subject to accepted minimum reliability standards is the principal objective of a vertically integrated utility since the rates that the utility charges its customers are set by an independent government regulatory body. Thus, maximizing profit can only be reached by minimizing cost. These rates are fixed by the regulating body following public hearings where both the utility and the consumers can argue for or against the level of the tariffs. One can view the public hearing process as a kind of auction where a price is set which is acceptable to both buyers and sellers; the buyer or customer is guaranteed not to have rates climb unreasonably, while the seller or utility is assured a fair rate of return on its investment.

At this stage, one might reasonably ask the following question. If the existing monopolistic structure is so just to all parties, why on earth should the move toward competition be so prevalent and powerful? Can both buyers and sellers profit from competition in the power industry? In this section, we will limit ourselves to stating that only time will ultimately tell since accurate predictions are certainly almost impossible to make, however, one can argue that there are two important factors in a competitive environment that could benefit at least some of the parties. The first factor has to do with new technology, in particular independent power producers whose highly efficient, low cost power plants may have not been installed without the new open access to the customer base. Secondly, the unbundling of utility services may result in more equitable tariffs being assigned to each individual task. These tariffs may more closely reflect the actual worth of the unbundled tasks compared to a single tariff which averages out the cost of many services.

Notice, however, that neither of these two arguments are directly related to the power systems operation efficiency. Throughout this chapter we consider dependence of total cost/price minimization on the way system services are provided and priced. The effects of how is this done on what customer eventually pays must be understood if one is ever to make the case for useful "unbundling" of services. We conclude, after much soul searching throughout this chapter, that it is impossible to make the case for unbundling at this stage of industry debate, given that the charging mechanisms for system services are created in the overall rush to deregulate and without a real understanding of their implications.

Of course, other, often overlooked factors, which may argue against competition have to do with the public socio-economic service that an utility provides through investments in regions of the country that otherwise may be economically disadvantaged. The potential loss of jobs and increased system unreliability in remote regions are aspects that should certainly not be underestimated in any move toward the opening of the electricity business to competition.

Unbundling of functions within a corporation. The principal form of unbundling which takes place in a utility could be called corporate-unbundling. This contrasts with the unbundling of services which is discussed in this chapter. Corporate unbundling implies the segregation of generation, transmission and distribution into independent competitive commercial entities. This step need not imply divestiture into separate companies as long as there exist barriers that prevent conflicts of interest from arising.

If the original utility is large, its generation can be split up into a sufficiently large number of smaller independent competing generating companies or gencos to which new independent producers are added. In the case of the UK, for example, the Central Electricity Generating Board was split into four gencos in order to minimize the potential for excessive market power on the part of any single genco [13]. Such market power is viewed negatively by the regulators as it might be used to suppress competition from smaller producers and drive them into bankruptcy.

The transmission network is also subject to a form of "unbundling", however this notion as well as its application becomes more nebulous. It can be argued that a transmission network is a natural monopoly since power flows cannot be directly controlled (at least not as easily as generation). If the transmission network is split into a number of physical parts, e.g. transmission lines, transformers and substations, and these are then sold separately to a number of shareholders, it is difficult to devise a mechanism for these transmission companies or transcos to compete fairly. This is so since the power flowing due to a contract between a generator and a load cannot be guaranteed to flow through a specific transco, flowing instead though the entire network, following the laws of physics. In fact, some radial transmission lines or congested subnetworks may be the only possible path for certain contracts, making these subnetworks

POWER SYSTEMS OPERATION: OLD VS. NEW   19

pure monopolies. Generally, therefore, transcos are treated as monopolies and, as such, are subject to regulation of the transmission tariffs they can charge for network access.

The other issue regarding the corporate unbundling of the transmission network has to do with the service that such companies provide, namely transmission transfer capability [8, 9], that is, the ability of the network to transfer power between one or a group of sellers to another single or group of buyers. There are two main types of transfer capability defined with respect to a network, total or TTC and available or ATC. The first refers to the maximum possible capability, while the second is the available transfer capability relative to the current operating point. The quantity ATC is a service that can in theory be reserved at a price as either a firm or a non-firm contract. Depending on the level of congestion in a network, the amount of available transfer capability may be at a premium in which case it may not be possible to accommodate all possible contracts through the network. In this event, some mechanism or independent unbiased entity may have to intervene in the competitive process to allocate transmission capacity in a manner acceptable to all parties. One such entity is the so-called ISO or independent system operator whose role, among others, is to allocate ATC during congestion.

The last component of the corporate unbundling is the creation of independent distribution companies or discos whose role is to provide low-voltage, normally radial service to individual industrial, commercial or residential customers. The physical disco is, as with the transmission company, a monopoly, since the connected loads have no other path through which to receive power.

Service unbundling. Some of the main services that can be unbundled for separate tariffication are: the generation of real power, real-time load following and transmission loss generation, stand-by spinning reserve, short-term spinning reserve for security purposes, load-frequency control, Var generation and interruptible loads. These will be discussed in more detail in the next section.

Again here, although in theory these services can be operated by physically distinct and independently operated enterprises, the level and, in certain cases, the location of these services need to be coordinated. This is required because of the strong physical coupling and physical restrictions among all these quantities. Market forces alone cannot ensure that this coordination will take place. In particular, the need to maintain power system security (that is, normal system operation in terms of load supply, frequency and voltage stability) under a number of specific contingencies is just as vital under competition as it is under the traditional monopoly.

Power exchange and independent system operator. Above, we argued for the establishment of an independent entity whose main role is to coordinate the various services purchased in the open market. Such an entity is called the Independent System Operator or ISO. An important question can be posed concerning the ISO, namely, what is the ISO independent of? One interpretation is that the ISO, through its security control actions, should not unfairly preclude one or more market participants from fulfilling a contract arrived at through the existing market negotiation process. Fairness, however, does not mean equal treatment, rather, it has to do with the ISO implementing its actions according to a set of rules previously agreed upon by all market participants.[2] Thus, as long as the ISO follows these rules, it is possible that some contracts may be curtailed or modified by uneven amounts if the security of the network so requires. Examples of such actions may be found in the operation of the England and Wales Power Pool (EWPP) where the system operator has the authority to alter the generation contracts if these would create a network flow congestion or violate some other network constraint.

The Power Exchange or PX, in general, is the body which implements the agreed upon mechanism by which all market participants trade electricity. This mechanism generally takes the form of an auction whereby many sellers and buyers submit bids to supply or receive power. An auctioneer, then selects a subset of the bids according to a set of rules. If the auctioneer is buying power from the generators on behalf of the pool, as in the EWPP, then the auctioneer's job is to minimize the pool purchase cost. If the auctioneer, as in the Alberta pool, receives both selling and buying bids, then the auctioneer's function is to match the aggregate generation price curve to the aggregate load price curve as closely as possible. If the structure is one of private bilateral contracts, then the two partners (or brokers on their behalf) carry out the role of the auctioneer by matching individual buyers and sellers.

The ISO and the PX could be separate bodies, as it is the case in California [2], the former concerned with maintaining system security without knowledge of contract prices, while the PX's main job is to facilitate the trading process. In some instances, however, the two functions are carried by the same body, for example in the UK, or are proposed to be done this way, such as in New York and New England.

Pool and bilateral market structures. There exist a number of approved and proposed competitive structures for the trading of electricity [14].

- (1) Wholesale competitive generators bid to supply power to a single pool. Load-serving companies buy wholesale power from the pool at a regulated price and resell it to the retail loads.

- (2) Wholesale competitive generators bid to supply power to a single pool, while load-serving companies then compete to buy wholesale power from the pool and resell it to the retail loads.

- (3) Combinations of (1) and (2) with bilateral wholesale contracts between generators and load-serving entities.

- (4) Combinations of all previous plus contracts between all entities and retail loads.

Trading entities. It is useful to define a number of commercial entities in a competitive power system structure. These are:

- (i) Generator-serving entities. The goal of these is to trade on behalf of one or a group of generators.

- (ii) Load-serving entities. Their goal is to trade on behalf of the loads.

- (iii) Pure trading entities or marketers. The basic function of the evolving power exchanges and marketers is to facilitate certain level of aggregation for trading power on a complex system consisting of thousands of physical loads and hundreds of suppliers. The role of these is to trade on behalf of any generator, load or other trading entities. One can think of a pool as one type of trading entity.

Transactions networks

Under competition, generation, loads and flows are not the only system decision variables. Instead, to these, one must add the contracts between trading entities. Such contracts are defined by the market conditions and, together with the bus loads, they can be considered as the system inputs driving the power system.

Transaction networks is a useful general tool for representing the information flow among trading entities as well as the contracts among them under any competitive structure. Figure 2.1 shows the most general form of a transactions network where the arrows represent matrices of transactions between the various types of trading entities, namely G for generator-serving entities, D for load or demand-serving entities and E for pure trading entities or marketers. We note that marketers, generators and load-serving entities can also trade with each other as shown by the arrows starting and ending in the same circle. The quantities GG,GD, and so on, represent submatrices of bilateral contracts among the different types of entities.

CLASSIFICATION OF OPERATION TASKS IN TODAY'S POWER INDUSTRY

The operation of today's power industry has as main purpose the provision of sufficient generation and other controllable equipment (capacitors, reactors, flexible AC transmission system (FACTS)) at the interconnected system level. The operation of a power system can be classified according to the following five tasks [17]:

- *Task 1:* Meet the given predicted time-varying demand at minimum operating cost.

- *Task 2:* Compensate for transmission losses (real and reactive) that occur in the system as the predicted demand is supplied.

- *Task 3:* Meet various operating constraints (such as thermal or stability constraints on transmission lines, voltages at both demand and supply buses).

22 POWER SYSTEMS RESTRUCTURING

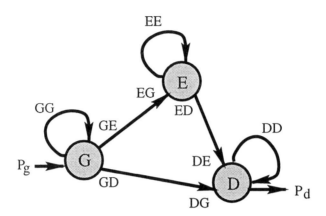

Figure 2.1 A general transactions network

- *Task 4:* Provide flexible generation in real-time to balance deviations from the anticipated demand, as they occur.

- *Task 5:* Provide stand-by network resources (generation, Vars) in case any single outage occurs on the system ($n - 1$ security criterion).

The result of meeting these operation tasks is the continuity of high-quality power supplied at the end user (measured in frequency and voltage), despite the user's variations from the scheduled demand and despite unexpected, major changes in the equipment status.

To meet anticipated demand, at present, economic dispatch and unit scheduling (task 1), loss compensation (task 2), and the satisfaction of the static operating constraints (task 3) are integral services provided by *all* generating units. Small random deviations from anticipated demand are tracked by automatic generation control (AGC) (task 4). Typically, a large system has only a handful of AGC units directly dedicated to systemwide regulation in response to small load random variations.

At present, operation planning is done at a systemwide level with the single (*bundled*) objective of performing all five tasks at the least possible total cost. Consequently, the price of electricity as seen by each customer reflects the average cost of designing the system and doing *all five* operation tasks.

A typical hierarchical information/control structure in operating today's large power systems is shown in Figure 2.2 and Figures 2.3–2.5. All power plants participate in unit commitment (tertiary level), while only some, relatively flexible plants are regulating units in each control area (secondary control). Almost all power plants participate in primary control by means of governors and excitation control.

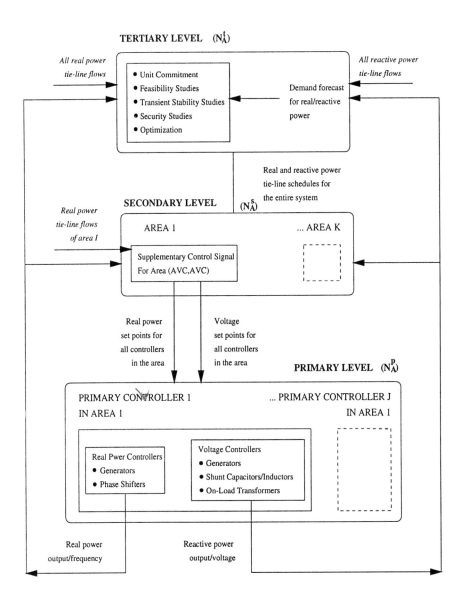

Figure 2.2 Hierarchical measurement/control structure of systems control services [20]

TEMPORAL DECOMPOSITION OF PROCESSES WITHIN THE REAL-TIME OPERATION

To prepare the ground for analyzing the intertemporal aspects of power system operation and the costs of balancing supply and demand in the new industry, we briefly review in this section principles of system operation by decision and

24   POWER SYSTEMS RESTRUCTURING

control in the present industry. Real-time operation is based on both temporal and spatial decomposition of a single dynamic process into subprocesses. Under certain assumptions which ensure that this decomposition is valid, it becomes possible to simplify the overall operating complexity and rely on hierarchical principles to control these subprocesses at a decentralized level with very little systemwide coordination. Ingenious decentralized control schemes, such as AGC, for example, are easily interpreted in light of such a hierarchical control layout.[3]

More importantly, understanding these basic principles of today's operation and control enables us to pose the need for possible different operation and control paradigms as the industry restructures. As one may expect, the concepts and specific implementations underlying today's operation of complex power systems are quite involved. In this section only the rudimentary aspects are reviewed which we consider as necessary for understanding why and how operation and control may change in the future.

Normal operation in today's industry

A power system is considered to be in normal operation when all equipment (generation and transmission) is functional as expected, and the only major changes are result of load dynamics. Therefore, power systems dynamics in normal conditions are driven primarily by changes in demand. Typical demand curves are shown in Figures 2.3–2.5. It can be seen from these figures

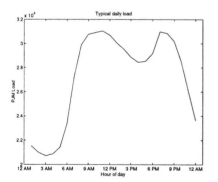

Figure 2.3 A typical daily load curve (PJM system)

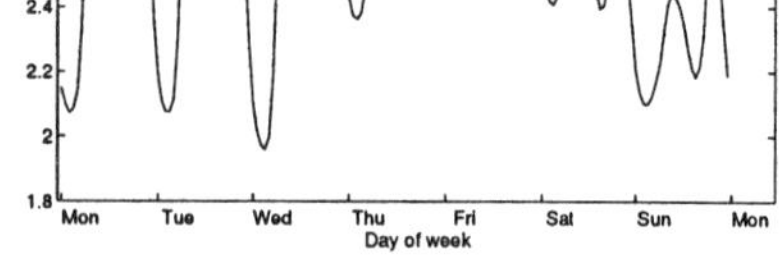

Figure 2.4 A typical weekly load curve (PJM system)

that load dynamics exhibit several rates of response, ranging from very fast random variations (order of seconds), through hourly, daily, weekly and seasonal patterns of larger deviations.

In today's U.S. industry real-time generation/demand balancing is done *both* in an open-loop manner for the predicted demand, as well as by the closed-loop regulation of frequency deviations from nominal that are caused by the actual

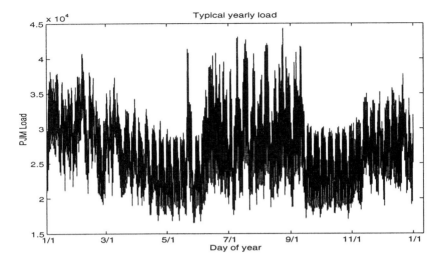

Figure 2.5 A typical yearly load curve (PJM system)

generation/demand imbalance at each control area level. At the tertiary level of the hierarchical organization shown in Figure 2.2 generation-based scheduling is open loop, while the secondary level control is closed-loop and automated. Variations of this can be found throughout the world; for example, the UK system does *not* have automated frequency regulation, and, moreover, since the entire system is effectively a single control area, tertiary and secondary level merge into a single level.

Open-loop generation scheduling. Economic dispatch and unit commitment are functions that generally incorporate some measure of anticipation in change. The economic dispatch function generally evolves over a 5-30 minute period and is used to schedule generation outputs $\hat{P}_{gi}(t)$ of the power plants in service each 5 minutes to follow hourly system load trend $\hat{P}_{d,sys}(t)$.[4]

In the present system the coordinator must also project which units should be operated in subsequent periods. Unit commitment programs are used to fulfill this function. Operating capacity additions must be planned in advance given the fact that it may take from 1 to 10 hours to restart all available steam plants, for example. Other conditions, such as hydro storage facilities, may require that weekly concerns be included (for example, start the week with full storage, end the week with no storage, and pump the storage full over the weekend). For purposes of analysis in this chapter, we consider a day ahead unit commitment decision for each day as a discrete time process taking place each hour.

At present, energy management systems and various utility control centers respond to their anticipated demand and perform economic dispatch and unit commitment. The units under the jurisdiction of a particular EMS or a control center generally schedule their own units for the *preagreed upon* power exchange with the neighboring companies. Tie-line flow scheduling among different companies is done bilaterally without any overall systemwide coordination[5].

Closed-loop frequency regulation. Assuming that the generation schedules determined in the open-loop manner are implemented in real time, the variations between the *actual* and *predicted* demand still create a time-varying supply/demand mismatch in real time. The main objective of the generating units participating at the secondary level is to regulate this mismatch by using available generation resources in real time. The power output of each unit is adjusted to respond to the supply/demand mismatch[6].

At the fastest time frame associated with system operation (less than several seconds) power plants respond in a fully automated, decentralized way to the supply/demand mismatch seen in local frequency deviation of each generator $F_{gi}(t)$ from its set value $F_{gi}^{ref}(t)$[7]. Also, the local voltage is controlled using excitation systems; their function is to stabilize terminal generator voltage $V_{gi}(t)$ to its set value $V_{gi}^{ref}(t)$. This lowest level of the hierarchical control scheme is sometimes referred to as the *primary control level* [20, 21]. Load also has a self stabilizing effect since power consumed decreases as frequency and voltage decrease.[8]

Response over a time period .5-10 minutes to a supply/demand mismatch of the entire control area is provided by the so-called regulating units. In the United States this function is also automated, and is activated in response to the actual mismatches. The regulating units automatically provide energy within a 10-minute period; this is done by regulating the set point deviations $F_{gi}^{ref}[kT_s]$ each $kT_s = 2$ seconds or so, for $k = 1, 2, \cdots$ so that total supply/demand deviations at each control area meet present NERC recommendations, such as the A_1 and A_2 criteria [23]. This automated regulation of supply/demand deviations around their scheduled values is sometimes referred to as the *secondary level control [20, 21]*.[9] Both primary and secondary generation control are activated in response to actual conditions.

These standard systems control functions are summarized as follows:

Normal State

Time Frame	Function
2-3 seconds	Inertia, loads, excitation systems
7-10 seconds	Governors
.5-10 minutes	Regulation (AGC)
5-30 minutes	Economic dispatch, manual control
1-10 hours	Unit commitment, restarting/shutting off units

Price of electricity versus total cost of performing tasks 1-5 in today's industry: Two part tariffs

In today's industry the only emphasis has been on minimizing total cost of meeting demand. No explicit "unbundling" into subprocesses reflecting cost of particular operating tasks is carried out. The price of electricity reflects (or is supposed to) both fixed and O&M costs of performing all five operating tasks.

The costs are averaged over time and geographically distributed customers, more or less on a pro rata load share basis. In other words, electricity is *the same product* at any time and any geographical location within a given power utility. There is no temporal nor locational unbundling of the value of electricity in today's industry.

While only the average price is derived and charged, it is relevant to recognize a two part tariff-based reasoning underlying the rates structure in today's industry; both fixed cost and the O&M recovery are guaranteed. Since much discussion in what follows is concerned with the market outcomes measured in terms of total cost as well as in terms of price of electricity, and, particularly, with the fact that optimizing for one does not lead to the same result as when optimizing for the other, we observe here that in today's industry costs of major uncertainties (demand fluctuations, fuel cost, availability of power plants) are averaged out over all customers. We are not aware of any utility study that has attempted to correlate the two.

CLASSIFICATION OF OPERATION TASKS IN THE COMPETITIVE POWER INDUSTRY

The industry changes taking place today are primarily reflected in the execution of task 1, that is, in making this task competitive instead of coordinated. This problem, which is one of balancing supply and demand, has been studied extensively in the context of many industries and the pros and cons of competitive supply/demand markets, and their implications on the economic efficiency are directly applicable to task 1. Many important subproblems related to creating an efficient market for basic supply/demand in a deregulated electric power industry, such as the issues of stranded generation cost and market power, must be resolved, however they are not the main subject of this chapter.

Here, we assume instead, that a competitive supply/demand market exists, and that it could take various forms and degrees of coordination. We refer to the trading entities in such an arbitrary industry structure as market participants (MPs) [17, 20, 24]. In section above, we have defined several representative industry structures for performing task 1.

In this chapter, we illustrate some possible approaches to market-based performing of tasks 2-5, for example transmission loss compensation, tertiary level regulation of tie-line flow constraints and frequency regulation. Resulting new hierarchies are generally different for these new solutions relative to the hierarchy in today's power systems operation shown in Figure 2.2.

The *fundamental problem* is that tasks 2-5 are dependent on how task 1 is done since their main objective is to balance the system in case this is not done by means of the main (supply/demand) market. Also, power quantities traded at the main market may change once the charges for services 2-5 are known. This interdependency could be solved in more than one way, e.g.,

1. Create resources for meeting all tasks 1-5 in a coordinated way, retain all technical services as they are, and introduce coordinated mechanisms for creating market price for such bundled technical services. The only cost

28 POWER SYSTEMS RESTRUCTURING

allocation (unbundling) that takes place in this scenario is according to the ownership of these resources. Various forms of proposed poolcos are centered around this scenario.

2. Allow for an arbitrary mechanism for task 1, perform some of the tasks 2-5 at the end user level according to prespecified technical performance, and some (minimal) subset of services 2-5 at the interconnected system level. The truly interconnected operating services [25] could be created ahead of time in a competitive manner, and used in a coordinated or decentralized way in real-time.

If step 2 is implemented, a charge to the MPs for providing system services to the primary electricity market must be established. Moreover, the price-charging mechanisms for these services may differ depending on how efficiently these are provided in real-time. If it were not for tasks 2-5, financial and physical processes concerned with the basic procurement of supply to meet anticipated demand could be fully decoupled, like in other industries. This is relatively easy to do, but it would have detrimental consequences of not giving incentives for provision of tasks 2-5 in a way necessary to benefit the consumer.

MEETING PREDICTED DEMAND IN TODAY'S INDUSTRY (TASK 1)

The main objectives of real-time operation in today's industry are to (1) schedule maintenance of power plants, (2) turn them on and off and (3) change generation outputs as required to supply the predicted system demand $\hat{P}_{d,sys}(t)$ shown in Figures 2.3-2.5 at least total O&M cost. Under the normal operating conditions each power company generally schedules its own plants for some assumed power exchange $\hat{T}_{net,i}$ with the neighboring systems (see Figure 2.6). Power exchange with the neighbors is only modified in real time for sharing reserves under unexpected contingencies.

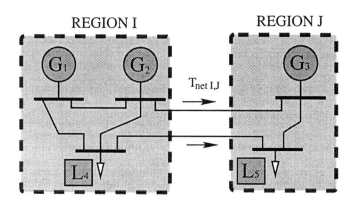

Figure 2.6 A control area

Since generators cannot instantly turn on and produce power, unit commitment must be planned *in advance* so that enough generation is always available to meet system demand with an adequate reserve margin in the event of generation or transmission line outages or in case demand exceeds the expected amount. The units could only be chosen so as to optimize the *expected total cost over a long term horizon*, rather than actual (unknown) total cost. Critical parameters that need to be used for computing turn on and turn off times for various plants are the start-up cost, rate of response, must-run time, etc.

Moreover, as it can be seen from Figures 2.3-2.5, typical demand varies on daily, weekly, monthly and yearly basis. It is generally as predictable as the weather is.[10] In order to meet tasks 1-3, at present a system operator relies primarily on static network and generation modeling tools such as load flow studies, economic dispatch simulations, or (optimal power flow) OPF analyses [61]. Such models assume that the demand is known, which is only true within the accuracy of the forecasted variables (load, unit outage statistics). In Appendix 2.1 we review the basis theoretical background for creating algorithms used in today's industry for performing task 1.

MEETING DEMAND IN THE NEW INDUSTRY: PRIMARY ELECTRICITY MARKET (TASK 1)

In the new industry, the scheduling of power sold and purchased by various market participants is done according to a general transactions network setup shown in Figure 2.1. The "dynamics" of *asynchronous* schedules (transactions) determined in a general bilateral setup (transactions are established sequentially in time, and for various durations desired) and a *synchronized* unit commitment-based schedule (each day, at 9pm, for example, *all* units are scheduled for the next day) are quite different.

In the new industry, the predicted system demand $\hat{P}_{d,sys}(t)$ will be supplied competitively by many producers. One could identify several different mechanisms for providing generation to where it is needed. These are,

- Physical bilateral transactions which take place when individual (or groups of) suppliers self-schedule power $\hat{P}_{gi}$ to directly supply a large load $\hat{P}_{dj}$.[11] Physical bilateral transactions are typically long-term and both price $\hat{p}$ and quantities $\hat{P}_{gi}/\hat{P}_{dj}$ committed are based on the predicted market conditions, but cannot change as the market conditions evolve. These are by definition *nontradeable* and are exersized according to the one time contractual terms.

- The tradeable bilateral transactions are based on long-term contracts between two (or more) parties i and j to purchase power $\hat{P}_{dj}$ and sell power $\hat{P}_{gi}$ at certain price $\hat{p}$ that is based on market predictions. However, in contrast to the physical bilateral transactions, both parties i and j are allowed to trade at least portion of the committed sale/purchase through a short-term day ahead spot market, if such market exists, or in a bilateral way with some other market participants.[12] Clearly, these transactions are no longer interpretable as identical physical transactions as defined by the initial contract between parties i and j since the physical injections generally take place at a different bus k than at buses i and j.

- Real time spot market transactions are based on providing bids into a coordinated power exchange/ISO[13] day or week ahead and a coordinator determining which plants are used and at which price.[14]

Without loss of generality, all three types of transactions can be viewed as transactions between some two entities i and j and can be denoted by specifying $\hat{DG}_{ij}$, the power obligated to the entity j by an entity i.[15]

This conglomerate of trading power directly among various parties as well as into a coordinating spot market forms a *primary electricity market*. As most of the readers may be aware, the debate concerning which types of these transactions should be allowed has been quite divided between the proponents of entirely bilateral (self-scheduled as well as tradeable) transactions [28] and the proponents of solely coordinated power scheduling by an ISO [30].

Contract specifications. While in today's industry the supply functions of specific power plants are known to the system operator executing unit commitment/economic dispatch functions, how much is made known and to whom as power is traded between different entities depends on the market mechanisms in place.[16] Generally, participants in bilateral contracts are not required to make their supply functions public. Their objective is to optimize own expected profit/benefit by finding right parties to trade power with, and by directly negotiating terms of trade only with those parties.

The contract specifications are *potentially* different when establishing bilateral contracts than when bidding to sell power into short-term spot markets. Typical bilateral bids will specify quantity $\hat{P}_{gi}/\hat{P}_{dj}$ and price $\hat{p}$ at which power will be traded independent of real-time electricity market price. Bids into a spot market are characterizable similarly as the basic supply functions and are not necessarily identical with the cost functions, particularly when significant market power exists.[17]

The basic role of a short-term spot market. The ultimate role of an ISO remains to ensure that total generation meets demand. An ISO does not own generation itself and can no longer be held responsible for supplying total demand most efficiently. Also, if the price of electricity is very high and some consumers cannot pay for it, an ISO cannot be held responsible for supplying such load unconditionally. Generally, there is no guarantee (i.e., it is nobody's obligation) that the demand will be completely met. There is a definite possibility that $\hat{P}_{g,sys} = \hat{P}_{d,sys} - \sum_j \sum_i \hat{DG}_{ij}$ is nonzero in real time (recall that $\hat{DG}_{ij}$ stands for a bilateral transaction defining load supplied at j by a generator located at bus i). A real-time spot market can be viewed as a possible mechanism for providing this additional generation and it is assumed to be necessary for balancing generation and predicted demand on a day ahead basis.

In Appendix 2.2, we review a theoretical background and the basic computer algorithms for decision making when selling and purchasing power in the new industry. Market participants have to decide if they wish to sell and purchase power at all for the predicted market conditions, and, once they are committed, decide in real time how much to use. These two decision making processes serve the purpose of *partial* system generation/demand balancing in real-time. Some portion of predicted system demand $\hat{P}_{d,sys}$ will be served through a day ahead spot market, for example. Decisions to enter long-term bilateral contracts or to purchase/supply power through a real-time spot market will depend to a large extent on a type of market participants; some market participants are natural candidates for long-term agreements (such as nuclear power plants and inflexible consumers), some units are more naturally suited to follow the demand needs and respond to the real-time price opportunities. In what follows we assume existence of both types of market participants. Such a mix of contractual events is shown in Figure 2.7. One could refer to the decentralized

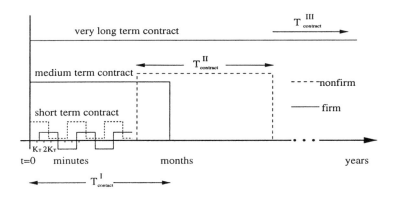

Figure 2.7 Contractual processes of interest

decision making by market participants to sell and purchase power as a decentralized unit commitment/economic dispatch process. In Appendix 2.2 the basic aspects of this process are reviewed.

Spot market clearing process. Power on the spot market is traded on hourly basis. Before the start of each hour generators enter bids specifying the quantity of power they are offering and the price they are demanding. A generator may divide the power he intends to sell into many smaller bids, so that he can effectively offer a bid curve which reflects his marginal cost curve, if he so desires. The predicted spot market demand $\hat{P}_{g,sys}$ is generally estimated by the spot market coordinator (PX or ISO). Once a spot market demand has been determined, the generator bids are stacked starting with the lowest price. The point at which the stack of bids intersects the cumulative load specifies the clearing price, defined as the price demanded by the most expensive bid

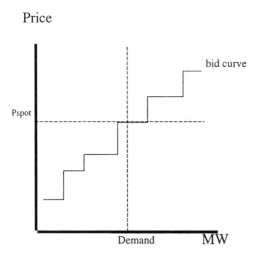

Figure 2.8 Spot market clearing curve

accepted. This price will be awarded to all accepted bids. The process of determining the clearing price is illustrated in Figure 2.8.[18]

Possible reasons for decentralized unit commitment

The conclusions drawn based on the analyses in Appendices 2.1 and 2.2 that there is effectively no difference in market outcomes when performing coordinated or decentralized unit commitment must be scrutinized very carefully. Two major causes of potential differences are,

1. Complex supply functions, and

2. Stochastic nature of the problem.

There has been quite a bit written recently on this topic [34, 35] and it is beyond the scope of this chapter to reproduce all the arguments made for decentralized unit commitment. The arguments are based on the claims that two critical unit commitment-related issues, the presence of nonconvex cost functions and uncertain conditions, are more naturally managed by the market participants themselves than by some coordinating body. These two problems are briefly discussed next.

Nonconvex optimization problems. This problem is basically created by the cost curves not being concave and is a well known obstacle when designing unit commitment software for applying to the real life power plants. While in most idealized analysis we assume these curves $C_i(P_{gi})$ to be smooth continuous quadratic functions, these are quite complex (discontinuous and nonconcave) for some critical technologies, including hydro- and CCGT-plants [13]. This can contribute to the nonexistence of a solution to the problem, or its nonuniqueness.

Dealing with market uncertainties. The theoretical equivalence between the simplest coordinated and decentralized unit commitment/economic dispatch summarized in Appendix 2.2 is often used as *the argument* that a competitive and coordinated markets would lead to the same economic outcomes. However, if the predicted price $\hat{p}$ made by the individual decision makers is considerably different from the actual spot market price p, a solution to the decentralized unit commitment will generally result in considerably different profit than the expected.

A decentralized decision making process for trading power both bilaterally and into a spot market is generally accompanied by various financial instruments designed to either enable the market participants to take short term opportunities and/or to minimize the risks related to the unavoidable uncertainties. There exist two qualitatively different types of decision makers: (1) risk takers and (2) risk-averse players. Depending on the type of approach taken, the tools for dealing with market uncertainties are different. As soon as the demand (even its predicted component) is assumed not to be known ahead of time, risk averse suppliers would engage in longer-term, firm bilateral contracts which guarantee that their power would be sold; more risk-prone suppliers will respond to the real-time spot market and make a higher profit for the same amount of power sold.

Of particular interest in this chapter is how is the fact that an electric power system is never at an equilibrium accounted for in various industry structures and what are the possible effects of specific market structures on the quality of technical performance, price of electricity and the total cost of supply meeting demand (system efficiency). As in today's industry, these objectives are often *not the same* and careful consideration of the economic and technical interplay must be carried out prior to deciding on the "best" industry structure of a given physical system. It is in this context that the industry structure begins to effect market outcomes.

Risk takers (market speculation). A successful risk taker must engage in a multistage decision making to deal with uncertainties presented to him by the market price fluctuations, as well as by the need to hedge with other generators for ensuring reliable delivery according to the contractual agreements.

A basic tool for successful risk taking when selling and buying electricity is effectively a decentralized stochastic unit commitment. To illustrate potential impact of uncertainties in decentralized unit commitment decision making on the outcome, consider a simplified situation with a single generator owner which sells electricity with hour-long contracts at the spot price of the market day ahead. This problem in its most general form is a *stochastic* dynamic programming problem with only one random variable $\hat{p}$ and one random output $\hat{P}_{gi}$ [36]. The owner must make unit commitment decisions *before* knowing the spot price of the next day. To start with, the decision making day ahead is the problem of finding optimal set of unit commitment decisions $\underline{u}_i = [u_{i1} \; u_{i2} \cdots u_{i,24}]^T$ by a power supplier i so that,

$$\max_{\underline{u}_i}\{\Sigma_{k=1}^{24} \hat{p}_k \hat{P}_{gi}[k] - u_i C_i(\hat{P}_{gi}[k])\} \tag{2.1}$$

the total expected profit over the next 24 hours is maximized for the expected market price vector $[\hat{p}_1 \; \hat{p}_2 \cdots \hat{p}_{24}]^T$. Here $\hat{p}_k$ denotes the expected spot market price each hour k over the next 24 hours. In a deterministic decentralized unit commitment one assumes that prices $\hat{p}_k$ at various stages during the 24 hour horizon are *not correlated*.

In a stochastic decentralized unit commitment, a decision maker engages in a relatively complex dynamic programming (DP)-based multistage decision making. The computation of the optimal decision that takes into consideration correlation between prices and different stages is effectively made according to the general formula (2.44) in Appendix 2.1: Starting at the last time stage, the optimal cost and optimal decision is calculated for every possible state. Next, the optimal cost to go and decision are calculated for every state in the next to last stage, using the optimal cost information from the last stage. This process continues until the first (current) time stage is reached.

Because the state includes price, the number of state variables is at least two (that is, plant status and price). The price model assumed is either a random walk or a mean-reverting process [58, 37]. A stochastic decision making is the basis for profitability in competitive industries, and serious work, primarily by marketers, is under way to apply this approach to selling and buying power [39].

Much of profits made by the marketers are based on a probabilistic decision making [38]. The difference between predicted and actual prices gives tremendous opportunity to build up actual profits, but it also involves risks if it is not done with sufficient knowledge.[19]

Risk-averse decision makers (hedging). Many bilateral contracts are generally equipped with various financial instruments typical of competitive markets for dealing with market price and other uncertainties, such as forward markets. The most often considered mechanism for doing this in the electricity markets is the so-called *contract for differences (CFD)* in which a firm bilateral transaction is defined by the *preagreed upon* power sold $\hat{P}_{gi}$, power purchased $\hat{P}_{dj}$ and the price $\hat{p}$. The CFD[20] basically states that the *actual (not expected) profit* to be made by a supplier of P_{gi} will be,

$$\begin{aligned} \pi_i &= P_{gi}p - (p - \hat{p})\hat{P}_{gi} \\ &= p(\hat{P}_{gi} - Pgi) + \hat{P}_{gi}\hat{p} \end{aligned} \tag{2.2}$$

and the benefit of a consumer,

$$\begin{aligned} b_j (\equiv -\pi_j) &= -P_{dj}p + (p - \hat{p})\hat{P}_{dj} \\ &= -p(P_{dj} - \hat{P}_{dj}) - \hat{p}\hat{P}_{dj} \end{aligned} \tag{2.3}$$

In real time, generator sells P_{gi} to the spot market at the market price p (totalling revenue pP_{gi}), load purchases P_{dj} on the spot market at the market price p (totalling negative revenue pP_{dj}); the CFD covers $(p-\hat{p})\hat{P}_{gi}$ and $(p-\hat{p})\hat{P}_{dj}$. The rest of power $(\hat{P}_{gi} - P_{gi})$ and $(\hat{P}_{dj} - P_{dj})$ is provided bilaterally by the market participants themselves.

In the simplest case, when $\hat{P}_{gi} = \hat{P}_{dj}$, formulae above can be rewritten as

$$\pi_i = \hat{p}\hat{P}_{gi} + p(P_{dj} - \hat{P}_{dj})$$
$$b_j = -\hat{p}\hat{P}_{dj} - p(P_{dj} - \hat{P}_{dj}) \qquad (2.4)$$

It follows from equation (2.4) that the uncertainty imposed by the market (both quantity traded P_{gi} and price p being different from the power quantity and price defined by the bilateral contract $\hat{P}_{gi}$ and $\hat{p}$) induces only *second order effects* when entering CFDs because the second term (the only uncertain term) is a product of price and *difference* between the actual and predicted power quantities. This is one well known way of minimizing risk in competitive markets.

Uniqueness of the electricity markets. In summarizing the above, we conclude that in today's industry minimal O&M cost is achieved through the *simultaneous* scheduling of all available resources. In the new industry, however, systemwide resource optimization is attempted by each resource optimizing its own profit/benefit and *adjusting* itself over time to what others are doing (reflected in the electricity price). In what follows we show that under relatively mild conditions the two processes should converge over time to the same result. The market rules will definitely effect the *rates* at which a decentralized physical scheduling approaches minimal total cost defined by the coordinated unit commitment and economic dispatch. Consequently, the electricity price will be nonuniform; at any chosen time, one will observe locational price differences caused by the delays in decentralized markets.

This issue of market delays is not as essential in many other competitive industries, whose supply/demand imbalance is handled through inventories. The electricity market is in many ways a *real-time market*, possibly the only one of this sort. The electricity must be provided in real-time rather than stored. This implies that different financial mechanisms currently used in other industries will have considerably different dynamics over time. For example, it is often assumed that most of the financial options are rarely exercised. This premise is based on the assumption that the expected value of an option is always higher than its exercised value, see Figure 2.9 [40]. This means that a commodity is rarely physically traded and that most of the financial instruments do not have a physical counterpart.

It is very difficult to apply the same assumptions to the electricity market. Instead, *most* of the financial transactions are exercised, after trading for profit/benefit reasons. As an example, a tradeable bilateral transaction is partly exercised by the parties entering the bilateral contract, and partly by (physically) purchasing/selling power into a spot-market. Balancing power

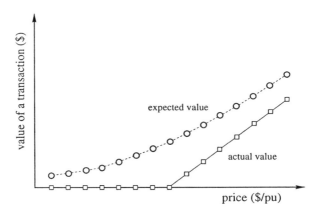

Figure 2.9 Value of an option

even in a bilateral tradeable transaction means physical delivery by someone. These general issues concerning influence of the financial mechanisms in place on the market performance of the electricity markets have only been studied to a very limited extent so far [41, 42, 44, 45]. In the next section we offer simple simulations illustrating this dependence at its basic level.

Note: One possible ambiguity between financial and physical deliveries arises in contracts for energy, rather than contracts for capacity; as long as market participants deliver on the average a predefined capacity, the contract is met. This may have some implications on frequency regulation (to be discussed in a later section of this chapter concerned with AGC for the new industry –task 4).

Intertemporal aspects of performing task 1

While in this chapter we are not primarily concerned with the financial decision making, it is not possible to completely separate the physical and financial processes from each other. In what follows, we analyze four basic industry structures:

1. Mandatory day ahead power exchange, without bilateral transactions;

2. Optional power exchange, with both bilateral and day ahead spot-market transactions being strictly physical (that is, *nontradeable*);

3. Optional power exchange, with the option to trade contractual responsibilities with other market participants.

4. Optional power exchange, with bilateral and forward day ahead and longer-term spot markets.

It was argued in the Appendix 2.2 that when *assuming known demand and neglecting start-up, must-run time, shut-down costs of power plants*, both entirely coordinated and the decentralized unit commitment/economic dispatch would result in an identical total cost. The conclusion would then be fairly straightforward: *Any electricity market structure* would have the same outcome, optimizing total cost would be equivalent to price minimization and no issues would be left.

In this section, however, we suggest that prices seen by customers, despite the same total operating cost, will depend on a type of accounting in place. In particular, because of the unique features of the electricity markets described earlier, prices will depend if contracts are forward markets (ex ante charges that do not change as real time prices vary), or if the contracts are based on ex post, after the fact, accounting of what actually has taken place in real time.

Another critical complication comes from how is fixed cost accounted for. In a competitive industry, there is generally no guaranteed capital cost recovery and prices are based on expected market conditions. As described earlier, in today's industry the accounting is after the fact proportional to the power consumed at a flat rate that is based on the guaranteed cost recovery. Here we only analyze the impact of ex ante versus ex post pricing on the electricity prices as seen by the consumers. An entire chapter in this text by Graves et al is devoted to the issue of fixed cost recovery by means of one-part or two part tariffs.

We illustrate next possible differences in economic outcomes when having mandatory poolco and optional poolco with physical bilateral transactions only. These differences are easily interpreted in terms of ex ante and ex post pricing effects.

Ex post and ex ante electricity pricing. The real-time price nonuniformity is created by two different pricing mechanisms depending on a type of contract: Long-term firm bilateral contracts are mainly based on a so-called *ex ante* electricity pricing; the price at which electricity is sold over the contract duration is *preagreed upon* at the beginning of the contract between the parties involved and it does not change as the real time spot price varies.

The spot market participants are usually compensated *ex post*, that is *after the fact*, say at the end of each day in a day ahead spot market setup.

Primarily because bilateral contracts are paid *ahead of time*, and spot market participants at the *end of each day*, at any instant of time electricity prices at different locations are different. The main reason for nonuniform electricity prices at different locations can be attributed to the fact that the primary market is not at its theoretical (economic) equilibrium.[21] While generation and demand balance, the prices at which this is done are not optimized according to the economic dispatch.

Bilateral transactions are established at different times and priced ex ante at prices $\hat{p}$ generally different than actual real-time spot prices p. Long-term bilateral contracts are not likely to reflect the actual market price, simply because the real-time long-term price cannot be known with high certainty. In addition to the price fluctuations caused by deviations in demand from predicted, the actual market price of electricity will depend on the activities of other bilateral

38   POWER SYSTEMS RESTRUCTURING

market participants. The closer $\hat{P}_{g,sys}$ to zero, the lower real-time spot market price, however this data is not easily predictable.[22]

The most successful market participants will be the ones with tools for *projecting* the so-called *residual demand*, that is the difference between the predicted system demand $\hat{P}_{d,sys}$ and the generation provided by all other market participants.

An example of ex ante and ex post pricing. To illustrate possible differences, consider a simple electric power system consisting of two generators and a single load shown in Figure 2.10.

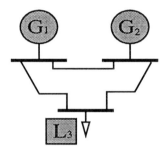

Figure 2.10   A three bus example

We illustrate how for a variety of scenaria customer is likely to see a different price of electricity.[23]

Ex post poolco set prices

- **Case (a) Inelactic demand**

    Shown in Figure 2.13 is the optimal generation scheduling which would correspond to a mandatory poolco: *All* generation is required to bid into the pool, and scheduling is done to minimize the total generation cost every 15 minutes over the four hour time horizon of interest in order to meet the predicted demand $\hat{P}_{sys} = \hat{P}_{L3}$ shown in Figure 2.11. In Figure 2.12 the corresponding price of electricity, the same at each bus, is shown.

- **Case (b) Elastic demand**

    When the demand is modeled as price elastic with its utility function represented as $U(P_{L3}) = 214.166 P_{L3} - 10 P_{L3}^2$ for deviations in demand of ±5% around the nominal load trajectory shown in Figure 2.11 and ∞ outside this range, the optimal demand obtained using generalized economic dispatch (see Appendix 2.2) is shown in Figure 2.14. The optimal generation outputs (assuming $G1$ to be a slack bus) are shown in Figure 2.15. Market spot price in this case is shown in Figure 2.16. It can be seen from these figures that the spot price is lower when demand responds to the price of electricity than in the case when a fixed demand must be met independent of the electricity price.

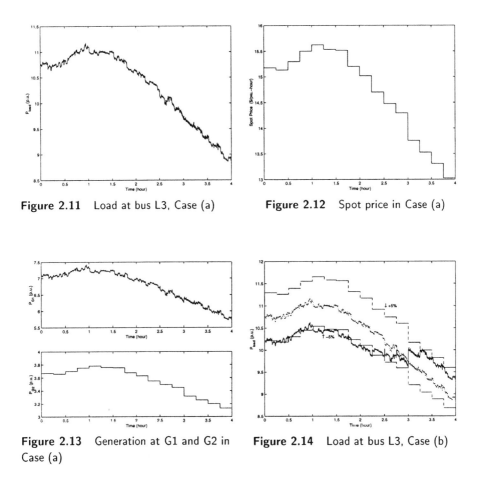

Figure 2.11 Load at bus L3, Case (a)

Figure 2.12 Spot price in Case (a)

Figure 2.13 Generation at G1 and G2 in Case (a)

Figure 2.14 Load at bus L3, Case (b)

Case (c) Mix of 4 hour bilateral (ex ante) and spot market (ex post) pricing

Next, the same demand is assumed to be supplied through a long-term bilateral contract at the price of electricity $10\$/p.u. - hour$ for supplying 5 p.u. from $G1$ to $L3$. The rest of power is scheduled using conventional economic dispatch every 15 minutes for meeting the remaining load. A comparison of the total electricity bill over the 4 hours as seen by the customer assuming inelastic demand is shown in Figure 2.17. Similar comparison for the case of elastic load is shown in Figure 2.18.

Effect of generation constraints. The same scenario is repeated next assuming that maximum generation $P_{G1}^{max} = 6.5 p.u.$ and $P_{G2}^{max} = 6.5 p.u.$. It can be seen, as expected, that the less expensive generation at $G1$ is used to its full capacity during the first 2.6 hours when the demand is at its peak, and that more expensive generation by $G2$ is produced to meet the remaining demand.

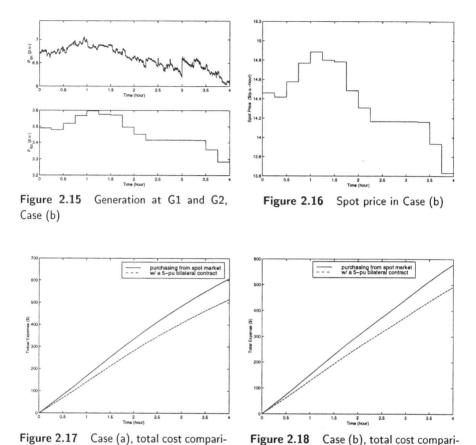

Figure 2.15 Generation at G1 and G2, Case (b)

Figure 2.16 Spot price in Case (b)

Figure 2.17 Case (a), total cost comparison with and without a bilateral transaction

Figure 2.18 Case (b), total cost comparison with and without a bilateral transaction

Shown in Figure 2.19 is the generation used in this case, and in Figure 2.20 the spot price.

The overall spot price is considerably higher than in an otherwise identical case without generation constraints. Optimal generation assuming elastic load is shown in Figure 2.21 and the corresponding spot price in Figure 2.22. As expected, the highest electricity price is when the demand is inelastic and when the capacity of inexpensive generation is constrained; this is followed by the case when demand is inelastic and the generation constraints are accounted for.

This example shows that, in addition to the popular debate centered around analyzing electricity market equilibria characteristics as a function of industry structure, the *intertemporal effects* (technical and financial) must be carefully taken into consideration since a power system is never in a stationary operation. Even when there are no obvious stability/synchronization problems, the system trajectory constantly evolves as illustrated early on in this chapter. While it is difficult to separate the two issues (delays in market activities and the

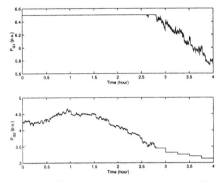

Figure 2.19 Generation of G1 and G2 with 6.5 p.u. generation constraints

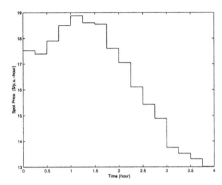

Figure 2.20 Spot price changes with generation constraints

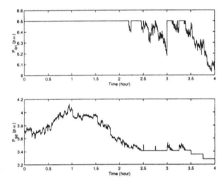

Figure 2.21 Generation of G1 and G2 with 6.5 p.u. generation constraints (elastic load case)

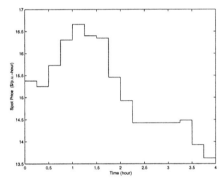

Figure 2.22 Spot price changes with generation constraints (elastic load case)

intertemporal physical dependencies), we stress the sensitivity of the results (technical performance, price and total cost) on the industry structure chosen.

The question of an electricity market equilibrium. It may be appropriate to address the issue of "market clearing" in the context of the examples shown here. What is that one means by this concept in various cases illustrated here? Cases shown here are mutually different.

One way to address this question is to recall basic models used for representing price dynamics in competitive markets [46, 43] which are based on the premise that the rate of change of market price is a function of market demand excess. Using this definition, it is clear that in cases of interest here the system is away from its equilibrium due to the continuous presence of the real-time demand excess (nonzero $\hat{P}_{g,sys}$). Moreover, according to this definition, the system is *never at its equilibrium*. Demand excess varies continuously at different rates and generation generally lags demand changes, therefore demand excess is always present.

To get around these issues, one could define an optimum operating trajectory as the *moving equilibria* associated with the preagreed upon longer time interval T for assumed known demand pattern and generation obtained using deterministic unit commitment/economic dispatch.[24] It is a straightforward optimization exercise to show that assuming smooth concave cost curves, one can never do better in terms of total cost over the entire period T using decentralized optimization techniques. Cumulative deviations from this trajectory would, therefore, be a measure of suboptimality caused by market structures. This difference over time $T \to \infty$ will generally be smaller if adjustable longer term financial mechanisms, like tradeable bilateral transactions or longer-term forward spot markets, are part of the industry structure in place.

We observe that in the industry structures not equipped with tradeable bilateral transaction mechanisms or longer-term spot forward markets, there would always exist a cumulative difference between the performance of an idealized coordinated market and decentralized markets of this type. It is essential to allow for *self-adjusting financial mechanisms*, such as tradeable bilateral transactions, in order to induce convergence of a decentralized market performance to its theoretical equilibrium. Simply put, the long-term bilateral transactions will not actively consider adjusting to the market conditions, if they are not given an opportunity to do this. We introduce next theoretical background for understanding *closed-loop market dynamics* assuming no restrictions on the market, that is an option to trade.[25]

Optimal primary market: tradeable transactions

Each generator in the network is given a choice of how much of his production capacity to allocate for sale to which of the available markets.[26] In order to make an informed decision, a generator needs a means of evaluating expected profits in each scenario. It also needs to account for the risk factor associated with reserving his capacity for sale on a short-term market. In this section we derive the expressions for the expected profits on each of the markets, and show the conditions that must be met for the overall process to lead to market clearing. Once we have modeled how generators enter the market, we can proceed to show how physical disturbances translate into deviations in market price.

Assumptions concerning behavior of power producers. The following assumptions are made in modeling the behavior of power producers:

1. Generators are entirely financial decision makers. They will not act in the interest of system security or performance unless they are given clear financial incentives to do so.

2. Generators have no market power. They are price takers in all markets (no gaming).

POWER SYSTEMS OPERATION: OLD VS. NEW 43

3. Demand is inelastic to fast price changes on the spot market.

4. All power producers have smooth quadratic cost curves, and consequently linearly increasing marginal costs.[27]

For simplicity, the cost is modeled as

$$C_i(P_{gi}) = aP_{gi}^2 + b \qquad (2.5)$$

$$\frac{dC_i}{dP_{gi}} = 2aP_{gi} \qquad (2.6)$$

$$(2.7)$$

Profits under long-term contracts. When a generator enters into a long-term contract, it is obliged to sell power at a rate given by the contract curve and at a prespecified price. This eliminates any risk of not being able to sell its power, but also removes the ability to take advantage of short term price peaks on the spot market. Since generators are assumed to be price takers, the profit associated with selling on the long-term market is easy to calculate. Faced with a long-term projected price p_{LT}, each generator i will set his total power output P_{gi} so that his marginal cost is equal to p_{LT}.

$$p_{LT} = 2aP_{gi} \qquad (2.8)$$

Using this constraint, the total profits of the generator in terms of long-term price is,

$$\pi_{LT} = p_{LT}^2 / 4a \qquad (2.9)$$

Since the long term price is known in advance, there is no risk involved in this contract, and we can view (2.9) as a *guaranteed profit*.

Profits under short term contracts. Consider next the case of the producer that decides to reserve all its production capacity for sale on the spot market. Again we assume that the generator is a price taker, who at each hour $[k]$ will see a new spot market price $(p_{ST}[k])$. At this price the market will absorb any amount of power $P_{gi,ST}[k]$ a supplier can generate. As in the long-term case, the producer will maximize his profits by setting his power output $P_{gi,ST}[k]$ to a level such that the marginal cost of generation is equal to the current spot market price. For each discrete spot market interval the profit of the power producer will then be given by,

$$\pi_{ST}[k] = p_{ST}^2[k]/4a \qquad (2.10)$$

If the spot market price was pre-determined, the producer could simply sum up the projected profits over each discrete interval $[k]$, compare this to the profit on the long term market (2.9), and thus decide where to place production capacity.

In reality, however, the spot market involves a great deal of uncertainty. Countries that have undergone deregulation have experienced a considerable increase in price volatility on the short-term market [7]. In order to allocate production capacity between the long term and spot market, the producer needs to generate an estimate of its expected profit on each market. To achieve this, we will model the spot market price as a random variable $\hat{p}_{ST}[k]$, with expected value $U_S[k]$, and variance $\sigma_S^2[k]$. Using equation (2.10) we can now express the expected profit in terms of the characteristics of this random variable,

$$\pi_{ST}[k] = E[\hat{p}_{ST}^2]/4a = U_S[k]^2/4a + \sigma_S^2/4a \qquad (2.11)$$

This expression tells us a great deal about the effect of price volatility on the spot market; if we compare equation (2.11) to our expression for profits on the long-term market (2.9), we find that they have a similar form. Indeed if we set the expected price level $U_S[k]$ on the spot market equal to the actual long term market price, we find that the expressions for profits on the markets are identical with the exception of the term $\sigma_S^2/4a$ on the spot market. This factor is a direct result of the *profit being a nonlinear function of price*, in this case a simple quadratic function. A marginal increase in spot market price will, therefore, create a large increase in overall profits, while an equivalent decrease in price will cause a smaller decrease in profits. As a result, an increase in the price volatility (i.e. larger σ_S^2) will result in greater expected profits on the spot market, as predicted by (2.11).

In an industry where participants choose between investing only on the long-term market or only on the spot market, the market equilibrium will be reached when expected profits are equal on both markets. Since price variance is always positive, this can only occur if the expected value of the spot market price is below the actual long-term price. This price differential can be expressed directly as a function of spot market *price volatility*,

$$\hat{p}_{ST}^2 = U_S^2 + \sigma_S^2 \qquad (2.12)$$

The result predicted in (2.12) seems counter-intuitive. It is important to realize that the above model does not take into consideration that most generators are likely to be risk adverse. If we would include this behavior in our modeling we would have to add a negative risk correction term to the right hand side of equation (2.12). As it stands, the model simply reflects the effect of passing an uncertain price signal through a nonlinear system.

Mixed strategy solutions: Tradeable contracts. The above analysis will allow us to find equilibrium prices under the condition that each generator uses a pure strategy of selling only on the long-term market or only on the spot market. In reality there is nothing to prevent a producer from dividing his output between the two markets. In order to specify the long and short term supply curves, we first have to determine under which conditions is it profitable for a producer to be selling on both markets. Consider the following example. A producer with marginal cost $MC = 2aP_{gi}$ sees a long-term price p_{LT}. It therefore commits a capacity of $P_{gi,LT} = p_{LT}/2a$ to the long-term market, setting long term price equal to its long term marginal cost. During the course of the long-term contracts, the producer notices that the short-term price increases above the level of his current marginal cost. It can now increase its profits by selling power on the spot market until the marginal cost of production is equal to the spot market price. The same is true for all generators who sell power on the long-term markets.

The spot market supply curve. We will now proceed to derive the supply curve for the spot market of a simple generic system. Assume our system contains a total of N generators with identical cost curves.[28] Each generator has a marginal cost curve of $MC = 2aP_g$. Further assume that a subset of M generators decides to reserve all their capacity for sale on the spot market. The remaining (N-M) generators will sell power according to the mixed strategy described above. We derive the shape of the supply curve by considering two separate situations,

1. For $\hat{p}_{ST}[k] < p_{LT}$, the spot market will be supplied only by the subset of M generators. Under these conditions the supply curve is given by,

$$\hat{p}_{ST} = 2aP_g/M \qquad (2.13)$$

2. For $\hat{p}_{ST}[k] > p_{LT}$, all generators in the system will supply power to the spot market. This will reduce the slope of the supply curve by the factor of N/(N-M). Combining this change of slope with the curve described in (2.13), the total supply curve for the spot market takes on the form,

$$\begin{aligned}\hat{p}_{ST}[k] &= 2aP_g/M & \text{for } \hat{p}_{ST}[k] < p_{LT} \\ \hat{p}_{ST}[k] &= (1 - M/N)p_{LT} + 2aP_g/N & \text{for } \hat{p}_{ST}[k] > p_{LT}\end{aligned} \qquad (2.14)$$

The resulting shape of the supply curve is depicted in Figure 2.23. It can be seen from this equation that we are dealing with a *two-slope supply curve*. The break point coincides with the price level where generators committed to long-term contract begin to enter the spot market.

Effect of supply curve on price volatility. The reason we have gone through such length in deriving the structure of the supply curve is that it is this curve that represents the *link between the physical and economic processes* modeled above. In the short term, the system is driven by physical disturbances in the form of generator/demand mismatches. Such disturbances translate directly into spot market demand. The shape of the supply curve tells us how this demand will cause movements in spot market price. In effect, *the supply curve is a transfer relation between the physical and economic disturbances on the system.*

Let us use a simple example to illustrate the effects of the supply curve on price volatility and generators profit levels. Assume the supply curve is of the form described in equation (2.14), and the physical disturbance ($\hat{P}_{d,sys}$) is a

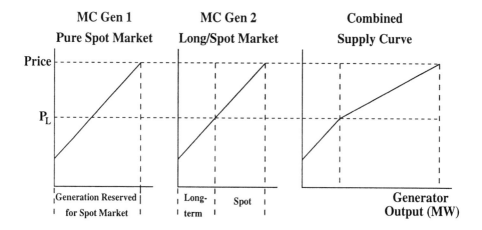

Figure 2.23 Two-part spot market supply curve

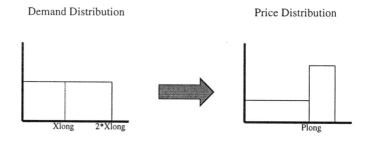

Figure 2.24 Distributed spot price

random variable evenly distributed between zero and $(M/a)p_{LT}$. Figure 2.24 shows how the resulting distribution of spot market price is weighted by the supply curve.[29] Notice how the distribution of the disturbance was selected so that it fell symmetrically around the breaking point of the supply curve. When we examine the resulting distribution of the spot market price we find that it no longer displays this symmetry. While spot price is equally likely to be above or below the long-term price level p_{LT}, the deviation range is significantly smaller for price levels above p_{LT}. On a system-wide level, this means that price is more volatile in the lower price ranges, and that one is less likely to experience extreme price peaks. It follows that by not restricting balancing generation to a few selected generators, one could avoid having high price volatility in the upper price ranges, avoiding unreasonable peaks in spot market price that could be extremely destructive for the end consumer.

If we examine these results from the perspective of the individual generator, we find that the spot price curve has a significant impact on how the producer allocates resources between the long and short-term markets. In modeling the profits on the spot market, we found that due to a *nonlinear relationship* between spot market price $\hat{p}_{ST}[k]$ and producer profits $\hat{\pi}_{ST}[k]$, the expected profits actually increased as the price becomes more volatile, see (2.11).

The shape of the supply curve however is telling us that even if the demand to the spot market is extremely volatile, this will not necessarily translate into high price peaks. The incorporation of long term contracts for generators into the spot market therefore has a distinctly negative effect on the projected profits of the purely short term producers. Reduced profits on the spot market will cause generators to re-evaluate their allocation decisions, causing more producers to enter into long term contracts. This will change the shape of the spot market supply curve by moving the breaking point to the left, increasing the slope of the first segment.

A self adjusting market clearing process. Since the price level of the new supply curve is higher than the old for any given demand, the generators which chose to remain with the pure spot market strategy will see an increase in their profits. Market participants will continue to reevaluate their strategies until an equilibrium is reached where expected profit levels for both strategies are equal.

This equilibrium will shift as new generators enter the market, or as the characteristics of the load changes. The process by which the market responds to such changing conditions is as follows:

1. The power producers decide how to allocate their generation capacity between the spot market and long term market. Their decisions are based on a known long-term price p_{LT} and an estimate of how the spot market price $\hat{p}_{ST}[k]$ is going to behave. This in turn determines the shape of the spot market supply curve.

2. Fast fluctuations in load causes power imbalances on the system which translate into demand for short term balancing power. This power must be purchased on the spot market.

3. The change in demand for power on the spot market translates into a fluctuation in the spot market price $\hat{p}_{ST}[k]$ away from the projected value. The magnitude of the price change, given a deviation in demand, will depend on the shape of the spot market supply curve.

4. Increased volatility in spot market prices will increase the profit incurred by generators investing their capacity in this market. If the volatility remains high during the course of the long-term contract period, more

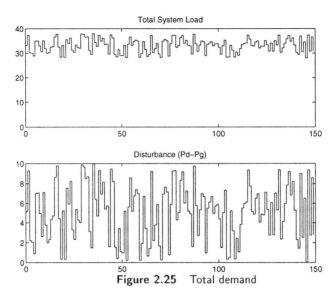

Figure 2.25 Total demand

generators will enter the spot market during the next period, and we are back at step 1.

Thus we have closed the loop and shown how the system can *adjust itself* until it reaches a stable equilibrium.

Simulations of a market clearing process. We now demonstrate the *adaptive behavior of market* participants by simulating a small power system. The system consists of five identical generators, and is driven by the stochastic system demand $\hat{P}_{d,sys}$.

Generator Characteristics Each generator has a total production capacity of 10 units of power. The cost curve is given by $C = P_g^2 + 4P_g + 2$, yielding a marginal cost curve of $MC = 2P_g + 4$.

Load Characteristics The load consists of a fixed base portion of 28 units, and a stochastic portion with probability density function evenly distributed between zero and ten for each discrete step. The base portion is supplied by long term contracts which are renewed every 50 time steps. The stochastic portion represents hard to predict load variations for which the system operator must purchase balancing power on the spot market. The total load and its stochastic portion are shown in Figure 2.25.

Analysis of generator behavior for each time sequence.

- **Sequence A: Time 0-50** We have chosen an arbitrary starting point from which the market can evolve. Generators one through four have committed generation to the long term market, each supplying a load of seven units. The long term market price, given by the marginal cost of

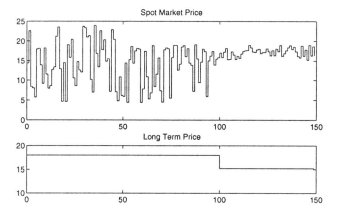

Figure 2.26 Spot market price dynamics

generation, is 18. The fifth generator has chosen to participate only in the spot market.

- **Sequence B: Time 50-100** Long term contracts are allocated as in A. Generators one through four now decide to offer their excess generation for sale on the spot market when spot market price exceeds long term prices. Generator five behaves as in A.

- **Sequence C: Time 100-150** Faced with diminishing profits, generator five decides to enter the long term market. This decreases the long term demand seen by each of the other generators. Each now supplies a load of 5.6 units, driving the long term market price down to 15.2. The spot market is now supplied solely through excess capacity not used to fulfill long term obligations.

Analysis of the simulation results. If we examine the price plots from the simulated system, we find several interesting trends. First if we look at the plot of spot market price shown in Figure 2.26, we find that the market *adapts itself to reduce price volatility.* As we move from sequence A to B, we remove the price peaks on the spot market. This is because we have moved from a steep single slope supply curve to a two slope supply curve. The long-term price is also shown in this figure. As we enter sequence C, the downward volatility of the spot market price disappears. This is a result of all generators participating in the long term market. There will, therefore, be no one willing to supply the spot market when price is below long term price levels. In addition to reducing price volatility this trend will, of course, raise the average spot market price. This effect however is overpowered by the simultaneous decrease in the long term price, so that we see a drop in the overall per unit price of power. The trend of the market is toward a decrease in average price and a reduction of price volatility, as shown in Figure 2.27. If we examine the overall cost of

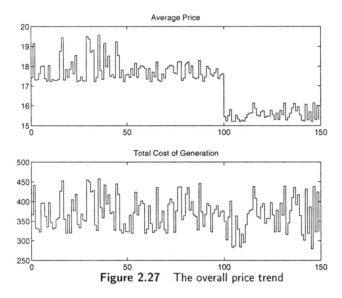

Figure 2.27 The overall price trend

production, we find a similar trend. The total cost of production for sequence A is 1,873. It drops to 1,848 for sequence B, and falls further to 1,777 for sequence C.

In summary, these results of our simulation exemplify what has long been preached by the economists. A truly competitive market will optimize the use of its resources so as to drive down market price. The challenge in shaping the future of the American power market will be to remove any barriers to competition so that liquid electricity markets characterized by this self-adjusting process are created.

OPERATION TASKS 2-5

The emphasis of the remainder of this chapter is on open questions and solutions related to tasks 2-5, particularly on,

1. Possible new paradigms for engineering these services;

2. Possible charging mechanisms; and

3. The impact of specific engineering and financial solutions on the technical and market performance.

We deal with these general questions as we review objectives of tasks 2-5. We start this in the next section by revisiting task 2.

TRANSMISSION LOSS COMPENSATION (TASK 2)

The implementation of various transactions, as of any other physical power injections, leads to an undesired loss of real and reactive power inside the transmission system equipment. The real power loss is typically between 2% and 5% of total generation injected into the grid. The effect of the transmission loss on total cost of generation needed to supply given demand is easily computed, as reviewed in the Appendix 2.3.

Reactive power loss, although much larger than real power loss, is not compensated using the same criteria in each utility; power companies use shunt capacitors, under load tap changing transformers and other reactive power means in addition to generators to compensate for the reactive power loss. Because the variable (operating) cost of a generator is not explicitly dependent on its reactive power output, the operating cost of reactive power compensation is not as straightforward to estimate as it is the case with the real power loss [67, 74]. The cost of other (non generation-based) compensation means is primarily related to maintenance and fixed cost.

Consequences of not compensating for transmission loss. If the real power loss is not compensated in real time, system frequency will deviate from its nominal value. In very large systems, such as the U.S. interconnection, it is not extremely critical to compensate for real power loss with high accuracy in order for the operation to be technically acceptable; the sensitivity of system frequency with respect to a 2% power imbalance is relatively insignificant [108].[30] Still, in today's industry the real power transmission loss is made up in real-time operation as frequently as generation is scheduled and with as high accuracy as the loss is being estimated. This is done mainly for economic reasons when scheduling generation to meet the predicted demand.

On the other hand, a threshold within which the reactive power is balanced in real time has not been viewed to be as critical for operation nor for efficiency as it is viewed to be for the real power balance, because the only consequence of insufficient reactive power loss compensation in real-time operation is seen in bus voltage deviations around their nominal values. As long as they remain within the acceptable operating limits, there is no immediate operating problem. The effects of inadequate reactive power loss compensation may be more pronounced in a long run since large voltage deviations may result in faster deterioration of the system equipment.

It follows from this that the transmission loss compensation is an electricity efficiency issue more so than a technical issue. To develop possible ways of efficient loss compensation in the new industry, we should consider abandoning the idea that it is the ISO who is solely responsible for transmission loss compensation. Instead, we should begin to consider transmission loss service as a market-based service. In the remainder of this section, we review several concepts for developing such markets.

52 POWER SYSTEMS RESTRUCTURING

Real power transmission loss compensation in the new industry

As we review approaches proposed up to date by the ISO's in forming, it quickly becomes clear that the proposed solutions are simplified estimates of transmission charges. On the other hand, if one were to consider highly theoretical estimates of very accurate loss compensation, this may require costly and complex real-time information (metering) and computer software. Thus, it is important to assess all possible options by relating accuracy of the specific method and the underlying cost of metering, data processing and real-time computing. In what follows an initial assessment in this direction is presented.

Industry proposals. To avoid complexity, several ISO's in forming have proposed formulae for system use reflecting estimated pro rata load share. These proposals are in response to FERC's suggested pro forma tariffs which in the case of transmission loss simply say that for each kWhour of system use, the transmission loss compensation cost would be certain fixed dollar amount [78]. The implementation of even the simplest pro rata load share formula requires measurements of the real power consumed by loads. This would require new metering installations, since at present only the energy consumed is measured.

The load pro rata share-like formulae are generally less accurate than even the approximate B coefficient-based formulae[31] since they do not reflect locational nor temporal aspects of transmission system use. The approximate B coefficient-based formulae evolved as popular means of estimating total transmission loss prior to times when large scale computers were used in the power industry.

Some other ISO's in forming that plan to use nodal pricing based on the original theory of spot pricing [61], such as the NY ISO, will effectively base charges for transmission loss compensation on computations identical to formula (2.58) in Appendix 2.3 also known in today's industry. However, at present there are several obstacles to transmission loss unbundling according to formula (2.58). In addition to the computational complexity of running optimal power flows for a system consisting of thousands of buses, the load is not being measured in sufficient detail to apply such computations according to the formula in real time. In NY, for example, the load is estimated based on historical data and it does not correspond to the real-time demand. In order to know the real-time demand and apply this method of charging, one must install measuring equipment at the retail level.[32]

Given this overall status of the proposed charging mechanisms for transmission loss compensation, which range from being overly simplified through theoretically idealized, but difficult to implement, it is obvious that we must analyze the tradeoff of these techniques and also investigate conceptually new approaches.

Possible market-based approaches to transmission loss compensation.
In the new industry, losses can be compensated in many different ways which generally depend on the structure of the primary electricity market in place.

In the case of a mandatory poolco, it is fairly obvious that an ISO would be in charge of this function. Since the price is a *bundled* price reflecting the price of supplying and delivering power, the unbundling of charges into different components is not done.

In some more flexible electricity markets that allow for self-scheduling bilateral transactions, one could consider transmission loss compensation by many entities as the transactions are implemented. Moreover, since, as it was pointed out above, the transmission loss compensation is not necessarily critical for basic technical performance of a power system, the type of transmission loss compensation and charging mechanism chosen will effect mainly the overall efficiency of the primary market and prices of electricity at different locations throughout the system.

In the new industry one could consider at least three qualitatively different ways of compensating for the transmission loss created as different transactions are implemented,

1. Each market participant produces locally power necessary to compensate for the transmission loss created by its own transaction.

2. A market participant pays an additional charge for the real power loss compensation to some other market participant who compensates his loss in a bilateral market-based way (not through an ISO).

3. Finally, the most discussed option is to have an ISO procure additional generation and compensate for total transmission loss based on minimizing the cost of loss compensation, and charge the users responsible for creating losses accordingly.

While the third approach is usually implied in the present industry restructuring debate, the first and second approach lend themselves more naturally to the competitive electricity market; transmission loss created by a specific market participant is compensated (physically and financially) in a decentralized way by system users and this eliminates the need for the ISO's coordinated loss compensation and charging.

Transmission loss and market inefficiency. As transmission opens to the system users, it is necessary to rethink the physical and financial mechanisms of transmission loss compensation. Viewed by the market participants, transmission loss is effectively an externality to the primary electricity market concerned with supplying power to the predicted demand (task 1 above) that adds new burden to the electricity price and leads to deviations in market prices from the ideal market price defined in Appendix 2.2.

Possible objectives of meaningful provision and pricing for transmission loss would be to either (1) induce minimal deviations from the ideal market efficiency, (2) minimize the impact on the price of electricity, or (3) take into consideration impact of losses in a compatible way to all.[33] We conjecture here that the three criteria are likely to lead to a similar result only when bus injections are close to those defined assuming perfect market conditions.

It is critical to observe that computing transmission loss in different primary market structures will result in completely different total loss and it will be compensated in the real time operation by different market participants. Consider, for example, the following three scenaria:

- Actual quantities traded determined by the market (like in California), ISO only runs load flow and checks for their feasibility;

- ISO schedules generation to meet specified (predicted) demand using economic dispatch that accounts for transmission loss (like in NY and NE ISO's); and,

- ISO schedules both generation and demand using generalized economic dispatch with the objective function defined in the Appendix 2.2 (like in Alberta).

In the first scenario, the load flow computations will show the amount of real power generated by the distributed AGC units (!) playing the role of a slack bus and compensating total transmission loss. In the second scenario generators will produce suboptimal outputs relative to what would be determined by the perfect market conditions, and the load will pay for it. In the last scenario, both generation and demand are adjusted to compensate for losses in an optimal way as measured by the formula (2.59) in Appendix 2.2.

To begin to put these differences in context, we first recognize that charges for performing tasks 2-5 generally lead to deviations of electricity prices from what is determined during the primary market process. We have concluded in Appendix 2.2 that under perfect market conditions at the (economic) market equilibrium the price of electricity is the same everywhere.

Usage-based loss compensation and market efficiency. The question of providing and charging for transmission losses can be discussed either assuming perfect market conditions, or recognizing from the beginning that a power system will never be in such a stationary operating mode. Conclusions concerning the most effective implementation of tasks 2-5 strongly depend on which of these two assumptions are made.

Because of this, one must proceed carefully with interpreting implications of using particular formulae for transmission loss charges on market deviations from its theoretical optimality defined in equation (2.48) of Appendix 2.2. It directly follows from comparing (2.48) and (2.58) that the theoretical impact of transmission losses on the electricity prices as seen by market participants is reflected in the ratio $1/(1 - \frac{dP_{loss}}{dP_{gi}})$. We observe here that this ratio changes as market inputs (bus injection vector P_i) vary; only at the economic equilibrium of the primary market this ratio represents the measure of locational inefficiencies caused by transmission loss compensation. Therefore, it is theoretically meaningful to apply usage-based transmission loss allocation formulae (such as in [68]) as a means of reflecting market efficiency deviations caused by transmission loss only at the economic dispatch-determined operating point. At any other operating point at which market clears (not necessarily through an economic dispatch) line flows-based transmission loss allocation is not an adequate measure of inefficiencies created by the transmission losses. This theoretical point is critical to take into consideration as one attempts to argue that it is meaningful to apply usage-based formulae for system services, transmission loss in particular. Many other (non usage-based) methods are known in competitive economics for valuing effects of market externalities on the primary market [72].

Expressed in a jargon more familiar to an engineer, it is theoretically justifiable to charge for transmission losses according to use only when system injections are optimized, using an OPF-like method. It is not theoretically possible to say that if the estimated loss at bus i is larger for one set of injections than for some other set of power injections, this is necessarily a less efficient operating point as measured in terms of total cost minimization. Consequently, usage (line flows) based formulae penalizing market participants for creating transmission losses can only be directly justified when the market is cleared at its economic dispatch operating point.

Usage-based loss compensation and compatibility of charges. Since the primary electricity market constantly deviates from its "optimal" trajectory, the argument to charge usage-based for performing tasks 2-5 for efficiency quickly disappears.[34] The only argument left for some sort of usage based formulae is the compatibility of system services.

This again becomes a controversial issue concerning a *flow direction* with respect to which transmission loss is charged.[35] For example, it is well known that the total loss created by an arbitrary set of transactions depends on the *order* in which transactions are implemented.

To start with, losses are caused by the resistance of transmission lines and are mainly dissipated through heat which is given by $I^2 R$ where I is the current being transferred through the line and R is the line resistance. Depending on the reference direction, I can be either positive or negative. However, $I^2 R$ is always positive. It is possible (and often true) that

$$P_{loss}^{total} \leq P_{loss}^{trans1} + P_{loss}^{trans2} + \cdots + P_{loss}^{transN} \qquad (2.15)$$

because

$$I^{total} = I^{trans1} + \cdots + I^{transN} \qquad (2.16)$$

where some of I^{transi}'s ≤ 0 and some I^{transi}'s ≥ 0 but not because $P_{loss}^{total} = P_{loss}^{trans1} + P_{loss}^{trans2} + \cdots + P_{loss}^{transN}$ where some of loss components $P_{loss}^{transi} \leq 0$.

We use an example of a 5-bus network [75] to illustrate potentially drastic differences when computing transmission loss under what may appear to be minor assumptions. The matrix of bilateral contracts at the loads is given by,

$$DG = \begin{bmatrix} 65.7 & 14.1 & 203.7 & 203.8 & 280.4 \\ 115.1 & 155.8 & 249.3 & 10.4 & 16.1 \\ 158.9 & 201.3 & 2.3 & 115.0 & 20.0 \\ 125.3 & 206.0 & 176.7 & 279.1 & 253.9 \\ 158.1 & 27.6 & 196.2 & 124.8 & 210.4 \end{bmatrix} \qquad (2.17)$$

All bus power injections are result of these bilateral specifications only. Each element DG_{ij} in this matrix represents the power bought by demand at bus i from the generator at bus j. As explained systematically in the next chapter, when transmission loss is not taken into consideration, both the matrix of power delivered by generators GD and DG are the same matrices. However, to account for loss compensation, one must specify how the transmission loss is compensated. It can be shown that if the total loss is allocated according to the relative *magnitude* of the contract, a version of pro rata load share formula, one would charge relative charges according to the following,

$$L = \begin{bmatrix} 0.0 & 0.4 & 5.1 & 5.1 & 7.1 \\ 2.9 & 0.0 & 6.3 & 0.3 & 0.4 \\ 4.0 & 5.1 & 0.0 & 2.9 & 0.5 \\ 3.2 & 5.2 & 4.4 & 0.0 & 6.4 \\ 4.0 & 0.7 & 4.9 & 3.1 & 0.0 \end{bmatrix} \quad (2.18)$$

If the total loss is computed according to some sort of usage-based formula, the total loss computed as the sum of losses created by each individual transaction is 206.96 MW (assuming equality in formula (2.15) above) with losses computed using distribution factor formula for flow calculation, whereas calculated with all transactions present the total loss estimated is 47.50MW, and when using exact AC load flow for computing flows the total loss is 71.84 MW.

Observe that, as expected, computing and charging transactions as though they are the only inputs into the system, one at a time, generally *overestimates* the transmission losses created by all transactions applied simultaneously. The distribution factor-based formulae (including MW-mile method) are likely to *underestimate* total loss.

A reader interested in methods for bilateral markets presented in Chapter 3 will find it interesting that the same exact total loss is obtained using the new method. The solution of the matrix differential equation introduced in Chapter 3 yields in this case,

$$GD = \begin{bmatrix} 65.7 & 15.3 & 236.0 & 203.7 & 298.8 \\ 106.4 & 155.8 & 266.0 & 9.6 & 15.8 \\ 138.3 & 189.0 & 2.3 & 100.1 & 18.5 \\ 125.3 & 223.4 & 204.8 & 279.1 & 270.6 \\ 148.6 & 28.1 & 212.9 & 117.3 & 210.4 \end{bmatrix} \quad (2.19)$$

from which the loss allocation matrix is

$$GD - DG = \begin{bmatrix} 0.0 & 1.2 & 32.3 & -0.1 & 18.4 \\ -8.6 & 0.0 & 16.7 & -0.8 & -0.3 \\ -20.6 & -12.3 & 0.0 & -14.9 & -1.5 \\ 0.0 & 17.3 & 28.1 & 0.0 & 16.7 \\ -9.5 & 0.5 & 16.7 & -7.5 & 0.0 \end{bmatrix} \quad (2.20)$$

We note that some contracts are allocated a positive loss while others are allocated negative loss. These loss allocation amounts depend on the size and location of the contract relative to every other contract. Negative allocation implies that particular contract in the final analysis has contributed to the reduction of system losses because of the net counter flows.

However, one has to be careful with "negative" cost charges resulting in this new method. For example, consider a line ij with the current $I^{trans1} = 0.5001$ p.u. flowing from i to j. Superimposed to this flow is the current $I^{trans2} = 0.5000$ p.u. flowing form j to i. Ignoring the nonlinearity of the problem for the moment, the total loss on line ij caused by total current flow on the line is given by $P_{loss}^{total} = (I^{trans1} + I^{trans2})^2 R = (0.0001)^2 R$. $P_{loss}^{trans1} = (I^{trans1})^2 R = (0.5001)^2 R$ and $P_{loss}^{trans2} = (I^{trans2})^2 R = (0.5000)^2 R$. Using the method described in Chapter 3, the loss can be assigned to each transaction by $P_{loss}^{total} \approx P_{loss}^{trans1} - P_{loss}^{tarns2}$ thus charging transaction 1 and paying transaction 2 accordingly for their losses. Assuming the computation is done through a numerical method given .5% error margin, the operator cannot confidently charge transaction 1 and pay transaction 2 because in reality it could have gone the other way.

Zonal pricing. When analyzing the effect of input changes at several locations throughout the system one must ask the question of how far does the change at the particular bus propagate through the system and how does this change interact with the effects of other changes. Depending on the electrical distance between buses of interest, in some cases the effects will be fairly independent and in some cases they must be analyzed simultaneously. This should be kept in mind when attempting to suggest so-called zonal electricity pricing [82, 83]; some transactions jointly contribute to the losses and should contribute on a pro rata share basis to compensate for their joint effect, and others are separable. As simplifications, such as zonal charging, are considered instead of full locational unbundling one could use the notion of electrical distance as a possible approach to deciding which buses should belong to the same zone for electricity pricing.

It is very straightforward, at least in concept, to apply a version of clustering algorithm [84] to the basic distribution factors formula given in Appendix 2.3 to determine which line flows are affected the most by which system inputs. To start with, matrix D defined in equation (2.67) of Appendix 2.3 is generally a full matrix; if one neglects terms in matrix D smaller in absolute value than some predefined small ϵ, and re-enumerates rows and columns to obtain a block diagonal matrix, the question of prime interest is readily answered. Buses and lines belonging to the same block are strongly coupled and should be treated as a single zone (they are "electrically close"). The number of zones is determined by the number of blocks in the D matrix.

This approach provides basis for *reduced measurement structures and simplified usage-based charges* for transmission losses. Only total injection into the zone is relevant for determining the transmission loss created by the system users whose buses belong to this zone. Further allocation of charge among the buses within the zone is, most likely, best done on a simple pro rata load share basis.[36] This type of analysis could be useful for aggregating the system for purposes of having well defined groups of load serving entities and generation serving entities.

Most of the ideas for zonal pricing [82, 83], as a way of simplifying locational nodal pricing method, are based on grouping nodes according to ranges of nodal prices $\hat{p}_i$ for $i \in Z_k$; buses whose nodal prices are similar are considered to belong to the same zone Z_k.

A possible issue with such definition is that it is not only the nodal price $\hat{p}_i$, but also the power quantity $\hat{P}_i$ injected into the bus i of interest that determines relative contribution of bus i to the total cost in each zone. In other words, it makes sense to define an *equivalent zonal price* $\hat{p}_{Z_k}$ that satisfies

$$\hat{p}_{Z_k} \sum_{i \in Z_k} \hat{P}_i = \sum_{i \in Z_k} \hat{p}_i \hat{P}_i \qquad (2.21)$$

or,

$$\hat{p}_{Z_k} = \frac{\sum_{i \in Z_k} \hat{p}_i \hat{P}_i}{\sum_{i \in Z_k} \hat{P}_i} \qquad (2.22)$$

It follows from this formula that a zonal price is a *weighted sum* of locational nodal prices of the buses belonging to the zone.

In summary, one could introduce more than one notion of a zone depending on what the objective for its use is. If the objective is to measure market inefficiencies, formula (2.22) is more appropriate than equivalencing based on electrical distances and using distribution factors matrix D. On the other hand, if usage-based transmission pricing is adopted, defining zones based on electrical properties of the grid would be meaningful. Another technical detail to keep in mind is that when zones are defined based on electrical distances they overlap implying that zones cannot be treated in a fully separate way from each other [85, 86]. The implications of this fact when defining zones for pricing should be carefully considered.

Preliminary conclusions concerning loss compensation in the new industry. We conclude this section concerning transmission loss provision by observing the following:

- It is very difficult and complex to estimate transmission loss in real time and charge according to nodal prices defined in Appendix 2.4. Carrying out this complexity may not be justifiable in light of the fact that the primary market is not at its (economic) optimum to start with, and that charging nodal prices away from the optimum *has no meaning*.

- It is worthwhile investigating, instead, systematic ways of defining electrically distant zones and letting each aggregator/zone representative provide for the estimated transmission loss impact locally using algorithms as the one outlined in Appendix 2.4.

- Requiring loss compensation in kind (or paying to the zone representative who provides for the entire zone) on a pro rata load basis is fully justifiable since the electrical effects of all customers in a zone are strongly coupled.

POWER SYSTEMS OPERATION: OLD VS. NEW   59

- Treating each zone as though it has no effect on other zones is also justified by their very design based on a large electrical distance.[37] This would avoid unobservable processes of transmission loss compensation by AGC units by default when nothing else is done.

- Loss can generally be compensated either by other market participants who would be paid for this service, or by the market participants themselves. While the details of how this is done best in the new industry will be evolving over time, one must recognize the fact that a bilateral market participant ought to be free to compensate for its loss in kind, and not necessarily be obliged to pay an ISO for providing this for him. In Chapter 3 of this text, a specific loss allocation formula is derived that assumes distributed transmission loss compensation as well.

- The loss unbundling questions are relatively new since in today's industry bundled service is provided at the average price, and the discussion provided here suggests caution as one proceeds with evaluating any proposed method for loss compensation (physical and financial).

OPERATING WITHIN TECHNICAL CONSTRAINTS (TASK 3)

It is well known that an unconstrained primary market solution may not be implementable because it may lead to violations of line flow constraints or bus voltages limits (generally referred to as the security constraints) to unacceptable levels. This will have impacts on the profits/benefits of market participants.

The objectives of a transactions management system are often viewed differently by a system provider than by the system users.

Objectives of a system provider

A system provider (an ISO) in charge of maintaining system integrity as market-driven bus power injections vary is primarily concerned with maintaining secure system conditions in its area and the coordination of its real-time operation with the neighboring systems so that the operation of a large interconnection comprising several ISOs also remains secure.[38] When some of the desired transactions are not implementable in real time, the ISO will need standby tools for systematic adjustment of proposed transactions to ensure that security constraints are met.

The role of an ISO in facilitating primary market efficiency is less clearly defined. To a large extent, it depends on the primary market structure in place and on explicit incentives given to the ISO to develop computer software for assessing and posting the most likely and most critical system limitations.[39]

Objectives of a system user

In contrast to the ISOs, a system user (or a group of system users requesting simultaneous access), sees system security requirement as yet another uncertainty in its basic decision making concerning supply/demand trades in the primary electricity market. Depending on how the transactions are managed, some market participants may either not be given permission to use the system at certain times or they may be charged for effecting security margins.

The main objectives of system users are (1) to make sure that they are served equitably and (2) to minimize uncertainties imposed by the system constraints. These uncertainties can be reduced either by a system provider actively posting most likely technical constraints (including probability of their occurrence) [11] and/or by purchasing financial insurance.

Objectives of transmission service pricing

In the current restructuring debate much confusion has arisen concerning transmission service pricing. This issue must be pursued very carefully with the clearly defined objectives of a system provider under open access. For example, if the main objective of an ISO is to ensure that the operation is secure while the market efficiency will be determined by the primary market processes, then the only justifiable transmission charge is the so-called access fee necessary to recover the fixed cost of the transmission equipment. Charging mechanisms for this are fairly well understood [70]. In this case an ISO needs tools for systematic adjustment of transactions based solely on engineering criteria.[40]

A more difficult scenario is the one in which an ISO is expected to facilitate market efficiency in addition to making sure that the system remains intact. An ISO could do this in several ways, the most obvious being public posting of system conditions and of the projected critical constraints. However, since an ISO is not necessarily in charge of generation scheduling and, therefore, is not likely to have full access to economic data, the ISO must develop tools for "learning" the market trends based on the past and the ongoing real-time market activities.

Methods for projecting system bottlenecks based on market activities are potentially useful for all types of primary market structures. However, in a poolco type primary market structure in which PX and ISO are the same entity, market inefficiencies created by transmission system constraints are reflected in the nonuniform locational nodal prices $\hat{p}_i$. As a consequence of these unequal nodal prices, a so-called merchandise surplus [87] is created which is the difference between what consumers pay and what the suppliers are paid. This raises the issue of congestion pricing. Opinions concerning merchandise surplus allocation are diverse, such as having (1) proposals to use these funds as insurance against system uncertainties for those system users who hold unconditional transmission rights to either use the system or be compensated for not using it [29] or (2) proposals to use this money toward enhancing the system in the future to eliminate the most likely major obstacles to facilitating unconstrained market transactions [31].

Notice that in entirely bilateral markets it is not necessary to charge for market inefficiencies, as long as adjustments of transactions are well defined and compatible for all. However, a mix of bilateral and spot markets creates difficulties in applying methods based solely on nodal transmission pricing; since

POWER SYSTEMS OPERATION: OLD VS. NEW   61

spot market transactions are scheduled routinely for efficiency, while bilateral transactions are adjusted only as the last resort when secure operation cannot be achieved without curtailing some of them, the two types of transactions are not managed in a compatible way. To avoid this inconsistency, it is necessary to "unbundle" the primary market decision process from the transactions management for security.[41] This leaves an ISO with the only option of requiring the transactions to adjust according to technical criteria. Transactions adjustments could be done either by an ISO, or in response to a charge for effecting security margins.

The three methods described below illustrate some possible variations for managing transactions to ensure secure operation; method (1) is applicable for transmission pricing in a mandatory poolco [29], method (2) requires adjustments of power quantities facilitated by the system, without any charges related to system constraints [90], and method (3) introduces a real-time transmission price feedback to the system users which helps the users adjust their requests as the security margins vary [17, 18, 19].

Assessment and posting of available transfer capability (ATC)

The need for systematic assessment of available transfers is well recognized and efforts are under way, primarily by NERC, to integrate seemingly unrelated industry efforts (such as interchange scheduling, distribution factor calculations, transaction information systems, system security information, transmission reservation and scheduling and, most recently, OASIS) into a systematic Transactions Management System (TMS) of the future. These activities have intensified over the past two years in response to the FERC Orders 888 and 889 that mandate open access transmission in the United States [78].

Various notions of available transfer capability. One of the obstacles is the inconsistency of methods used for evaluating ATC in various parts of the United States. In the Appendix 2.5 of this chapter several notions of the available transfer capability (ATC) are presented as a function of the primary market structure [9]. For example, one could speak of a *bilateral transfer capability*, a *bus transfer capability*, and others. This distinction was reflected at one point of the restructuring debate in the United States when the pros and cons for so-called *network-* and *point-to-point* service were analyzed. Nevertheless, most of the efforts for computing this quantity are based on an implicit assumption that all transactions are scheduled simultaneously by an ISO [79]. The multidimensional nature of the ATC has been well recognized in several references [8, 9, 81] and it is important to keep it in mind as "the ATC" of a particular transmission serving entity is defined, computed and posted. It is, therefore, important that power transfer capability be considered in the context of its specific goal, or a set of goals and the primary market industry structures in place. These goals are not additive in developing a composite goal, since the uses are not all applied simultaneously.

A brief treatment of security and its relation to transfer capability in Appendices 2.5 and 2.6 of this chapter points out that the transfer capability is not only a function of thermal line flow limits. Instead, bus voltage constraints and system dynamics are also essential ingredients of a secure system operation.

The role of voltage constraints in assessing ATC. Despite the present tendency to compute ATC based on estimating real power transfer limitations independent of the voltage changes created, it is likely that major bottlenecks to power transfers over far electrical distances will be voltage-support related. This will eventually be recognized, and it will require development of systematic real-time voltage adjustment algorithms to facilitate market needs. While such algorithms exist (see, for example [101, 20]), they will not be actively used before an economic signal is in place to reflect the value of voltage support [103].

The role of system stability constraints in assessing ATC. Depending on the physical characteristics of a transmission system, some parts of the system are likely to be constrained by the possibility of transient and small-signal instabilities while attempting to facilitate market transactions. It is well-known that such limitations exist at present in various parts of the US transmission grid, for example the East-West intertie problem in NY, several bottlenecks in the South-North transfers, voltage instability problems in the western part of the US system, to mention just a few. These limitations could have equal effects on market outcomes as when transactions have to be denied because of thermal or (steady state) voltage security limits. To make issues even more involved, many of these truly dynamic constraints often depend on a type of primary control in place on both generation and transmission equipment. Typical controllers on power plants are governors, excitation systems and power system stabilizers. Typical direct-flow controllers on the transmission system are much talked about Flexible AC Transmission Systems (FACTS). Physical storage is likely to begin to play a major role in overcoming fast dynamic instabilities as system users adjust to the slower signals given to them by an ISO.

It is worthwhile observing that the present restructuring debate has underestimated the importance of these devices for serving the market needs. The ATC is typically computed without much systematic consideration for voltage and stability problems. This is understandable, since accurate methods for assessing this type of problems are not feasible in real time operation. This puts even more importance of developing methods for assessing most likely problems, by means of system equivalencing techniques [104]. It becomes critical to give incentives for sufficiently robust system control design to eliminate voltage and stability related constraints [105]. None of this is addressed by anyone actively. If anything, developments in this area are left open to possible abuses of some system users for their own gains on the market. The chapter by Gorski and DeMarco in this text addresses possible malicious design of governors to disable competitors from staying on the system. At least one report can be found concerning similar market issues when using FACTS technologies [95].

Compatibility of transmission service. When an individual system user (or a group of users) approaches a transmission system provider requesting certain transmission service, the ATC estimated by a transmission provider generally depends on the bus power injections resulting from implementing other transactions and also on the location and timing of the transaction requested.

A system provider being charged with a nondiscriminatory service provision to all (compatible service) could easily appear as making arbitrary decisions when adjusting nonfeasible transactions. It is for this reason mainly that ATC should be defined and used carefully in the context of the primary market structure in place. Computing ATC in a mandatory poolco is different than computing ATC to serve bilateral transactions in a sequential way as they request access to the grid.[42]

In active primary markets decisions to adjust quantities traded and, therefore, requests for their implementation are likely to evolve at different rates and in an asynchronous way (recall Figure 2.7 in this chapter), making the order and amounts by which particular transactions are adjusted an open question. Since the ATC depends on *the order* in which transactions are approved and a system provider has no control over when they get requested, it is necessary to design transactions management systems so that *no preference* is given to one over the other transaction. One way of doing this is to periodically provide all transactions with a charge for effecting security margins, or, equivalently, to adjust them according to their relative contribution to the reduction in security margins.[43]

Review of the proposed transactions management systems for secure operation

There have been several qualitatively different approaches proposed for incorporating ISO and user concerns into a transactions management system that schedules operation within technically acceptable operating constraints. These are,

1. Introduce locational based marginal costs (LBMC) [29, 30] (or, nodal prices [61]) to bundle the charges induced by system constraints in the primary electricity market price p of meeting task 1 (method 1).

    These nodal prices are computed in a real-time spot market setup by performing generalized constrained economic dispatch (or optimal power flow [61]).

2. Facilitate primary market requests, as they arrive to the ISO, by either fully approving them or initially curtailing them as the system constraint is violated and defining technical conditions under which further primary market trades can be done [90] (method 2).

3. Provide a real-time charge to each system user according to their relative contribution to reduced security margins [17, 18, 20]. This charge is

provided not only when a constraint is active, but continuously as the operating conditions change. This allows system users to optimize their effective profits/benefits (after paying for system security) by adjusting their power quantities traded at the primary electricity market.

While these three approaches may appear to be very similar, their impact on market outcomes is potentially quite different. They range from paying charge for violating a transmission constraint only when the constraint is active (method 1), through being charged continuously for reducing security margins (method 3) and not paying any charge for violating transmission constraints, but taking the risk of being curtailed (method 2).[44]

The choice of transactions management system must be done very carefully, however. Given perfect market conditions, one could argue that all three methods, if implemented carefully, should lead to the same outcomes. Below we illustrate that when the primary market is away from its equilibrium, market rules in place for transactions management could make real difference in the outcomes of individual system users (or, groups of users). In what follows, we use simple power systems examples to illustrate these three methods.

Poolco-based congestion pricing (method 1). The so-called nodal pricing approach is based on the theory of spot pricing [61].

In the case of line flow constraints-related security charges, the transmission flow constraints are explicitly considered when performing generalized economic dispatch that is,

$$|T_{km}| \leq T_{km}^{max} \quad (2.23)$$

where T_{km} is the real power flow across lines connecting any two buses k and m. Because of line flow limits, the solution to this problem leads to different nodal prices $\hat{p}_i$ at each bus. The basic proposal is then to charge each market participant i price $\hat{p}_i$, hence according to their location in the network. As a result, loads located in remote areas where the transfer capability is limited will pay comparatively higher rates.

As an example, consider a simple three-bus network shown in Figure 2.10. Assume that, based on the bids, an ISO obtains the following cost/benefit functions[45]: $C_1(P_{G1}) = P_{G1}^2 + P_{G1} + .5$, $C_2(P_{G2}) = 2P_{G2}^2 + .5P_{G2} + 1$ and $U_1(P_{L1}) = 214.167 P_{L1} - 10 P_{L1}^2$. To start with, without any transmission line flow constraints, the generalized economic dispatch (Appendix 2.2) results in optimal power produced $P_{G1} = 6.58$ p.u, $P_{G2} = 3.42$ p.u. and power consumed $P_{L1} = -10$p.u.

Next, consider the same supply functions subject to the line flow limit $T_{G1-L1}^{max} = 5$ p.u. A simple load flow calculation shows that the real power line flow would be $P_{G1-L1} = 5.5$ p.u. for the above bus injections. System λ defined in (2.48) is $\lambda = 14.17$. However, since this solution violates the line flow constraint, the ISO would perform a price-bid based constrained economic dispatch and calculate the nodal prices. The result of a transmission constrained economic dispatch [96] obtains $P_{G1} = 5.41$p.u, $P_{G2} = 4.16$ p.u. and $P_{L1} = -9.583$ p.u. The corresponding nodal prices are $p_{G1} = 11.84$,

$p_{G2} = 17.16$ and $p_{L1} = 22.49$. The nodal price differences reflect the impact of the active transmission line constraint relative to the unconstrained market price $\lambda = 14.17$. The proposal in [29] is to use this approach for ex post transactions charging, computed at the end of each day, for example.

Multilateral markets (method 2). A qualitatively different approach to managing system constraints is to *unbundle* primary electricity activities (task 1) from the system constraints-imposed adjustments. This was proposed by Wu et al as a way of creating coordinated multilateral markets [90]. The basic idea is for the ISO to initially adjust transactions so that the line flow is at its limit, and to provide market participants with a set of equations that will have to be observed as further trading evolves. The coordination is achieved by the participants observing the technical constraints defined by the ISO. Market participants are left alone to decide on their transactions, without having to provide any economic data to the ISO. The technical constraints use the distribution factors-based line flows allocation formulae (2.67) in Appendix 2.4 that define how much each bus injection contributes to a line flow. For every line whose limit is reached, one such constraint is provided. This approach is qualitatively different from the nodal pricing-based transactions management since there is *no monetary charge* related to system constraints.

To illustrate this method, consider the same three bus system as above. Since the line flow T_{G1-L1} exceeds the operating limit T^{max}_{G1-L1}, the ISO first adjusts the transactions requested by the market to ensure that $T_{G1-L1} = T^{max}_{G1-L1}$ and imposes the following equality constraint,

$$T_{G1-L1} \approx \frac{2}{3}P_{G1} + \frac{1}{3}P_{G2} \qquad (2.24)$$

This is a *strictly technical information* which $G1$ and $G2$ must observe as they pursue further trades. In this case, if P_{G2} is increased at twice the rate of decrease in P_{G1}, then the flow in the previously congested line is unchanged. Therefore, $G1$ can sell additional power to the load if $G1$ *simultaneously* buys twice as much power from $G2$. The initial adjustment is arbitrary. To show sensitivity of profits/benefits of the individual market participants to the initial adjustment, consider two different initial adjustments: (1) Reduce both P_{G1} and P_{G2} by the amount of the excess line flow (.53p.u), and (2) Curtail only P_{G2} since this unit has a higher marginal cost; for the line flow to remain at its limit, $P'_{G2} = 1.83$ p.u, while $P'_{G1} = P_{G1} = 6.58$ p.u. Consequently, $P_{L1} = 8.42$ p.u.

Of course, there exist many other combinations of bilateral trades that are completely equivalent and satisfy the line flow limit. For example, $G2$ can increase its sales to the load if $G1$ sells half of that amount to $G2$. Alternatively, $G1$ can increase these sales to the load if $G2$ decreases its sales to the load by twice as much.

This trilateral (generally multilateral [90]) trade can be viewed as three coordinated bilateral markets: one, for exchanges between $G1$ and $L1$, another between $G1$ and $G2$, and a third between $G2$ and $L3$ [97]. The submarkets are coordinated because the price in one market affects the demand or supply in the other.

Generators $G2$ and load $L1$ make their decision to trade in the same way as before the congestion is reached, i.e., they follow the rule of their own profit/benefit maximization as given by equation (2.45) in Appendix 2.2. $G2$ continues to sell power to $G1$ as long as the price is greater than his marginal cost. The load's demand for power in the $G1 - L1$ (sub)market is determined by the same strategy: continue to buy power until the price is equal to his marginal utility.

The strategy for $G1$ is a bit more complicated, however. The profit for $G1$ in the trilateral market is function of the prices p_{G1-L1} and p_{G1-G2} in the two post-congested markets and the quantity that is sold after congestion is affected by this fact [97].

At the new equilibrium $G1$ sells 5.42 units of power and $G2$ will produce 4.17 p.u. The load purchases 9.58 units of power. The bilateral prices are $p_{G1-L1} = 22.5$ and $p_{G1-G2} = 17.17$. It can be shown that this equilibrium is reached *independently* from what the initial adjustment is.

However, total profits/benefits of $G1$, $G2$ and $L1$ are generally affected by the initial adjustment since the actual amounts sold at pre- and post-congested price are different.

Because the individual profits are affected by the initial adjustment rule used by the ISO, this seemingly free market approach is prone to the so-called *quantity control* (this could be contrasted to the nodal pricing-based approach which is sometimes qualified as a *price control based method* since the MPs are required to provide their economic information.)

Notice, however, that if the primary market accounting is done only at the end (and not at both precongested and congested prices), there is no difference between individual profits when applying methods (1)[29] and (2)[90]. Technical feedback reflecting system constraint (method 2) and price feedback (method 1) result in identical quantities and prices.

However, in an industry structure where bilateral transactions take place inside the secure operating regions without any charges for reducing security margins, trading will be done in two stages, for quantities traded prior to congestion, and at the end when an equilibrium is reached subject to an active line constraint. In this case, as long as an ISO has a direct control over how much can be sold, the profits are affected by the initial adjustments.

In contrast with methods (1) and (2) which are inactive before the security limit is reached, method (3) [17] provides a transmission price feedback reflecting relative contribution of individual market participants to the reduced system security as the system constraints are being approached. This should give sufficient warning to system users to adjust their transactions. Consequently, the probability of violating security limits is reduced considerably. This method is briefly described next.

Soft-constraints based transmission charges (method 3). One of the basic questions in the new industry is if the price feedback can be used to induce system security sufficiently fast in the real-time operation. We have already discovered through simulations of the primary electricity market that the *rate* at which the price feedback is used will determine how fast the self-adjusting process will evolve.

The basic idea of a real-time transmission price feedback for ensuring secure system operation is based on using the *error signal*

$$e(t) = [(T_{line\ 1}^{nom} - T_{line\ 1})\ (T_{line\ 2}^{nom} - T_{line\ 2}) \cdots (T_{line\ b}^{nom} - T_{line\ b})]^T \quad (2.25)$$

and creating vector of transmission service charges

$$TS(t) = [TS_{G1}\ TS_{G2} \cdots TS_{Gng}\ TS_{L1}\ TS_{L2} \cdots TS_{Lnl}]^T \quad (2.26)$$

where each TS is defined as a price feedback proportional to the error vector $e(t)$,

$$TS(t) = -diag[e(t)]diag[\alpha]e(t) \quad (2.27)$$

This price feedback design is *usage-based* in the sense that transmission service charge to a market participant at bus i is proportional to the relative contribution of this injection to the line flow violations; the larger α, the faster will system users respond to the security conditions. However, if the price feedback is too large, oscillations may occur. The conceptual use of real-time price feedback to regulate technical processes is known to potentially have a destabilizing effect [98]. To avoid these problems, one has to engage into systematic feedback design for regulating system conditions in response to the *expected ranges of system disturbances* (deviations in the primary market dynamics are disturbances to transmission provision tasks 2-5).

Rather than informing the user that he had to be suddenly curtailed, the user is given in real time operation a price feedback for effecting security margins. This price, of course, unavoidably, depends on all other system users and all users are expected to adjust to the system conditions over time by either paying for effecting system security and/or adjusting power quantities traded to avoid this charge.

We illustrate next on a 5-bus example shown in Figure 2.28 the feasibility of this approach.[46] The concept is rather simple: Each system user should pay for violating transmission constraints. In Appendix 2.7 a specific derivation of the general formula (2.27) for this specific 5-bus system is provided. This formula is based on the distribution factors matrix D defined in the Appendix 2.4, equation (2.67) and the transmission line constraints. The generalized distribution factors for basic use of line flow allocations do not have to be updated frequently as they do not vary considerably with the operating conditions. Instead, only the information about the proximity of critical line flows to their limits will need to be provided to market participants in real-time. Based on this information, a system user will be able to estimate the price for affecting system security.

Figure 2.28 shows the 5-bus system used to illustrate the idea of soft constraints based price feedback to avoid transmission constraints in real time.

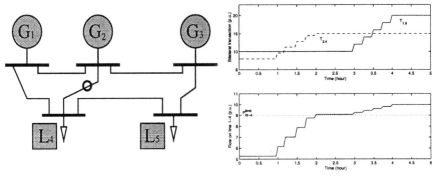

Figure 2.28 5 bus system

Figure 2.29 Transactions without charge no constraint

Assume there are only two bilateral contracts present on the system $T_{1,5}$, between $G1$ and $L5$, and $T_{2,4}$ between $G2$ and $L4$, respectively. To start with, the scheduled (nominal) values for $T_{1,5}$ and $T_{2,4}$ are 8 p.u. and 10 p.u, respectively. Over time, at the end of the first hour, at $t = 1$ hour, $T_{2,4}$ starts to deviate from its scheduled value in anticipation of an improved profit by $G2$ and benefit at $L4$. The quantity requested for approval by $T_{2,4}$ adjusts every 15 minutes and the maximum change is $\pm$ 20 % of its nominal value. Around $t = 2:15$ hour, $T_{2,4}$ settles to its optimal value $T_{2,4} = 15 p.u.$ At $t = 3$ hour, $T_{1,5}$ starts changing quantity traded and it finally settles to its optimal value $T_{1,5} = 20 p.u.$

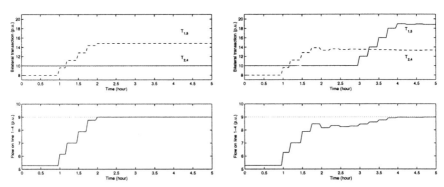

Figure 2.30 Case 1: obvious physical adjustment

Figure 2.31 Case 2: applying soft constraints

Accounting for transmission constraints. Now, assuming that the thermal limit of line 2-4 is 9 p.u, the line constraint will be obviously violated in absence of any price feedback reflecting proximity of the line flow to its limit. Shown in Figure 2.30 is the result of an obvious physical adjustment; transaction $T_{2,4}$ is curtailed to the value 9 p.u (case 1).

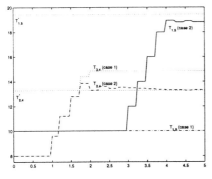

Figure 2.32 A comparison of cases 1 and 2

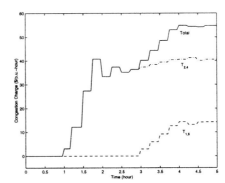

Figure 2.33 Congestion Charge

Shown in Figure 2.31 is an application of soft constraints-based price feedback described here. Each transaction is given a price signal based on its relative contribution to the line constraint. It can be seen that since the proximity to a line flow constraint depends on both transactions, $T_{2,4}$ will *self-adjust* as $T_{1,5}$ changes so they collectively result in an acceptable line flow.

In Figure 2.32 a comparison of cases 1 and 2 is summarized. This figure shows the result of case 2 actually approaching the optimal value $T^*_{1,5} = 19.43$ p.u. and $T^*_{2,4} = 3.28$ p.u when both transactions are scheduled simultaneously to minimize total cost subject to a given line constraint. This figure is an illustration of a *self-adjusting* process by primary market participants in response to the system constraints-based price feedback. The transmission charges for $T_{1,5}$ and $T_{2,4}$ are shown in Figure 2.33.

These simulations indicate the basic preaching of competitive economics: Given the right price feedback, the primary market will adjust in a decentralized way to the near theoretically optimal conditions *without* requiring coordinated primary market. System users are likely to learn the value of their transaction being implemented and compare this value with what they are charged for. If the difference between the two is big, this gives an incentive to the system user to adjust to the situation by seeking longer-term solutions (trade different power quantities, invest into system enhancements, purchase some sort of insurance against system uncertainties). *In theory*, assuming perfect market conditions, one will have exactly the same situation as when comparing performance of real-time spot markets and their competitive market equivalents, i.e., there will not be much difference between the two. Recall that the long-term market adjustments are in response to the observation of the real-time electricity prices. Following an identical thinking, one can argue that if an adequate real-time signal concerning system security status is given to the market participants, they will respond to it and converge to the market conditions in which constraints are directly coordinated with the primary market processes.

Method 3 recognized dependence of ATC on the activities of all system users, and, consequently, that the status of the transmission system as a function of market activities must be communicated and updated in real-time, and that the system users should adjust their primary market activities to account for their impact on system conditions. Nodal spot pricing, as well as multilateral trading subject to active system constraints gives the signal to the system user later than necessary for the user to adjust and avoid creating the constraint.

In summary, it is possible to manage transactions within system security constraints in more than one way. If market delays and intertemporal effects are neglected, all three methods lead to an identical market outcome. Further studies are needed to investigate market performance away from the perfect market conditions. Both theoretical and experimental studies are needed in this direction.

Dealing with system constraints-related market uncertainties

System constraints are basically an externality to the primary electricity markets. Their effect could be reduced considerably by the system providers posting projected system constraints and/or by developing financial instruments for hedging against these uncertainties. Very little has been done so far in developing tools by the system providers to inform market participants concerning probabilities of technical limitations. If these were made known, system users could account for the risks related to system provision when making market decisions. An initial effort in this direction was reported in [11]; however, rather than assuming that all market transactions are equally probable, as assumed in this work, it is necessary to develop *conditional probabilities* about the system constraints in terms of most likely market events.

Since a probabilistic information about system security status is not made readily available, market participants are considering possible means of hedging against system related uncertainties. One possible approach proposed by FERC in [92] suggests creation of primary and secondary markets for purchasing "transmission rights". Taking this path brings up many unresolved questions, see [15, 16]. More work is needed to come to some resolution concerning the credibility of this idea.

The second idea is to establish transmission congestion contracts (TCC) between market participants and a transmission system provider [29]. This idea was originally conceived as part of the nodal pricing for transmission congestion, in particular the controversy surrounding the question of who should be paid the so-called merchandise surplus (MS) resulting from unequal nodal prices [87, 88]. This surplus is defined as the difference between money paid by the demand and the money paid to the power suppliers. This surplus is often large as the constraints are violated.

This issue has been discussed quite actively. One possible solution would be to collect this surplus for future system enhancements to avoid congestion. A peak-load pricing for transmission was recently proposed in [31, 32] as one possible method for implementing this idea. The second approach to using merchandise surplus is by introducing financial instruments for hedging against system constraints-induced market uncertainties. This is briefly described next.

Transmission congestion contracts (TCCs). One possible approach to distributing the merchandising surplus suggested is to use it to compensate market participants who own *transmission congestion contracts (TCCs)* [29] to inject into the system a prespecified amount of power in a point-to-point fashion. If TCCs are not feasible in a real-time operation, the owners of these transmission rights would be compensated ex post for not being able to exercise their transaction in such a way that they are indifferent to the fact if they have produced power themselves, or have been compensated for not selling it. The TCCs should be viewed as a strictly financial instrument for dealing with transfer capability-related uncertainties, much in the same way as the CFDs (described earlier in this chapter) are used to deal with the primary market uncertainties. They are effectively CFDs between the system provider and the system users who own these contracts.

As an illustration of a TCC, consider the same three bus example and an existing TCC to transfer 7.5 p.u. from bus 1 to bus 3. This could represent a financial right of $G1$ to sell 7.5p.u. to load $L1$. Consider than the following actual dispatch of $P_{G1} = P_{L1} = 7.5$ p.u and $P_{G2} = 0$ p.u, which results in line flows $T_{13} = 5p.u. = T_{13}^{max}$, and $T_{23} = 2.5$ p.u. The corresponding nodal prices are $p_1 = 16.$, $p_2 = 0$ and $p_3 = 64..$ Since the congestion is not binding yet, the capacity right of 7.5 p.u. is implementable at the production cost of 7.5×16, equaling \$120. Transmission cost is \$0, congestion rent is \$0. Now, after computing the OPF for this case, we obtain nodal prices and quantities $P_{G1} = 5.76$p.u, $P_{G2} = 3.74$p.u and $P_{L1} = -9.5$ p.u, and $p_1 = 12.52$, $p_2 = 15.44$ and $p_3 = 24.17$. To derive the effect of transmission constraint, take the difference between unconstrained and constrained prices $t_1 = 13.5 - 12.53 = .97$, $t_3 = 24.17 - 24.17 = 0$, congestion payment to bus 1 is $(.97-0)5.76 = 5.59\$$, the congestion rent is $(.97 - 0)(7.5 - 5.76) = -1.736$, production cost equals $12.53 \times 5.76 = 72.23\$$, and the purchase at bus 3 costs $24.17 \times (7.5 - 5.7645) = 41.94$). The algebraic sum of the congestion payment, rent, production cost and the payment by the load amount to \$118.02, which is almost \$120.00 (within the numerical error).

The initialization of transmission congestion rights in electricity markets which are far from being liquid and the impact of this on the overall market inefficiency pose open questions. The unattractive features are the requirements to submit economic data into a centralized trading entity (such as PX), and, moreover, potential arbitrage caused by the inability to differentiate between an inefficiency caused by a system user gaming for profit or by an active technical constraint.

BALANCING SUPPLY AND DEMAND IN REAL TIME (TASK 4)

In addition to scheduling for predicted demand, it is necessary to balance actual supply/demand. As we have described in the introductory material of this chapter, in today's industry any real-time supply/demand mismatch at slower time scales is regulated by means of an automated closed-loop scheme known as the AGC [26].[47] In a competitive power industry one should be cautious when

72 POWER SYSTEMS RESTRUCTURING

defining objectives of AGC. An ISO, responsible for supply quality, will have to project the expected ranges of supply/demand mismatch for which regulating generation is needed. This is not so easy to do since real time mismatches are generally the result of (1) deviations of power contracted from power delivered, (2) transmission losses and (3) random demand fluctuations. While none of these are easy to project, conventional AGC is only intended to regulate system frequency in response to random fluctuations in system demand $P_{d,sys}$ from its predicted value $\hat{P}_{d,sys}$.

Some gaming in the primary electricity market is likely to take place, in particular if contracts are defined as contracts for energy and not for power. When the spot price is high, producers will deliver less power precommitted in long-term energy contracts and use the remaining capacity for short-term profits; they will do the opposite at times of low electricity price. They still meet their energy contract this way [106]. This strategy is likely to increase the real time mismatch created under (1). Moreover, if transmission loss is not estimated and compensated accurately in real time, this mismatch will also contribute to the overall burden on the new AGC (component 2).

In addition, a well known self-stabilizing effect of loads may change considerably with the increase of price responsive demand: power consumed will not necessarily decrease as frequency decreases. Magnitudes and rates of demand fluctuations around the demand specified in the primary electricity market will be nonuniform for various types of users. Arc furnaces, for example, are more likely to create large demand fluctuations, while many residential customers are not significant contributors to the real power supply/demand mismatch. This is why [99, 33] proposed a standard request for service to an ISO by a user located at bus i should define power $P_i(t)$ to be purchased/sold in the primary market as a function of time, and also an expected band of deviations $\pm \Delta P_{LT,i}(t)$. This is needed since a consumer cannot be expected to forecast his consumption exactly and since a generator will be unable to track fast load fluctuations in real time. A contract needs to specify a band around the contract curve inside which load and generation are allowed to deviate without being penalized, as shown in Figure 2.34. Each system user (supplier and/or consumer) should be required to register this type of contract with the ISO in charge of its control area(s). The ISO will, in turn, use this information to determine the relative charge for AGC for each system user. Basically, the tighter the bound, the smaller the AGC charge. This is a technically sound criterion, since the bound represents the maximum mismatch between load and generation. In case such a mismatch occurs, the ISO has to procure regulating generation. The AGC charges paid by the ISO to the units participating in AGC will be recovered by the AGC charges to system users. By setting the AGC fees in an individualized manner, the ISO gives generators an incentive to minimize the width of their contract bounds. Setting the bounds too narrow could, however, result in significant penalties if the load deviates outside the allowed margins. If AGC fees and penalties are set accordingly, the width of the bounds specified in the contract should provide an accurate measure of the actual volatility of the load. This information can in turn be used by the ISO

POWER SYSTEMS OPERATION: OLD VS. NEW 73

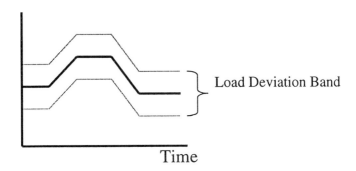

Figure 2.34 Contract specification

to estimate the maximum cumulative disturbance on his system. Using this estimate he can then decide how much regulating generation to procure for future periods.

With a contract specification of this type, it is possible to design a very simple market for frequency regulation in the new industry [99]. This is briefly summarized next.

Market for regulation

This market offers an alternative to the ISO for balancing supply and demand in real time. Bids for participating in frequency regulation are defined as an obligation to alter their power output $P_i(t)$ in response to system frequency deviation $\Delta F(t) = F_{ref} - F(t)$. For example a bid could take on a form,

$$P_i(t) = \overline{P} + k\Delta F(t) \quad \text{for } \Delta F(t) < \Delta F_{max} \quad (2.28)$$
$$P_i(t) = \overline{P} + k\delta F_{max} \quad \text{for } \Delta F(t) > \Delta F_{max} \quad (2.29)$$

where $\overline{P}$ is the nominal output, as shown in Figure 2.35.

The system operator effectively pays a regulating unit to maintain a regulating reserve. By deciding how many of these bids to accept, the system operator determines the amount of the regulating reserve for its area. This decision will be closely linked to the maximum anticipated disturbance defined by the by the sum of the bounds in the long term market (Figure 2.34). The system operator may choose the accept enough bids to match its regulating reserve with the sum of the long term bounds, or he may rely on the spot market to pick up any additional imbalance on its system. The decision on how much regulating reserve to maintain has a profound effect on the price volatility of

74 POWER SYSTEMS RESTRUCTURING

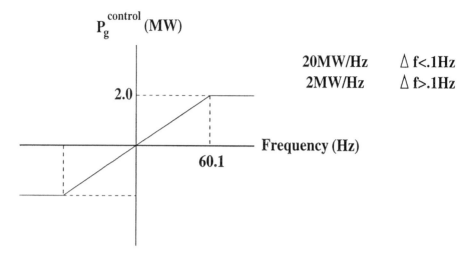

Figure 2.35 Contract for regulation

the spot market, as described in [33]. For detailed treatment of this topic, see [106, 99, 33, 107, 108].

PROVIDING FOR RELIABILITY UNDER OPEN ACCESS (TASK 5)

The operating tasks 1-4 described so far in this chapter are functions related to system operation during normal conditions. Sufficient standby capacity must be made available to respond to the loss of the largest power unit. This is provided by the operating reserve capacity, which must be available within 10 minutes. This operating reserve is required by NERC in today's industry. The unit commitment is done taking into consideration these reserve requirements.

The effect of the reserve requirement on the price of electricity is significant, since today's utilities overbuild generation to meet this level of reliability. The net result is that all customers have very reliable power at a higher average price than if this requirement were not imposed.

The reliability requirements are likely to become more relaxed in the new industry if many customers decide to accept lower reliability of service in exchange for lower price of electricity, i.e., if they are willing to be interrupted more often than at present [89].

Moreover, in today's industry the LOLP (Loss of Load Probability) criterion is generally difficult to meet without having at least sufficient standby reserve as the capacity of the largest power plant. As the technological developments have made it possible to have smaller scale power plants, at first sight it appears that to meet the same LOLP criterion one may need less reserve. This question is analyzed next.

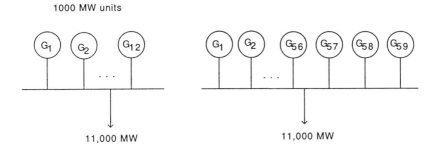

Figure 2.36 Example of the impact of size of generating units on system reliability

Impact of reduced generator capacity on system reliability

Under competitive power systems a trend has emerged toward smaller capacity units away from the large mega-projects. Here we briefly examine the impact of this trend on the reliability of the system through example shown in Figure 2.36. We suppose that the original system has a load of 11,000 MW and is being supplied by 12 generating units, each with a capacity of 1000 MW for a reserve of 1000 MW. It is then assumed that the load is supplied by N smaller generators, each with a capacity of 200 MW. The number N will vary but we begin with $N = 56$, which is the minimum number needed to meet the load and provide a reserve margin. It is assumed that the generating units in both cases are unavailable 2% of the time. The results of this illustrative example are shown in the table below,

	System load = $11,000 MW$			
Case	Unit Size (MW)	No. Units	Reserve (MW)	LOLP (%)
A	1000	12	1000	2.31
B	200	56	200	30.87
C	200	57	400	10.59
D	200	58	600	2.73
E	200	59	800	0.65

System reliability is measured in terms of the LOLP or Loss of Load Probability which is the probability that the generating capacity will not be sufficient to meet the load. In cases A and B, for example, some load will have to be shed if two or more units become unavailable. In case C, loss of load will take place if three or more units are unavailable, while in case E, loss-of-load will occur if five or more units become unavailable.

As is observed in the LOLP column, in case A, loss of load occurs 2.31% of the time, that is, some load shedding takes place. For this case the reserve is 1000 MW. In cases B through C, we consider the situation with smaller unit sizes but with considerably more units so as to be able to meet the load of 11,000 MW. With only a reserve of 200 MW as in case B, the LOLP is considerably higher than with case A, that is, 30.87%, a value which would be unacceptably high. Thus, providing a reserve, as in case A, equal to the size of the largest unit, leads to unacceptably high LOLP even though the loss of any single 200 and 1000 MW has equal probability. This behavior can be explained by the fact that in the smaller capacity case there are many more units and consequently the likelihood that more than one unit will be unavailable is more significant than in the case with only 12 units.

We note from the table that as the number of units increases in cases B through D, so does the reserve with a consequent decrease in LOLP. In case D, where the reserve is 800 MW the LOLP is .65% compared to the 1000 MW reserve in case A, where the LOLP is 2.31%.

This basic analysis seems to indicate that the trend in a competitive industry toward smaller units can help improve the system reliability even with reduced reserve margins (provided that a minimum level of reserve is maintained). Another factor which may strengthen this conclusion is the possibility that the unit unavailabilities for smaller newer units will also improve.

For completeness, we include the formula for LOLP used to derive the values in the table. We let U be the unit unavailability and $1-U$ the unit availability. The number of units is installed is denoted by n, while m is the maximum number that can become unavailable (for the given unit capacity) in order to be able to meet or exceed the given load. Then,

$$LOLP = 100[(1-U)^n + nU(1-U)^{n-1} + \frac{n(n-1)}{2}U^2(1-U)^{n-2} + \cdots$$
$$+ \frac{n(n-1)\cdots(n-m)}{m(m-1)\cdots 1}U^m(1-U)^{n-m}] \qquad (2.30)$$

Possible new notions of reliability provision and its value in the new industry

It is clear based on this analysis that the relative size and the number of plants on the system effect the LOLP at which a customer will be served. In electricity markets that are not entirely coordinated (mandatory poolcos), LOLP as a single (systemwide) quantity begins to lose its importance. Moreover, reliability of service will have to be separated into reliability of power delivery (transmission and distribution service) from the reliability on the supply/demand side.

Power supply/demand reliability

Reliable provision of power supply and consumption is conceptually a primary market question. Consequently, the ways in which it is accomplished will greatly depend on the primary electricity market structure, i.e., on the mechanisms by which task 1 is met. Reliable supply/demand is *no* longer an ISO's responsibility in an arbitrary electricity market structure.

For example, in a mandatory poolco with no active demand participation reliability remains ISO's responsibility. He must procure this service either on a market basis, or reserves will be required much in the same way as in today's industry. The question of reliability requirements being provided on a market basis is discussed in [15]. It is not reasonable to expect nonuniform reliability of service to customers under such arrangement. For theory and examples of the unit commitment *switching curve law* including consideration for reserve constraints, see [12].

A PX/ISO arrangement is a natural next step for locational differentiation in supply/demand reliability patterns. Independent producers selling power into the PX could make bilateral arrangements with other suppliers for providing committed power into the PX at the times when they can not do this.

Similarly, bilateral arrangements can be made among two (or more) parties i and j under a specified $LOLP_{ij}$. This $LOLP_{ij}$ is generally different than the systemwide LOLP presently imposed by NERC. Nonuniform reliability service as a function of price of meeting the basic task 1 is a straightforward concept in this bilateral industry structure. For some initial developments in this area, see [62].

Reliable delivery of power

As the reliability of supply meeting demand takes on a new meaning for different electricity market structures, reliable power delivery remains in the domain of a transmission provider. It is clear that system users whose $LOLP_{ij}$'s are not as strict as of the others would be willing to tolerate less reliable power delivery, as well.

The very concept of nonuniform LOLP at the supply/demand side creates many fundamental questions for developing and implementing reliable transmission system delivery. The technical concepts for doing this in a systematic manner are not in place at present. If this avenue is to be taken, much fundamental research must be done first.

Moreover, while it makes sense that transmission service has different value to system users with nonuniform LOLP's, it is difficult to implement these concepts assuming that the transmission system remains a natural monopoly. A new regulatory framework that enables performance-based transmission service is needed before one engages into what amounts to peak-load pricing for transmission service [32].

In closing, it is very likely that reliability requirements in the new industry will change particularly concerning mechanisms and valuing of supply/demand reliability. This is likely to have unforeseen impacts on electricity prices throughout the system. To facilitate this obvious market need, much basic research is needed for developing methods of keeping the transmission system fully functional. In a long run, reliable power delivery at a price will evolve also. This is a much more complicated concept and it will take significant technical and regulatory developments to bring to life. A mind twisting question is a development of formula similar to (2.30) above for valuing reliability of transmission lines to system users with nonuniform reliability needs.

CONCLUSIONS

Within each large interconnection of the power grid in North America the actual mechanisms of providing power to a customer are likely to be nonuniform, at least for some time to come. In some parts day ahead electricity spot markets are beginning to be implemented, in some other parts bilateral trades are dominant with the system operators standing by to facilitate their technical implementation.

On the power supply side, many will *actively* respond to the market opportunities, either by aggressively finding the "best" customer to serve, or by responding in real time to the short-term opportunities to make profit.

In a competitive power industry many market participants desire to have full control over their own units concerning the on-off status and the actual power produced. Consequently, the basic process of supply meeting a time-varying demand (operating task 1) must be analyzed keeping in mind that the contractual/financial management (electricity pricing) and the physical energy trades and their costs are potentially quite different. Both processes must be recognized as actively evolving and effecting market outcomes, although often the financial processes are slower than the real-time technical processes of scheduling power. One could generally identify two distinct subprocesses, (1) financial, selling and/or purchasing power, and (2) physical, real-time implementation of the transactions requested based on (1).

The arguments pro and against a particular type of primary market structure are often based on the vested interests of various parties promoting a specific approach. It is important, however, to provide a systematic treatment of the issues and not be biased. In this chapter, we have attempted to analyze critical issues.

The pros and cons for one type of market versus the other are usually debated in context of the implications on tasks 2-5, so-called transmission congestion, in particular, but not in the context of task 1. To fill up this gap, we have, instead, taken a look first at the primary market performance (operating task 1) as a function of the industry structure in place. Reasons for the primary electricity market posing a unique challenge to the state-of-the-art theory of competitive economics are summarized first. The price of electricity as seen by a consumer may be quite uncorrelated with what is obtained when the objective is to minimize the total cost of supply meeting demand. The differences are shown to be generally a function of the electricity pricing methods in place, in particular the *ex ante* or *ex post* pricing.

We next introduce basics of possible electricity market dynamics in which *all three* types of transactions (bilateral physical, bilateral tradeable, spot market trades) are allowed. We show that under the reasonable assumption of no gaming, this market structure is self-adjusting to both physical disturbances (supply/demand mismatches) and to the price changes. In other words, despite unique physical features of the primary electricity market, given sufficient flexibility for decentralized unit commitment/economic dispatch, the market will clear. Moreover, the economic outcome over long term horizons in self-adjusting markets with sufficient flexibility for participants to respond to the changing market conditions in real time should not depend if charging is done at the beginning or the end of a contract. An example from [33] is used to illustrate the basic self-adjusting mechanism of a market comprising all three types of transactions.

Given that tasks 2-5 make the power industry very distinct, one is faced with several fundamental questions. To start with, it is not clear a priori that all of these tasks must be performed according to the present hierarchical structure; it is possible that, as in task 1, some of the functions in 2- 5 could be performed by MPs provided that there is an established market for such services. This question is particularly relevant for tasks performed by power plants since *all* energy-related services could be procured through competitive markets, in much the same way as the primary market exists for procuring generation to meet predicted demand. At least in concept, there is no difference between the imbalance created because not sufficient generation was scheduled ahead of time (at the tertiary level) and the imbalance that needs to be regulated in an automated way in response to the real-time mismatch. If the generation at the tertiary level is made competitive, the regulating generation should also be procured in a competitive way. This does not imply that this generation is necessarily operated in an uncoordinated way by the operators of individual power plants; technical design for decentralized regulation is an open question at present.

For example, if some of the tasks 2-5 are done by MPs, it is critical to specify the level of technical performance to be met by the individual market participants. An additional question arises when attempting to facilitate *tradeable market transactions*; the idea of some other party partially delivering power by party at bus k instead of party at bus i to fulfill contractual obligation between parties i and j if the market conditions are right leads to the need to make a distinction between physical and economic transactions. Only physical transactions are relevant when performing tasks 2-5 in such setup; however, if a transaction is tradeable or not makes a big difference in the economic outcomes of the market. Similar ideas could be pursued in potential markets for performing tasks 2-5; the idea of tradeable responsibility for compensating transmission loss created by a market participant, or for violating transmission limits is bound to emerge, much in the same way as the responsibilities for meeting environmental constraints are viewed as potentially tradeable [4].

These distinctions between physical and commercial processes lead to truly novel, difficult engineering questions not attempted in the past. On the other hand, the electricity markets are likely to function as expected if least technical restrictions are put on the economic-based decision making. Possibly the biggest challenge to engineers is to rethink the ways tasks 1-5 are done in today's industry with the ultimate objective of having meaningful markets for performing these tasks.

80   POWER SYSTEMS RESTRUCTURING

The tendency has been so far in the new industry to analyze the financial processes (such as procurements of services to meet tasks 1-5) independently from their technical implementations. In this chapter we do this partially, but then return to the questions concerning necessary feedback from financial to technical processes, and vice versa, for providing right incentives to maintain high quality performance in an efficient way. Much of the discussion in this chapter struggles with the question of how this should be done.

Furthermore, if the case is made that some of these services must remain coordinated in real-time at the system level, one must decide how the resources for performing these services should be planned, created and used in real-time. Equally important are questions related to the price-charging mechanisms for the services that remain coordinated.

Most importantly, total cost of supply meeting demand (efficiency) and the price of power paid by the consumer are not always directly correlated. We have suggested that for the two to be related one generally needs a real-time technical or price feedback for performing tasks 2-5 to the market participants.

Appendix 2.1 Theoretical and algorithmic background for meeting task 1 in today's industry

Conventional economic dispatch

The economic dispatch program is routinely used for changing the power generated every 5-15 minutes to fit the anticipated demand in near real-time.

Mathematically, this is the problem of minimizing the total generation cost

$$\min_{P_g}(\sum_{i=1}^{ng} C_i(P_{gi})) \tag{2.31}$$

such that total generation equals total load,

$$\sum_{i=1}^{ng} P_{gi} = \sum_{i=1}^{nd} P_{di} \tag{2.32}$$

This basic version of unconstrained economic dispatch finds a solution to this optimization problem for a system of arbitrary size. A necessary condition for solving this basic economic dispatch problem is,

$$\frac{dC_1}{dP_{g1}} = \cdots = \frac{dC_{ng}}{dP_{ng}} = \lambda \tag{2.33}$$

This condition defines the least generation cost for meeting given demand. The term λ is known as the short-run marginal cost (SRMC) and at the optimum of (2.31) all unit marginal costs are equal to it.

Notice that conventional economic dispatch assumes demand to be a given input for which generation use is optimized under the constraint that total generation meets total demand (equation (2.32)). It is straightforward to generalize this formulation to account for price-elastic demand. This generalization is stated in the Appendix 2.2 concerned with task 1 in the new industry.

The simplest unit commitment

The basic unit commitment problem (without start-up costs or minimum up/down time constraints) is as follows,

$$\min_{u,P_g} \sum_{i=1}^{n_g} u_i C_i(P_{gi}) \qquad (2.34)$$

subject to

$$\sum_1^{n_g} P_{gi} = P_d \qquad (2.35)$$

where u_i equals 0 or 1 depending on whether the unit is off or on. Following the Lagrangian relaxation method, one first forms the Lagrangian function,

$$L(u, P_g, \lambda) = \sum_{i=1}^{n_g} u_i(C_i(P_{gi}) - \lambda P_{gi}) + \lambda P_d \qquad (2.36)$$

By minimizing (2.36) over P_g first, one obtains the conventional economic dispatch equal incremental condition, that is,

$$\frac{dC_i}{dP_{gi}} = \lambda, \quad \forall i \qquad (2.37)$$

which permits one to solve for P_{gi} in terms of λ, the system incremental cost, or,

$$P_{gi} = P_{gi}(\lambda) \qquad (2.38)$$

Equation (2.38) is then substituted into (2.36) so that the Lagrangian becomes,

$$L(u, \lambda) = \sum_{i=1}^{n_g} u_i(C_i(P_{gi}(\lambda)) - \lambda P_{gi}(\lambda)) + \lambda P_d \qquad (2.39)$$

Finally, the Lagrangian method minimizes $L(u, \lambda)$ with respect to u giving the *switching curve law*

$$u_i = \begin{cases} 0 & \text{if } C_i - \lambda P_{gi} > 0 \\ 1 & \text{if } C_i - \lambda P_{gi} < 0 \end{cases}$$

that is, the unit is off if the average cost $\frac{C_i}{P_{gi}} > \lambda$ and on otherwise. Essentially, this says that the unit should be off if the average unit cost is too high, i.e., higher than the system incremental cost. This simple rule is used to decide if the unit should be turned on and off. Once on, a conventional economic dispatch is used to adjust to demand changes if these are monitored more frequently. Observe that the same λ is obtained when simply doing conventional economic dispatch, namely that both unit commitment and economic dispatch result in the same marginal cost when the demand is assumed known and no intertemporal effects in starting and running power plants is accounted for.

Deterministic unit commitment that accounts for various cost components

In its very general form, the unit commitment problem may be written as follows [36],

$$u_k^* = arg\min_{u_k} E_{w_k}(\sum_{k=1}^{N} C_k) \qquad (2.40)$$

that is, a unit commitment is the problem of computing decisions u_k^* which minimize the total *expected* cost over some future horizon while satisfying all constraints. Here E_{w_k} stands for the expected value with respect to the random variable w_k. For unit commitment, w_k is typically real power demand $P_{d,sys}(t)$, as well as the availability of generators and transmission lines. Unit commitment decisions are made at periodic intervals k. The time period between decisions is referred here as a stage. The future horizon of interest is N stages long.

Because of the enormous computational complexity of the stochastic unit commitment, many simplifications are made in practice. The possibility of generator and line failures is typically handled by providing adequate reserve margin [47, 48, 49], so that all loads are served with very high probability.

A simplified version of the general unit commitment problem formulation results when the optimization is performed assuming that the *demand is equal to the forecasted value* [47, 50]. In this case a random load $P_{d,sys}(t)$ is replaced by its predicted value $\hat{P}_{d,sys}(t)$. Even under this assumption, the real life unit commitment problem is a complicated problem with a large number of possible decisions [36]. In particular, the rate of response of units, start up and shut down costs, must run times, etc, resulting in a complicated production cost function. The total cost incurred during stage k, denoted as C_k, is,

$$C_k = \sum_{i=1}^{ng}(\int_0^{h_k} C_{gi}(P_{gi}(t))dt + u_k(i)I(x_k(i) < 0)S_i + (1 - u_k(i))I(x_k(i) > 0)T_i)$$

$$(2.41)$$

Here I is a conditional statement and it has value of 1 if the statement is true and 0 if it is false. h_k stands for the number of hours in stage k, S_i and T_i are start-up and shutdown costs for generator i, respectively. x_k is the state variable (in the simplest case, the only state is the on/off status of plants as the decision is being made for the next stage).

Given a predicted system demand over the future horizon of interest T (T is a week or longer in Figures 2.3–2.5), typically used unit commitment software computes the turn on and turn off times for various plants taking into consideration the start-up cost, rate of response of plants, must-run time, etc. Very few software packages take into consideration transmission constraints.

Stochastic unit commitment problem

Many unit commitment methods in use are deterministic with respect to the load, meaning that the optimization is performed assuming that the actual demand $P_{d,sys}$ is equal to its forecasted value $\hat{P}_{d,sys}$ [47, 50]. It can be shown that a schedule obtained by replacing a randomly varying demand by its expected value would be a suboptimal method, sometimes being very far from optimal, when applied to the stochastic unit commitment problem formulation given in equation (2.40) above. Because of this, some power companies have developed unit commitment methods that allow for a probabilistic load distribution [49, 51].

We stress in the later part of this chapter that a probabilistic (multistage) decision making is possibly the most critical part of becoming a successful market participant in the new power industry. While much of such decision making will be made by the marketers, it is important to understand the necessary decision making process of this type by the ISOs themselves (or by the power exchanges, if the two functions are separated). An effective day ahead, or week ahead, coordinated forward spot market is in many ways equivalent to using a stochastic unit commitment in today's industry [36]. The seemingly unnecessary distinctions in formulating unit commitment problem must be understood to appreciate issues such as causes of real time electricity price volatility and inefficiency in the new industry.[48]

The stochastic unit commitment problem is the problem of finding a sequence of optimal unit commitment decisions (for example, for the next 24 hours) $\underline{u}[1]$ $\underline{u}[2] \cdots \underline{u}[24]$ where $\underline{u}[k] = [u_1[k] \; u_2[k] \cdots u_{ng}[k]]^T$ is a decision vector for all generating units at stage $[k]$ so that,

$$\min_{\underline{u}[1] \; \underline{u}[2]\cdots \underline{u}[24]} E_{P_d[1] \; \cdots P_d[24]} \sum_{k=1}^{24} \sum_{i=1}^{ng} C_i[k] \qquad (2.42)$$

Mathematically, the stochastic unit commitment problem can be expressed as a dynamic programming (DP) problem including control inputs, system states, and uncertain random variables [52, 47, 36]. Time is broken down into a series of stages, and a control decision is made at the beginning of each stage. The system can be described by the following equations,

$$x_{k+1} = f_k(x_k, u_k, w_k) \qquad (2.43)$$

where $k = 0, 1, \cdots$, x_k is the state vector at time k, u_k is the control input at time k and w_k is a random disturbance. The state transition equation (2.43) defines how the state changes from one stage to the next [47]. At each stage, there is a cost to be paid. The problem is to determine a control policy u_k that minimizes the cost (or, maximizes the reward).

In the case of finite number of stages N, at each stage k there is cost $g_k(x_k, u_k, w_k)$ incurred. Additionally, there is a terminal cost $g_N(x_N)$ which depends on the final value of the state vector. The objective of the problem is to find the control policy that minimizes the total expected cost over N stages, known as the *optimal policy*. Dynamic programming is a method for finding such policy, and it is expressed mathematically as [52],

$$\begin{aligned} J_N(x_N) &= g_N(x_N) \\ J_k(x_k) &= \min_{u_k \in U_k(x_k)} E_{w_k}[g_k(x_k, u_k, w_k) + J_{k+1}(f_k(x_k, u_k, w_k))] \end{aligned} \qquad (2.44)$$

84   POWER SYSTEMS RESTRUCTURING

where $J_k(x_k)$ denotes the optimal expected cost when beginning at stage k. An optimal policy u_k^* is obtained so that it attains the minimization in equation (2.44) for each x_k and k.[49]

While this formulation is complex and generally not solvable exactly in real time for large electric power systems, potential benefits from such *multistage decision making* must be kept in mind. This fact was recognized some time ago in [59] and recommendations were made to perform so-called dynamic dispatch that incorporates active load prediction.

The most critical distinction between the simplified unit commitment methods in today's industry and the solution obtained using a DP approach comes from the assumption that the events at various stages are not correlated [61]. This assumption cannot be made with any significant uncertainties present (demand forecast, generation and transmission lines availability, etc.)

Appendix 2.2: Theoretical background and algorithms for meeting task 1 in the new industry

Decentralized economic dispatch

When competitive bilateral transactions take place, each party has as its main objective to maximize its profit,

$$\max_{P_i} \pi_i(P_i) \tag{2.45}$$

where $\pi_i = pP_i - C_i(P_i)$ stands for the profit made by the market participant i through some sort of trading process, given (known) p. If the market participant is a generator then

$$\pi_i(P_{gi}) = pP_{gi} - C_i(P_{gi}) \tag{2.46}$$

while if it is a load,

$$\pi_i(P_{di}) = -(pP_{di} - C_i(P_{di})) \tag{2.47}$$

In the above equations, p is the price paid, C_i is the cost function. Thus, under perfect conditions, when the market converges to a single electricity price for both sellers and buyers, p, one can maximize π_i to yield

$$\frac{dC_{g1}}{dP_{g1}} = \cdots = \frac{dC_{ng}}{dP_{ng}} = \frac{dC_{d1}}{dP_{d1}} = \cdots = \frac{dC_{nd}}{dP_{nd}} = p \tag{2.48}$$

This is simply obtained by each market participant optimizing own profit/benefit for the assumed (exogenous) market price p. The process of bilateral decisions will stabilize at the systemwide economic equilibrium under a perfect information exchange among all mps.

Thus, since the necessary conditions (2.33) and (2.48) are identical under perfect market conditions both the presently used unconstrained economic dispatch and a competitive market process should lead to the same power quantities traded, and to the same total social welfare optimum. Most importantly, in this case the optimal electricity price is reached under the same conditions as the social welfare is maximized. The performance objectives of the individual market participants (price) and the total operating cost are equivalent.

Decentralized unit commitment

Consider first a simplified situation with a single generator owner which sells electricity into a day ahead spot market. This problem is much simpler than the stochastic unit commitment problem in today's industry, as there is only one generator to consider, and all of the random disturbances are presumed to be reflected by the price at which power is sold[50]. The generator owner is assumed to be a price taker in a competitive market place.[51]

The generation owner must make unit commitment decisions typically by certain time day ahead, *before* actually knowing the spot price of the next hour. After the spot price is known, the generator decides how much power to sell (dispatch) in order to maximize profit. The only control for the problem is $u_k(1)$ whether to turn on or off at stage k. The generation level $\hat{P}_{gi}$ may be regarded as a function of the control $u_k(1)$ and the expected price $\hat{p}_k$. If $u_k(1) = 0$, then $P_{gi} = 0$. If $u_k(1) = 1$, then $\hat{P}_{gi}$ at stage k is set to maximize the *expected* profit,

$$\hat{\pi}_k = \hat{p}_k \hat{P}_{gi} - C_i(\hat{P}_{gi}) \tag{2.49}$$

For a quadratic cost function $C_i(P_{gi}) = a_i P_{gi}^2 + b_i P_{gi} + c_i$ and not taking into consideration the generation limits power quantity that maximizes profit at stage k is easily found to be $\hat{P}_{gi} = \frac{\hat{p}_k - b_i}{2a_i}$.

Notice that here an assumption is made that $\hat{p}_k$ is an *exogenous* input to the decision making process, namely that the market prices are uncorrelated at different decision making stages. The same process gets computationally more involved when price is modeled as a state variable. The result, i.e., the overall profit should be higher with this additional variable. The problem, of course gets computationally more involved. For examples of this, see [62].

Theoretical equivalence of a coordinated and decentralized unit commitment

As the coordinated unit commitment of today's industry is being partially replaced by the decentralized scheduling in a bilateral way, an important question arises concerning similarities and differences of the two solutions.

In the simplest case of assuming deterministic (known) price, and ignoring start-up costs, must-run time constraints, etc. it can be shown that an individual decision maker would arrive at the *same average cost versus market price decision rule as the rule often used by a system operator scheduling plants in a coordinated way* reviewed in the Appendix 2.1 above. The proof for this goes as follows[52]:

Given a generator i with cost curve $C_i(P_{gi}) = a_i P_{gi}^2 + b_i P_{gi} + c_i$ and fixed cost C_{Fi}, its profit while on is,

$$\hat{\pi}_{on} = \hat{p}\hat{P}_{gi} - C_i(\hat{P}_{gi}) - C_{Fi} \tag{2.50}$$

The profit while off is,

$$\hat{\pi}_{off} = -C_{Fi} \tag{2.51}$$

After determining the value of P_{gi} that maximizes the "on" profit, the decision to turn on/off is made. The generator will turn on only if $\hat{\pi}_{on} > \hat{\pi}_{off}$, or

86   POWER SYSTEMS RESTRUCTURING

$\hat{p}\hat{P}_{gi} - C_i(\hat{P}_{gi}) > 0$ which is equivalent to,

$$\hat{p} > \frac{C_i(\hat{P}_{gi})}{\hat{P}_{gi}} \qquad (2.52)$$

which is the average cost rule used for coordinated unit commitment. Notice that generation limits do not affect this result. A generation limit may preclude a more profitable choice of $\hat{P}_{gi}$, but as long as the average cost rule is satisfied, the generator will turn on.

One can conclude based on this derivation that under perfect market assumptions and when neglecting minimum times, startup costs, etc a system operator would schedule the *same units* to what would be decided in a decentralized way by the individual power producers.

Unit commitment/economic dispatch at a spot market level

A bilateral trading process can be interpreted as a *decentralized* unit commitment that at least partially provides generation to supply system demand $\hat{P}_{d,sys}$. However, this process does not guarantee that the generation sold will meet the predicted system demand, and an additional unit commitment must be done in near-real time (on a day ahead spot market) to supply the remaining generation $\hat{P}_{g,sys}$.

This day ahead unit commitment at a spot market level is generally done in a coordinated way, except that it relies on price bids resembling supply functions which are not necessarily the same as cost functions. In that sense, the only difference with the unit commitment decision making is that this process is price- instead of cost-based.

For example, such combined unit commitment/economic dispatch in the UK poolco is presently done according to the average cost rule [63]; the only difference is that in a competitive poolco the decision is based on relative comparison of the average unit cost and the system price $\hat{p}$, instead of cost $\hat{\lambda}$.

Generalized economic dispatch. In the formulation of the economic dispatch given in (2.31) we assume that generation only is price responsive, while the demand is prespecified. It is straightforward to extend this formulation to account for demand price elasticity.

The generalized economic dispatch is then the problem of scheduling *both generation and demand* so that the total generation and demand cost is minimized as follows,

$$\min_{P_g, P_d} \left(\sum_{i=1}^{n_g} C_i(P_{gi} + \sum_{j=1}^{n_d} C_j(P_{dj}) \right) \qquad (2.53)$$

such that total generation equals total load (equation (2.32)). A necessary condition for solving this basic economic dispatch problem is,

$$\frac{dC_1}{dP_{g1}} = \cdots = \frac{dC_{ng}}{dP_{ng}} = \frac{dC_{d1}}{dP_{d1}} = \cdots = \frac{dC_{nd}}{dP_{nd}} \qquad (2.54)$$

Criterion (2.53) is used as a general measure of static efficiency in any competitive industry where it is often referred to as the (negative of) the *social welfare*. Both generation and demand cost functions $C_i(P_{gi})$ and $C_i(P_{di})$ are analogous and can be represented as a single function as shown in Figure 2.37 [87]. In

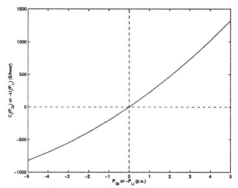

Figure 2.37 Simplified generation and demand cost function

addition, since both generation and load have hard upper and lower limits, the optimum must account for such restrictions. One way of accomplishing this is by assuming that at the limits the costs are infinitely high as depicted in Figures 2.38 and 2.39.

An important degenerate case of price elastic cost functions is the problem of a generalized economic dispatch subject to must run and must serve constraints for some generation plants and some loads, respectively, see [45, 44] for implications of this constraints on market outcomes.

Generalized unit commitment. A possible generalization of the conventional unit commitment formulation to include turning on and off price responsive loads is straightforward, at least in principle. This problem generalization is as follows,

$$\min_{u, P_g, P_d} \left(\sum_{i=1}^{n_g} u_i C_i(P_{gi}) + \sum_{j=1}^{n_d} u_j C_j(P_{dj}) \right) \tag{2.55}$$

subject to

$$\sum_{i=1}^{i=n_g} u_i P_{gi} = \sum_{j=1}^{j=n_d} u_j P_{dj} \tag{2.56}$$

where u_j is 0 or 1 depending on whether the load is on or off. The result is a *generalized switching curve law* for both power suppliers and price responsive demand; the switching law is identical for power plants as in the conventional unit commitment case, while the switching law for loads is

$$u_j = \begin{cases} 0 & \text{if } C_j - \lambda P_{dj} > 0 \\ 1 & \text{if } C_j - \lambda P_{dj} < 0 \end{cases}$$

that is, the load should be off if its average benefit is too low, i.e., lower than the price of electricity λ.

If the utility (benefit) functions of loads are made known to the operator making unit commitment decisions, this is implementable in a very similar way as it is done for the power plants at present.

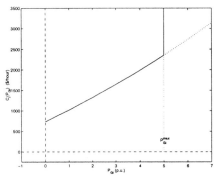

Figure 2.38 Generation cost function with capacity limits accounted for

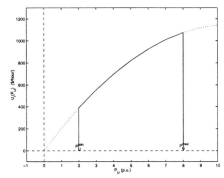

Figure 2.39 Demand utility function with capacity limits accounted for

Appendix 2.3: Methods of compensating and accounting for real and reactive power loss in today's industry

Most of the methods concerned with loss compensation are derived assuming real/reactive power decoupling. It is well known that this assumption is justified when the operating conditions are such that the power line flows are not very large, that is when the bus voltage phase angle differences are relatively small, and when bus voltages are very close to their nominal 1 p.u. values. When this is not the case, in particular as a consequence of large power transfers, it is not fully justified to assume that real power losses are only effected by the real power transfers, and the reactive power losses are effected by the voltage differences only.[53]

We first briefly review much discussed real power transmission loss compensation. This is followed by the less discussed methods of compensating and paying for the reactive power loss.

2.3.1 Conventional calculation of the real power loss

In today's industry the objective function of the economic dispatch with transmission loss accounted for is the same as given in equation (2.31) above, with the difference that the power balance constraint given in equation (2.32) takes on the form,

$$\sum_{i=1}^{ng} P_{gi} = \sum_{i=1}^{nd} P_{di} + P_{loss} \qquad (2.57)$$

The necessary optimality conditions result in,

$$\frac{dC_{g1}/dP_{g1}}{(1 - dP_{loss}/dP_{g1})} = \cdots = \frac{dC_{ng}/dP_{g,ng}}{(1 - dP_{loss}/dP_{g,ng})} = \lambda_{loss} \quad (2.58)$$

The nodal price $p_i = \frac{dC_{gi}}{dP_{gi}}$ at bus i is then computed $\forall i$ as,

$$p_i = \frac{dC_{gi}}{dP_{gi}} = \lambda_{loss}(1 - \frac{dP_{loss}}{dP_{gi}}) \quad (2.59)$$

The formula (2.59) for computing marginal costs of generators with thermal transmission loss accounted for is *exact*. It is straightforward to see that the *total cost inefficiency created by transmission losses* at each bus is reflected through the ratio $1/(1 - dP_{loss}/dP_{gi})$. Without the loss effect, the marginal cost of all generators is identical and equal to the ideal market price λ as defined in (2.61) Appendix 2.1. This ratio reflecting loss related inefficiency (sometimes referred to as a penalty factor) generally depends on the systemwide conditions, that is on the generation/demand into all system buses and not only into the individual bus locations.

In the past, prior to having standard load flow calculation packages, utilities have used approximate formulae known as the $\mathcal{B}$ coefficients [64]; the only information necessary for estimating total loss using the $\mathcal{B}$ coefficients is vector of net power injections $P = P_g - P_d$ into all buses (except the slack bus). At present, transmission loss is computed more accurately using results of load flow computations. These more accurate methods are reviewed next in context of computations for the new industry.

Appendix 2.4: Some possible loss compensation solutions for the new industry

Possible theoretical solutions for taking into consideration transmission loss charges are (1) generalized nodal prices and (2) usage-based formulae. Both of these methods are straightforward generalizations of the well known formulae in today's industry reviewed above, except for the fact that these formulae are not actively used for unbundling costs of various services, such as transmission loss compensation. These relations are reviewed next.

Generalized nodal prices To start with, if demand is assumed to be known, nodal prices which reflect effect of transmission loss on total generation cost inefficiency are identical to the formulae (2.59) above known in today's industry. While these formulae are not used for accounting purposes in today's industry, it is well known that these nodal prices reflect transmission loss-created inefficiencies.

These formulae are generalizable to account for price elastic demand, much in the same way as a generalized economic dispatch was posed in Appendix 2.2 as an extension of the conventional economic dispatch reviewed in Appendix 2.1. The only difference is that the equality constraint (2.32) is replaced by the equality constraint (2.57). The optimality conditions define nodal prices at all buses, loads and generators,

$$\frac{dC_{g1}/dP_{g1}}{(1-dP_{loss}/dP_{g1})} = \cdots = \frac{dC_{g,ng}/dP_{g,ng}}{(1-dP_{loss}/dP_{g,ng})} =$$
$$\frac{dC_{d1}/dP_{d1}}{(1-dP_{loss}/dP_{d1})} = \cdots = \frac{dC_{d,nd}/dP_{d,nd}}{(1-dP_{loss}/dP_{d,nd})} \quad (2.60)$$

The nodal price at bus i

$$p_i = \frac{dC_i}{dP_i} \quad (2.61)$$

reflects the *bundled* price of supply meeting demand *and* compensating for the share of transmission loss. These formulae are applicable if one wishes to differentiate locally the effect of losses at various buses in a mandatory poolco-like electricity market, with the only distinction that the optimization is based on supply function bids instead of on the actual cost curves.

Usage-based formulae for estimating transmission loss

It can be seen by comparing closed-form relations (2.54) and (2.60) that the nodal price at bus i deviates according to the ratio $1/(1 - \frac{dP_{loss}}{\partial P_i})$. It can be also shown that for most often operating conditions the function $\frac{dP_{loss}}{dP_i}$ is a *monotonically increasing function* of power injections, P_i [90, 74]; therefore, the larger injections, the larger flows created, the higher loss penalty factor. This provides an argument for some sort of line flow-dependent (usage-based) charges for transmission loss, that, at the same time, could be interpreted in terms of prices reflecting market inefficiencies created by transmission loss charges.

Several usage-based formulae for transmission loss compensation have been suggested recently. These formulae can be directly related to the methods used by the industry at present to *compute total transmission loss* for given power systems injections. We show next that the so-called distribution factors for flow allocation and transmission loss allocation in the new industry [68, 69] result in an identical total transmission loss as it would be estimated by the B coefficient formula for total loss estimation in today's industry when one does not wish to perform computer-based load flow calculations.

B **coefficients**   To start with, it is well known that thermal loss in an arbitrary transmission line connecting buses i and j, P_{Lij}, is

$$P_{Lij} = P_{ij} + P_{ji} \quad (2.62)$$

where P_{ij} and P_{ji} are real power line flows at the sending and the receiving ends of the line. Assuming voltages close to 1 p.u., and approximating $\cos \delta_{ij}$ by the first two terms in Taylor series expansion, an approximate line loss formula is $P_{Lij} = G_{ij}\delta_{ij}^2$ where G_{ij} is a line conductance. The total transmission loss is

simply
$$P_{loss} = \sum_i \sum_j P_{Lij} \qquad (2.63)$$

and is computed as,
$$P_{loss} = (M\delta)^T G(M\delta) \qquad (2.64)$$

where M is the line-bus incidence matrix of the system [65, 66] and G is a diagonal matrix whose diagonal terms are line conductances. The vector δ is usually approximated by the (lossless) DC load flow with,

$$P_g - P_d = B\delta \qquad (2.65)$$

where δ is a vector of voltage bus angles, except at the slack bus. As a result, the line losses become,

$$\begin{aligned} P_{loss} &= (P_g - P_d)^T B^{-1} M^T G M B^{-1}(P_g - P_d) \\ &= (P_g - P_d)^T \mathcal{B}(P_g - P_d) \end{aligned} \qquad (2.66)$$

Thus, just by knowing network topology and its parameters (line conductance matrix G and imaginary part of the bus admittance matrix B), one estimates total transmission loss using equation (2.66).

Distribution factors-based transmission loss formulae The vector of linearized line flows (losses neglected) is simply computed as

$$\begin{aligned} P_{flow} &= diag(B_{ij}) M\delta \\ &= diag(B_{ij}) M B^{-1}(P_g - P_d) \\ &= D(P_g - P_d) \end{aligned} \qquad (2.67)$$

It follows from (2.67),(2.65) and (2.66) that

$$P_{loss} = (diag(B_{ij})^{-1} P_{flow})^T G(diag(B_{ij})^{-1} P_{flow}) \qquad (2.68)$$

The total loss created by the bus injection vector $(P_g - P_d)$ is expressed in terms of the line flow vector P_{flow} in formula (2.68). It can be seen that the total loss is a quadratic (nonlinear) function of line flows. The accuracy at which the line flows P_{flow} are estimated, therefore, is more critical when estimating losses than for charges based on line flows themselves, such as usage-based open access fee, for example proposed in [68, 69].

Notice that a generalization of the distribution factor formula (2.67) is necessary to eliminate slack-bus induced assymetries when computing relative contributions of *all (including slack bus)* injections to a measured line flow, see [71].

Accuracy of estimating line flows The matrix D is known as the matrix of distribution factors and it determines the basic contributions of various bus

92 POWER SYSTEMS RESTRUCTURING

injections to the line flows. Any usage based formulae presently under consideration for estimating technical impact of various transactions, such as MW-mile [68, 69] are often based on this expression.[54]

A considerably more accurate formula for computing transmission loss is obtained when vector of voltage bus phase angles δ is computed using a nonlinear AC load flow.

The only difference is that δ is computed (instead of formula (2.67)) using,

$$\delta = [\frac{dP}{d\delta}]^{-1}(P_g - P_d) \quad (2.69)$$

where the term $[\frac{dP}{d\delta}]$ is a load flow Jacobian matrix evaluated at the load flow solution.

Transmission loss compensation in a poolco-like industry structure

As described in the main text, a range of methods of various accuracy is applicable to accounting for real power losses in a poolco industry structure. The charging methods recommended by FERC as well as by the ISO's in forming are directly intended for such an industry in which an ISO provides for transmission loss.[55] The most representative would be (1) the $\mathcal{B}$ coefficients/ distribution factors-based formulae, including MW-mile method and alike [38], (3) AC flows-based approximate calculations [80] and (4) new formulae described in Chapter 3.

Approximate formula (2.66) used in today's power industry for estimating total transmission loss can be directly applied for allocating the contribution of power injection into any bus i on total loss and the corresponding deviation in nodal price i needed to compensate for its contribution to the total loss using formula (2.59). For issues when using more accurate line flow allocation-based pricing of unbundled system services, see [80].

Transmission loss compensation in bilateral markets

The idea of a market participant compensating for its own transmission loss and not having to depend on an ISO for transmission loss provision is very attractive in competitive markets. Basically, the less coordination, the better. This poses an interesting question concerning the accuracy at which transmission loss caused by a known power injection into a bus can be estimated by the market participant himself. The first approach of this type was proposed in [90, 74] and it is based on the well known localized response property to a bus input change in a very large meshed electric power network.

An algorithm for estimating transmission loss created by an electrically distant system user (or an entire zone, by definition electrically distant) can be found in [74]. This algorithm assumes only knowledge of the net injection into the bus (or zone) for which transmission loss is being estimated $P_i = P_{gi} - P_{di}$, and knowledge of transmission network topology and parameters. The strictly technical information about the transmission should definitely be provided on systems such as OASIS [76]. In technical terms, the line-bus incidence matrix M, transmission system conductance G and susceptance Y matrices are assumed to be updated periodically and made a public knowledge through an OASIS-like system.

A possible algorithm for loss compensation by the end user The algorithm for loss estimation is extremely simple, and it consists of the following basic steps.

Step 1: The first step in estimating real power transmission losses is to re-enumerate the network using the tier based re-enumeration, in which tier I represents all buses directly connected to the bus where the injection is changed, tier II stands for buses directly connected to uses in tier I, etc, totaling N tiers. Based on this enumeration, the bus-admittance matrix B is partitioned into submatrices $B_{II}, B_{I,II}, B_{II,I}, B_{II,II}, ..., B_{N,N-1}, B_{N,N}$.

Step 2: Based on this bus re-enumeration, changes in nodal bus voltage angles are computed by a market participant himself, independently from the others. This is done using the forward/backward substitution based on formula (2.65) above as

$$\Delta \delta_I = (B_{I,I} - B_{I,II} \tilde{B}_{II,II}^{-1} B_{II,I}) \Delta P_1 \quad (2.70)$$

$$\Delta \delta_j = (-\tilde{B}_{jj}^{-1} B_{j(j-1)}) \Delta \delta_{j-1} \quad (2.71)$$

for $j = 2, ..., N$ where

$$\tilde{B}_{kk} = B_{kk} - B_{k(k+1)} \tilde{B}_{(k+1)(k+1)} B_{(k+1)k} \quad (2.72)$$

for $k = 2, ..., (N-1)$ and $\tilde{B}_{NN} = B_{NN}$ Once changes in nodal bus voltage angles is computed, the change in bus angle difference is computed using

$$\Delta \delta_{ij} = M \Delta \delta \quad (2.73)$$

Step 3: The next step in estimating real power transmission losses is to compute line by line change in real power losses introduced by a transaction using the formulation described as,

$$\Delta P_L(\Delta \delta_{ij,1}) = G_{ij}(2\delta_{ij} \Delta \delta_{ij,1} + \Delta \delta_{ij,1}^2) \quad (2.74)$$

where $\Delta \delta_{ij,1}$ is defined, in the same way as before, as the vector of change in angle differences due to a change in power injection at node 1.

Step 4: Each market participant will compensate for the losses introduced by its transaction according to formula (2.74) by producing this much additional power.

The idea is that each system user should do this. Consider, for example, a bilateral transaction between parties X and Y. Let us suppose there is a load X purchasing in the primary market 100 MW from the power supplier Y. According to the algorithm described here, load X will estimate losses created by a 100 MW power injection at its node. Let us call this losses ΔP_X. Then the load will use $(100 - \Delta P_X)$MW only. Generator Y, on the other hand, will have then to compute how much losses will he create and produce $(100 + \Delta P_Y)$MW.

94   POWER SYSTEMS RESTRUCTURING

This algorithm is easily implementable using a little microprocessor by each significant system user. This is an example of a "smart" meter needed in the new power industry. It is demonstrated in [74] that losses are significant only in lines connecting buses in the first two tiers around the location where the change is made.

Appendix 2.5: Alternative definitions of transmission capacity

More precise definitions of transmission capacity are clearly needed, one which can accommodate the existing regulated power exchanges as well as multilateral interchanges in the potential future open networks. Several definitions come to mind, each of which provides a different perspective into the kind of exchanges that the network is capable of. Four definitions are suggested here:

- **The system transmission capacity (STC)** is the *maximum total incremental generation* above a normal point of operation that can be securely[56] transmitted to the set of loads.

  The STC is essentially the conventional definition of the transmission capacity [25].

- **The bus transmission capacity (BTC)** is the maximum amount of power at a given network bus, incremental above a given operating point, that can be securely injected into or extracted from the network *regardless of its destination or origin*. The BTC can be calculated for each bus.

  The BTC defines the maximum possible generation (or load) at each bus, incremental from a given operating point, regardless of the bus or buses to which (from which) this power is delivered (received). The BTC defines an upper bound on the capacity to send or receive power to or from any other buses in the network regardless of the consequences that operation at this bus maximum may have on the capacity to send or receive power at other network buses. Normally, it will not be possible to operate at this maximum since it may sharply reduce the transmission capacity of other buses. Nevertheless, the BTC does provide a measure of transmission capacity unique to each bus and could be useful to assess the degree to which a bus can participate in power interchanges. For example, the BTC can serve as an upper bound on any potential interchange transaction to or from this bus.

- **The bilateral transmission capacity (BITC)** is the *maximum amount of power* at a given network bus, incremental above a given operating point, that can be securely injected into (extracted from) the network and *transmitted to (from) another specific bus*. The BITC can be calculated for each pair of generation and load buses.

  The BITC is a special case of the BTC assuming a purely bilateral exchange of power between two specified buses. Once again, the BITC establishes an upper bound on the possible power exchanges through the network between two buses regardless of the possible impact that operation at this maximum would have on other potential interchanges. In solving for the BITC, the load flow permits real power changes from the given operating point in only the sending (ΔP_i) and receiving buses (ΔP_j) so that,

$$\Delta P_i + \Delta P_j = \Delta P_{ij}^{loss} \tag{2.75}$$

where ΔP_{ij}^{loss} is the incremental transmission loss due to the transaction between buses i and j. Again, one important application of the BITC is a bound on any potential bilateral power exchanges between two buses. The BITC also can be generalized to multilateral exchanges involving sets of sending and receiving buses.

- **The uniform bus transmission capacity (UBTC)** is the *maximum amount* of power at a network bus, incremental above a given operating point, that can be securely injected into or extracted from the network assuming a *uniform allocation of transmission resources to each generation bus*. The UBTC can be calculated for each bus or for pairs of generator and load buses.

In a single regulated utility, the uniform allocation of resources to the generators is normally done through economic dispatch. In an open system, the motivation for the UBTC is to prevent any single generator from hoarding the available transmission capacity at the expense of other possible transactions. Under this approach, each generating bus would be assigned a "fair" percentage of the available transmission capacity which would be used to its maximum possible absolute power level according to the capability of the network and consistent with all other generation buses also trying to maximize their output. Variations of this definition include: (i) Leaving the loads free to find the best possible UBTC, (ii) Requiring specific multilateral exchanges between specified groups of generators and loads.

Appendix 2.6: Security regions and transmission capacity

The concept of security regions is very powerful in conceptualizing and, under certain conditions, in explicitly characterizing the multi-dimensional security margins of power networks in terms of controllable parameters. The security margins are crucial in defining any type of power interchange and its corresponding capacity. One can conceptualize security regions under steady-state or transient state or both transient and steady-state.

In its most general form, a security region, S, is a set of vector quantities, $\theta(t) = [x(t), y(t), d(t)]$, for each of which, the power network satisfies all steady-state and transient operational requirements under both the existing network topology and a set of contingency-degraded network topologies. In order to understand the most general concept of a security region consider that a power system can be characterized by a set of differential-algebraic equations of the form,

$$\frac{dx(t)}{dt} = f(x(t), y(t), d(t), t) \qquad (2.76)$$

and

$$g(x(t), y(t), d(t), t) = 0 \qquad (2.77)$$

with

$$f(x(0), y(0), d(0), 0) = 0 \qquad (2.78)$$

that is, where everything is in stable equilibrium at $t = 0$. The quantity x is the dynamic equations state vector (machine and load dynamics) while y is the load flow state vector (P and Q injections, voltage magnitudes and phase angles, power flows, phase-shifters, tap-changers). The vector d is a disturbance and parameter vector which, in its most general form, can be assumed to represent all bus load model parameters as well as all fault and contingency data, machine parameters and any other system parameter.

Equations (2.76), (2.77), and (2.78) model the load flow and the system dynamics for the existing as well as for all possible contingency network topologies and system dynamics. The disturbance vector, $d(t)$, describes all disturbances including faults, load changes and outages with respect to which the system must remain stable. In addition to the stability requirements, the system variables must satisfy a number of steady-state operational inequalities on line flows, voltage and frequency limits. Both the steady-state stability and the stability requirements can be expressed in the form[57],

$$h(x(t), y(t), d(t), t) \leq 0 \qquad (2.79)$$

Then the security region, S, can be defined as the set of initial states $\theta(0) = [x(0), y(0), d(0)]$ which satisfy (2.76)–(2.79) for the set of disturbances defined by $d(t)$, $t \geq 0$, namely,

$$S = \{\theta(0) | \frac{dx(t)}{dt} = f(\theta(t), t) \; ; \; h(\theta(t), t) \leq 0\} \qquad (2.80)$$

For illustrations of these formula on small system examples, see [9]. In order to find such a limit, even for this simple system, normally requires a sequence of simulations which repeatedly vary the generation level, P_m, until it becomes impossible to meet all security constraints. A general tool to mechanize the recursive simulation process and find arbitrary boundaries of the security region for arbitrary power systems is reported in [10].

Transmission capacity and security constraints. All of the four above mentioned definitions of transmission capacity can be systematically defined in terms of S as follows. Let TC be a generic term for any of the transmission capacity definitions. Thus, as defined above, TC is some type of incremental power injection and therefore a function of θ, $TC(\theta)$. Since TC is a maximum

power transfer within the requirements of the security region, TC can be defined by,
$$TC = \max_{\theta \in S} TC(\theta) \qquad (2.81)$$
For example, the function to be optimized in order to find the system transmission capacity, STC, is,
$$TC(\theta) = \sum_{j=1}^{ng} \Delta P_j^{gen} \qquad (2.82)$$
where ng is the number of generators and ΔP_j^{gen} is the incremental output of generator j from its existing operating point. In this case, the loads may be assumed to be free to vary with respect to the present operating point perhaps within some pre-specified limits or they may be required to vary uniformly. If the loads are allowed to be free, the value of STC will be higher than if they are forced to vary uniformly. For elaborate illustration of the relations between specific notion of transmission capacity and security regions, see [9].

Appendix 2.7: Usage-based transactions management formulae and self-adjusting of system users

For the five bus example, the transmission service charge TS reflecting proximity of the line flow P_{f2-4} to its limit is computed as,[58]

$$TS = \alpha(P_{f2-4} - P_{f2-4}^{nom})^2 \quad \text{if } P_{f2-4} > P_{f2-4}^{nom} \qquad (2.83)$$
$$ 0 \qquad \text{otherwise} \qquad (2.84)$$

and it is further distributed to transactions currently using the system according to the relative contributions of transactions $T_{1,5}$ and $T_{2,4}$ to this line flow (using distribution factors $\frac{\partial P_{f2-4}}{\partial T_{1,5}}$ and $\frac{\partial P_{f2-4}}{\partial T_{2,4}}$),

$$\begin{aligned}
\Delta P_{f2-4} &= \frac{\partial P_{f2-4}}{\partial T_{1,5}} \Delta T_{1,5} + \frac{\partial P_{f2-4}}{\partial T_{2,4}} \Delta T_{2,4} & (2.85) \\
&= 0.0912 \Delta T_{1,5} + 0.4360 \Delta T_{2,4} & (2.86) \\
&= \Delta P_{f2-4}^{T_{1,5}} + \Delta P_{f2-4}^{T_{2,4}} & (2.87)
\end{aligned}$$

Corresponding transmission charges for $T_{1,5}$ and $T_{2,4}$ are,

$$TS^{T_{1,5}} = \frac{\Delta P_{f2-4}^{T_{1,5}}}{\Delta P_{f2-4}} TS \qquad (2.88)$$

$$TS^{T_{2,4}} = \frac{\Delta P_{f2-4}^{T_{2,4}}}{\Delta P_{f2-4}} TS \qquad (2.89)$$

Self-adjusting formula to the usage-based transmission price

Each system user applies a very straightforward formula when deciding how to adjust its transaction request in response to changing system security constraints.

98   POWER SYSTEMS RESTRUCTURING

The decision making formula is effectively identical to the formula (2.45) used in decentralized economic dispatch/unit commitment described in the Appendix 2.2 earlier, *except* for the additional term reflecting its relative transmission service charge (2.89), that is, in case of a power producer i for example,

$$\max_{P_{gi}}\{pP_{gi} - C_i(P_{gi}) - \sum_k \sum_n TS^{kn} P_{gi}\} \qquad (2.90)$$

where TS^{kn} is the charge to P_{gi} for effecting security margin in a transmission line kn. This adjusting process is applied in the five bus example for the utility functions at the five buses as follows,

$$C_1(P_{G_1}) = P_{G_1}^2 + P_{G_1} + 0.5 \qquad (2.91)$$
$$C_2(P_{G_2}) = 2P_{G_2}^2 + 0.5P_{G_2} + 1 \qquad (2.92)$$

for generators 1 and 2, and for loads

$$U_4(P_{L_4}) = 360.5P_{L_4} - 10P_{L_4}^2 \qquad (2.93)$$
$$U_5(P_{L_5}) = 241P_{L_5} - 5P_{L_5}^2 \qquad (2.94)$$

Since $P_{G_1} = P_{L_5} = T_{1,5}$ and $P_{G_2} = P_{L_4} = T_{2,4}$, the cost/utility functions can be rewritten in terms of $T_{1,5}$ and $T_{2,4}$.

$$C_1(T_{1,5}) = T_{1,5}^2 + T_{1,5} + 0.5 \qquad (2.95)$$
$$C_2(T_{2,4}) = 2T_{2,4}^2 + 0.5T_{2,4} + 1 \qquad (2.96)$$
$$U_4(T_{2,4}) = 360.5T_{2,4} - 10T_{2,4}^2 \qquad (2.97)$$
$$U_5(T_{1,5}) = 241T_{1,5} - 5T_{1,5}^2 \qquad (2.98)$$

Acknowledgments

The first author appreciates continuous help of her graduate students Petter Skantze, Yong Tae Yoon and Chien-Ning Yu, as well as long discussions with Ziad Younes concerning market aspects of the new industry. This writing is made possible through a partial financial support of the M.I.T./ McGill Consortium on Transmission Provision and Pricing for the New Industry.

Notes

1. This new framework is described in depth in Chapter 3 of this book.

2. In many restructuring documents term "compatibility" is used instead of the term "fairness".

3. The basic assumption that must be met is that any lower level is stabilizable faster than the rate at which the higher level actions take place. For other, less obvious, conditions, see [3].

4. Symbol $(\hat{\ })$ is used to denote a predicted variable (demand, price), and not its actual value.

5. A possible automated coordination at the systemwide level for remaining within the tie-line flow constraints is referred to as the tertiary control level [21, 20].

6. The actual physical variable regulated at a generator-turbine-governor(G-T-G) set level is the frequency setting F_{gi}^{ref}. It is well known that when the system frequency F is nominal so-called droop characteristics determines the actual real power P_{gi} produced. This distinction is potentially relevant when introducing objectives of AGC (task 4) for the competitive industry. For purposes of task 1 discussion we assume that real power output P_{gi} is directly controlled.

7. While all power plants are equipped with governors in the present industry, serious concerns have been expressed concerning the bandwidth settings ϵ inside which the governor is not responding.

8. As demand becomes more price elastic in real time this natural demand response may be considerably different, which, in turn, may effect the overall system stability.

9. NERC has recommended recently somewhat more relaxed criteria for AGC [27].

10. System demand also changes from year to year as a function of many macroeconomics forces, but this is generally of concern only at the planning level.

11. If a transaction is intended for the supply to balance demand, $\hat{P}_{gi}$ and $\hat{P}_{dj}$ are identical. However, some of the market participants may provide for transmission loss compensation, security and reliability, in which case the two are not necessarily identical.

12. Different accounting procedures exist for tradeable transactions; these are similar to the financial mechanisms in conventional forward markets, such as contracts for differences.

13. In some evolving ISOs power exchange and ISO are two separate functions, as it is the case in California, and in some others they are one entity, as proposed in New York and New England.

14. An example of a week ahead forward market is the Nordic system [6].

15. For detailed modeling of basic power balance between these bilateral entities, the reader should see the next chapter in which it is explained that any market transaction is effectively a bilateral transaction, even when one of the entities does not physically produce or consume power.

16. The term supply function is used interchangeably with the term cost function.

17. See chapter by Younes et al for modeling of market power.

18. Variations of different auctions for generation exist, see [22].

19. This difference could be viewed as a *virtual storage* of an electricity market, and is different from the *physical storage* which is likely to ultimately play a major role in the new electricity markets.

20. Discussion on CFD contributed by Ziad Younes.

21. In this section we assume an ideal transmission system, lossless and without any physical constraints.

22. For an industry in transition the long-term bilateral markets are usually not liquid, an additional problem to worry about.

23. The simulations in this section are provided by Chien-Ning, M.I.T.

24. This setup is often referred to as historical data. Notice that the historical data do not account for load price elasticity.

25. The derivation provided is for an entirely bilateral long-term tradeable market and short-term hourly coordinated market. Variations of this can be analyzed; for example, a long-term forward electricity market advocated in the chapter by Graves et al can be analyzed using similar approach.

26. This section draws heavily on the work by Petter Skantze [33].

27. These assumptions are made to simplify analytic expressions used and are not essential. Assumption 3 is easily eliminated. However, the others are not straightforward to model analytically; experimental economics tools are often used to gain insights regarding the validity of these assumptions in specific markets [44, 45], also see chapter by Sheble in this book.

28. This assumption is made strictly for the simplicity of derivations, and is not required.

29. The notation X_{long} and P_{long} in Figure 2.24 stand for the long-term power traded $\hat{P}_{LT}$ and the long-term expected price $\hat{p}_{LT}$.

30. This general insensitivity of system frequency to the relatively small power mismatch will create much confusion when attempting to unbundle various system services to the market.

31. See Appendix 2.4 for a summary of these formulae.

32. This is not to say that it is impossible to install this equipment; for example, in the Nordic electricity market the demand is measured down to retail level. The tradeoff between the cost of equipment installed and efficiency gains must be considered very carefully, however.

33. To the best one can tell, the term "compatibility" used by FERC in the restructuring documents implies equal rules for charging utility owned generation/demand system users and the others. No further quantification is suggested.

34. This was illustrated under task 1 earlier in this chapter.

35. The discussion here and the numerical example below are provided by Yong Tae Yoon.

36. This simplified pricing method within a zone is also referred to as the postage stamp pricing [70].

37. This is not true for an arbitrarily defined zone, however.

38. For a brief summary of secure conditions, see Appendix 2.6 as well as many technical references.

39. An effort in this direction has already been initiated by requiring that the so-called available transfer capability (ATC) be posted by system operators in the OASIS system. OASIS stands for Open-Access Same-Time Information System, and is an Internet-based system in forming that provides information about wholesale electrical power transfers in the United States [76].

40. A version of method 3 described below could directly be used for this purpose, except that instead of charging system users for effecting security margins one could use the same signal for adjusting the transactions so that the system remains secure.

41. Recall that an ISO has no economic data about bilateral transactions and because of it is not in a position to schedule all transactions with the objective of system efficiency maximization.

42. For more mathematical treatment of this difference, see chapter 3.

43. This is qualitatively different from only curtailing the first transactions when a constraint is encountered, and leaving the others unchanged. This approach will definitely create system constraints-induced uncertainties even to the market participants already using the system.

44. The charge for effecting system constraints is often referred to as the transmission congestion pricing.

45. These are typically obtained using several data points offered in terms of quantity and price.

46. Simulaitons of this scenario done by Chien-Ning, M.I.T.

47. Mismatches at faster time scales (0.1Hz and higher) are regulated by control loops that are "inner" to the AGC loops.

48. The same issues exist in today's industry, except that these are masked from the power consumers since only average charge is stated.

49. Note that an optimal policy need not be unique.

50. For this formulation, we assume that the generator is capable of selling as much power as desired at the market equilibrium price p_k. For analyzing the impact of generation constraints on decision making and a numerical example illustrating intertemporal effects, see [36].

51. See Younes et al in this text for an analysis when this assumption is not met.

52. This proof provided by Eric Allen, M.I.T.

53. The decoupling assumption may become very questionable in the new industry because of the tendency to transfer large amounts of power for economic reasons.

54. This matrix is routinely used in many tight power pools at present for contingency screening.

55. This is in contrast to the idea of MPs adjusting their power produced/consumed to account for transmission.

56. Securely implies that all steady-state and transient operational and contingency conditions are met.

57. The inequalities are time-varying since the system variables have different limits during the transient and steady-state conditions.

58. In this case, we choose $\alpha = 4$ $/pu^3$-hour. The choice of P_{flow}^{nom} is critical in this method; an "anchor" point for which the open access is designed must be defined, and all inefficiencies and transmission charges for new system enhancements should be measured relative to the nominal system design.

References

[1] Security Coordinator Procedures, NERC, October 14-15, 1997.

[2] Gaebe, G.P., "California's Electric Industry Restructuring", Proc. of the EPRI Workshop on The Future of Power Delivery in the 21st Century, La Jolla, CA, November 18-20, 1997.

[3] Ilić, M., Zaborszky, J., *Analysis and Control of Electric Power Systems*, Wiley Interscience, Inc. (in press)

[4] Ellerman, D.A., Schmalensee, R., Joskow, P.L., Montero, J.P., Bailey, E.M., *Emmissions Trading Under the U.S. Acid Rain Program*, MIT Center for Energy and Environmental Policy research, 1997.

[5] "The organized markets in Nord Pool, the Nordic Power Exchange; The Day-Ahead Market", Nord Pool, May 1996.

[6] "The organized markets in Nord Pool, the Nordic Power Exchange; The Futures Market", Nord Pool, November 1995.

[7] Graves, F., "Antitrust Questions Raised by Peak Price Intensity and Market Volatility in Deregulated Power Markets", Stanford Energy Modeling Forum, September 1997.

[8] Proceedings of the Workshop on Available Transfer Capability, National Science Foundation, June 26-28, 1997.

[9] Ilić, M., Galiana, F., Fink, L., Bose, A., Mallet, P., Othman, H., "Transmission Capacity in Power Networks", International Journal of Electrical Power & Energy Systems, 1998 (also, Proceedings of the Power Systems Computation Conference, Dresden, Germany, August 1996.)

[10] Marceau, R.J., Malihot, R., Galiana, F.D., "A generalized shell for dynamic security analysis in operations planning", IEEE Trans. on Power Systems, vol. 8, pp. 1099-1106, August 1993.

[11] Cheng, J., Galiana, F.D., McGillis, D., "Studies of bilateral contracts with respect to steady state security in a deregulated environment", Proceedings of the 20th International Conference on Power Industry Computer Applications, pp. 31-36, May 11-16, Columbus, OH.

[12] Radinskaia, E., Galiana, F.D., "Generation scheduling and the switching curve law", the IEEE Transactions on Power Systems (submitted for review, December 1997.)

[13] Hunt, S., Shuttleworth, G., *Competition and Choice in Electricity*, John Wiley, 1996.

[14] Hunt, S., Shuttleworth, G. "Unlocking the grid", IEEE Spectrum, July 1996, pp. 20-25.

[15] Fink, L.H., Ilić, M.D., Galiana, F. D., "Transmission access rights and tariffs", Electric Power Systems Research 43 (1997) 197-206.

[16] Younes, Z., Ilić, M., "Physical and Financial Rights for Imperfect Electric Power Markets", M.I.T. LEES WP 97-005, May 1997.

[17] Ilić, M.D., Hyman, L., Allen, E., Cordero, R., Yu, C-N., "Interconnected System Operations and Expansion in a Changing Industry: Coordination vs. Competition", , pp. 307-355, in *The Virtual Utility: Accounting, Technology & Competitive Aspects of the Emerging Industry*, Shimon Awerbuch and Alistair Preston (coeditors), Kluwer Academic Publishers, 1997, series on Topics in Regulatory Economics and Policy Series.

[18] Hyman, L.S., Ilić, M., "Scarce resource, Real Business or Threat to Profitability?", Public Utilities Fortnightly, October 1, 1997.

[19] Ilić, M., Hyman, L.S., "Transmission Scarcity: Who Pays?", The electricity Journal, July 1997, pp. 38-49.

[20] Ilić, M.D., Liu, S.X., "Hierarchical Power Systems Control: Its Value in a Changing Electric Power Industry", Monograph in the Springer-Verlag London Limited Series, Advances in Industrial Control, March 1996.

[21] Paul, J.P., Leost, J.Y., Tesseron, J.M., "Survey of the secondary voltage control in France", IEEE Trans. on Power Systems, vol. PWRS-2, 1987.

[22] Bernard, J., Either, R., Mount, T., Schulze, W., Zimmerman, R., Gan, D., Murillo-Sanchez, C., Thomas, R., Schuler, R., "Experimental results for single period auctions", Proc. of the 31st Annual Hawaii International Conference on System Sciences, vol. III, 1998, pp.15-23.

[23] North American Electric Reliability Council, "NERC Policy on Generation Control and Performance", 1997.

[24] Ilić, M.D., Graves, F.C., Fink, L.H., DiCaprio, A., "A Framework for Operations in Competitive Open Access Environment", The Electricity Journal, March 1996, Special Issue on Restructuring of the Utility Industry.

[25] Interconnected Operations Services: Defining system requirements under open access, NERC working document, 1997.

[26] Cohn, N., "Research Opportunities in the Control of Bulk Power and Energy Transfers on Interconnected Systems", Special Report on Power System Planning and Operations: Future Problems and research Needs, EPRI EL-377-SR, February 1977.

[27] VanSlyck, L., "A brief Overview of Control Criteria Development", IEEE Winter Power Meeting, February 1997 (submitted).

[28] Fernando, C., Kleindorfer, P., Tabors, R., Pickel, F., Robinson, S., "Unbundling the US Electric Power Industry: A Blueprint for Change", TCA Co. Report, March 1995.

[29] Hogan, W., "Contract Networks for Electric Power Transmission", Journal of Regulatory Economics, vol. 4, 1992.

[30] Hogan, W., "To Pool or Not to Pool: A Distracting Debate", Public Utilities Fortnightly, January 1, 1995.

[31] Lecinq, B.S., *Peak-load Pricing for Transmission in a Deregulated Electric Utility Industry*, Master of Science in Technology and Public Policy Thesis, MIT, June 1996.

[32] Lecinq, B.S., Ilić, M., "Peak-load Pricing for Electric Power Transmission", Proceedings of the 30th Hawaii International Conference on Systems Sciences, January 1997, pp. 624-632.

[33] Skantze, P., *Closed-loop market dynamics in a deregulated electric power industry*, Master thesis, M.I.T., Department of Electrical Engineering and Computer Science, February 1998.

[34] Johnson, R., Oren, S., Svoboda, A., "Equity and Efficiency of Unit committment in Competitive Electricity Markets", Washington D.C, June 1-2, 1995.

[35] Li, C.A., Johnson, R.B., Svoboda, A.J., "A new unit commitment method", IEEE Trans. on Power Systems, vol. 12, pp. 113-119, Feb 1997.

[36] Allen, E., Ilić, M., "Stochastic Unit Committment in a Deregulated Utility Industry", Proc. of the North American Power Symposium, Univ of Wyoming, October 1997.

[37] Wolak, F.A., "Market Design and Price Behavior in Restructured Electricity Market", Stanford Energy Modeling Forum meeting, 1997. CAREFUL

[38] *The US Power Market: Restructuring and Risk Management* Risk Publications, London, July 1997.

[39] Kaminski,V., "The challenge of pricing and risk managing electricity derivatives", ibid, pp. 149-175.

[40] Graves, F., Overview of utility finance, M.I.T. lecture notes, 6.683, October 1997.

[41] Green, R., "The electricity contract market", Working paper, Cambridge University, Department of Applied Economics, UK, May 1996.

[42] Newberry,D.M., "Power markets and market power", The Energy Journal 16 (3) 41-66.

[43] Okuguchi, K., *Expectations and Stability in Oligopoly Models*, Springer-Verlag, Lecture Notes in Economics and Mathematical Systems, 1976.

[44] Backerman, S.R., Rassenti, S.J., Smith, V.L., "Efficiency and income shares in high demand energy networks: Who receives the congestion rents when a line is constrained?", The Univ. of Arizona Working paper, January 21, 1997.

[45] Smith, V.L., "Regulatory Reform in the Electric Power Industry", Regulation, 1, 1996, pp. 33-46.

[46] Samuelson, P.A., "The stability of equilibrium: Comparative Statics and Dynamics", Econometrica, p. 111, April 1941.

[47] Shaw, J., "A direct method for security-constrained unit commitment", IEEE Trans. on Power Systems, vol. 10, pp. 1329-1342, August 1995.

[48] Bertsekas, D., Lauer, G., Sandell, N., Posbergh, T., "Optimal short-term scheduling of large-scale power systems", IEEE Trans. on Automatic Control, vol. AC-28, pp. 1-11, Jan 1983.

[49] Takriti, S., Birge, J., Long, E., "A stochastic model for the unit committment problem", IEEE Trans. on Power Systems, vol. 11, pp. 1497-1506, August 1996.

[50] Erwin, S., et al, "Using an optimization software to lower overall electric production costs for Southern Company", Interfaces, vol. 21, pp. 27-41, Jan 1991.

[51] Hara, K., Kimura, M., Honda, N., " A method for planning unit committment and maintenance of thermal power systems", IEEE Trans. on Power Apparatus and Systems, vol. PAS-85, pp. 427-436, May 1966.

[52] Bertsekas, D., *Dynamic Programming and Optimal Control: Volume I*, Belmont, MA, Athena Scientific, Inc., 1995.

[53] Kaye, R.J., Outhred, H.R., "A theory of electricity tariff design for optimal operation and investment", IEEE Trans. on Power Systems, vol. 4, May 1989, pp. 606-613.

[54] Kaye, R.J., Outhred, H.R., Bannister, C.H., "Forward contracts for the operation of an electricity industry under spot pricing", IEEE Trans. on Power Systems, vol. 5, February 1990, pp. 46-52.

[55] Barta, S., Varaiya, P., "Stochastic models of price adjustment", Annals of Economic and Social Measurement, 5/3, 1976, pp. 267-281.

[56] Negishi, T., "The stability of a competitive economy: A survey article", Econometrica, October 1962, pp. 635-669.

[57] Siljak, D.D., *Large-Scale Dynamic Systems*, North-Holland, 1978 (chapter four on Economics: Competitive Equilibrium).

[58] Dixit, K.A., Pindyck, S.R., *Investment Under Uncertainty*, Princeton University Press, 1994.

[59] Fink,L.H., Erkman, I., " Economic Dispatch to Match Actual Data to the Actual Problem", Proc. EPRI 1987 Power Plant Performance and Monitoring and System Dispatch Conference, EPRI CS/EL-5251-SR, pp.4/23-43.

[60] Ross, D.W., Kim, S., "Dynamic Economic Dispatch of Generation", IEEE Trans. v.PAS-99, n.6, 1980; pp.2060-68.

[61] Schweppe, F.C., Caramanis, M.C., Tabors, R.D., Bohn, R.E., *Spot Pricing of Electricity*, Kluwer Academic Publishers, 1988.

[62] Allen, E.H., "Stochastic Unit Committment in a Restructured Electric Power Industry", M.I.T. PhD Thesis, May 1998.

[63] Gross, G., Finlay, D.J., "Optimal bidding strategy in competitive electricity market", Proc. of the Power System Computation Conference, p. 815-823, Dresden, Germany, August 1996.

[64] Kirchmayer, L.K., *Economic Operation of Power Systems*, John Wiley & Sons, Inc., New York, 1958.

[65] Brown, E.H., *Solution of Large Networks by Matrix Methods*, John-Wiley & Sons, Inc., 1985.

[66] Desoer, C., Kuh, E.S., *Basic Circuit Theory*, Mc-Graw Hill Book Co., 1969.

[67] Alvarado, F.L., Hu. Y., Ray, D., Stevenson, R., Cashman, E., "Engineering foundations for the determination of security costs", IEEE Trans. on Power Apparatus and Systems, vol. PWRS-6, pp. 1175-1182, August 1991.

[68] Shirmohammadi, D., et al, "Evaluation of Transmission Network Capacity Use for Wheeling Transactions", IEEE Trans. on Power Systems, vol. 4, October 1989.

[69] Head, W.J., "MAPP Transmission Service Charge", Mid-Continent Area Power Pool, Minneapolis, MN, 1996.

[70] Ilić, M., Yoon, Y.T., Zobian, A., Paravalos, M.E., "Toward regional transmission provision and its pricing in New England", Utilities Policy, Elsevier Science Ltd., 1997.

[71] Ng,W.Y., "Generalized generation distribution factors for power system security evaluations", IEEE Trans. on POwer Apparatus and Systmes, vol. PAS-100, pp. 1001-1005, March 1981.

[72] Laffont, J-J., Tirole, J., *A Theory of Incentives in Procurement and Regulation*, The MIT Press, 1993.

[73] Perez-Arriaga, I., "A conceptual model for pricing analysis of tranmsisison services", IEEE WM 1991.

[74] Ilić, M., Cordero, R., "On providing interconnected operations services by the end user: Case of transmission losses", Proc. of the NSF Workshop on Restructuring, Key West, FL, November 1996.

[75] Phelan, M., *Loss Allocation Under Open Access*, M.Eng. Thesis, McGill University, January 1998.

[76] *Industry Report to the Federal Energy Regualatory Commission on the Future of OASIS*, Prepared by the Commercial Practices Working Group and the OASIS How Working Group, October 31, 1997.

[77] Barker, J., Tenenbaum, B., Woolf, F., *Governance and regulation of Power Polls and System Operators: An Internal Comparison*, July 1997 Draft, Forthcoming Technical Paper, World Bank Industry and Energy Department.

[78] *Order No. 888*, Federal Energy Regualtory Commission, April 1996.

[79] *Available Transfer Capability Definitions and Determination*, North American Electric Reliability Council, June 1996.

[80] Zobian, A., Ilić, M., "Unbundling of transmission and ancillary services. Parts I and II", IEEE Trans. on Power Systems, , May 1997, pp. 539-558.

[81] Ilić, M., Yoon, Y.T., Zobian, A., "Available transmission capacity (ATC) and its value unde ropen access", IEEE Trans. on Power Systems, vol. 12, May 1997, pp. 636-641.

[82] Walton, S., Tabors, R., "Zonal Transmission Pricing: Methodology and Preliminary Results from the WSCC", Electricity Journal, November 1996, pp. 34-41.

[83] Hogan, W.,"Nodes and zones in electricity markets: Seeking simplified congestion pricing", IAEE Conference, San Francisco, September 1997.

[84] Zaborszky, J. et al, "A clustered dynamic model for a class of linear autonomous systems using simple ennumerative sorting", IEEE Trans. on Circuits and Systems, vol. CAS-29, Nov. 1982.

[85] Happ, H., *Diakoptics and Networks*, Academic Press, NY, 1971.

[86] Siljak, D.D., *Decentralized Control of Complex Systems*, Academic Press, Inc., 1991.

[87] Wu, F.F., Varaiya, P., Spiller, P., Oren, S., "Folk theorems nd transmission open access: proofs and counterexamples", POWER report PWR-23, U of California Energy Inst., Oct 1994.

[88] Oren, S., Spiller, P., Varaiya, P., Wu, F.F., "Nodal prices and transmission rights: a crititcal appraisal", The Electricity Journal, vol. 8, April 995, pp. 14-23.

[89] Tan, C-W., Varaiya, P., "Interruptible electric power service contracts", Journal of Economic Dynamics and Control 17(1993) 495-517. Norht-Holland.

[90] Wu, F.F., Varaiya, P., "Coordinated multilateral trades for electric power networks: Theory and Implementation", POWER PWR-031., June 1995.

[91] *Reliability Assessment* 1997-2006, NERC, October 1997.

[92] Federal Energy Regulatory Commission: Capacity reservation open access transmission tariffs; notice of proposed rulemaking, April 24, 1996 (Docket No. RM96-11-000).

[93] Guan, X., Li, R., Luh, P., "Optimization-Based Scheduling of Cascaded Hydro Power Systems with Discrete Constraints", Proc. of the 28th North American Power Symposium, M.I.T., November 1996.

[94] Gedra, T., Varaiya, P., "Markets and Pricing for Interruptible Electric Power", IEEE Trans. on Power Systems, vol. 8, February 1993, pp. 122-128.

[95] Baldick, R., Kahn, E., *Contract paths, phase-shifters and efficient electricity trade*, POWER report PWR-035, U of California Energy Inst., October 1995.

[96] Elaqua, A. J., Corey, S.L., "Security constrained dispatch in the New York Power Pool", IEEE paper no. 82 WM 084-2.

[97] Ilić, M., Allen, E., Younes, Z., "Providing for transmission in times of scarcity: An ISO cannot do it all", MIT LEES WP 97-003, May 1997.

[98] Schweppe, F.C., Tabors, R.D., Kirtley, J.L., "Homeostatic control for electric power usage", IEEE Spectrum, July 1982, pp. 44-48.

[99] Ilić, M., Skantze, P., Yu, C-N., "Market-based frequency regulation: Can it be done, why and how?", Proc. of the American Power Conference, Chicago, IL, April 1-3, 1997.

[100] Ward, C., Colloquium on economic provision of reactive power for system voltage control, The IEE Savoy Place, London WC2R 0BL, UK, October 9, 1996.

[101] Zobian, A., Ilić, M., "A steady-state voltage monitoring and control algorithm using localized least squares minimization of load voltage deviation", IEEE Trans. on Power Systems, 1995.

[102] Alvarado, F.L., Broehm, R., Kirsch, L., Panvini, A., "Retail Pricing of Reactive Power Service", Proc. of the EPRI Conference on Innovative Approaches to Electricity Pricing, March 27-29, La Jolla, CA, 1996.

[103] Banales, S., Ilić, M., "On the role and value of voltage support in a deregulated power industry", Proceedings of the North American Power Symposium, Malarmee, Wyoming, October 1997.

[104] Verhese, G.C., Perez-Arriaga, I.J., Schweppe, F.C., Tsai, K.W., "Selective Modal Analysis in Electric Power Systems", EPRI EL-2830, Research project 1764-8, Final Report, January 1983.

[105] Ilić, M., Hyman, L.S., "Incentives for Transmission Investments", Proceedings of the EPRI Workshop on Power delivery in the 21st Century, La Jolla, CA, November 18-21, 1997.

[106] Skantze, P., Ilić, M., "Price-driven model for system control under open access: short-term gaming for profit vs. system regulation", Proc. of the North American Power Symposium, M.I.T., Cambridge, MA, November 1996.

[107] Ilić, M., Yu, C-N., "Minimal System Regulation and Its Value in a Changing Industry", Proceedings of the IEEE Conference on Control Applications, Dearborn, MI, September, 1996.

[108] Ilić, M., Liu, X.S., Eidson, B.D., Yu, C-N., Allen, E.H., Skantze, P., *A Unified Approach to Real-Time Control, Accounting, and Supporting Policies for Effective Energy Management under Competition*, MIT-EL 97-001, January 1997.

3 FRAMEWORK AND METHODS FOR THE ANALYSIS OF BILATERAL TRANSACTIONS

Francisco D. Galiana* and Marija Ilić**

*Department of Electrical Engineering and Computer Engineering
McGill University
Montreal, Quebec, Canada

galiana@pele.ee.mcgill.ca

**Department of Electrical Engineering and Computer Science
Massachusetts Institute of Technology
Cambridge, MA 02139 U.S.A

ilic@mit.edu

INTRODUCTION

Under competition and open transmission access, bilateral electricity transactions, which are nothing more than contracts between sellers and buyers, have become the new decision variables in power system analysis. Even under the traditional structure, where all generators belong to one entity and all loads buy their power from that same entity, it can be argued that transactions are bilateral, namely, from the individual generators to the 'utility' and from the 'utility' to the individual loads. With competition, however, many more types of bilateral transactions are possible. Thus, individual generators can now sell power directly to loads or to a pool or to trading entities such as marketers which, in turn, can trade with loads or among each other. Electricity transactions can take on a variety of characteristics such as firm or non-firm, short or long-term, or they can make use of financial instruments such as futures and options. Furthermore, transactions can involve trades of electricity in the form of energy or instantaneous power or reserve, both against each other or, alternatively, in return for other sources of energy such as gas, coal or hydraulic potential. With time, additional innovative types of transactions will emerge, limited only by the ingenuity of the market and the likelihood of making a profit.

Transactions, although essentially financial agreements, when actually implemented, define physical quantities in the network such as loads, reserves and generation levels, consequently influencing real power flows, voltage levels, losses, operational costs and, eventually, system reliability. Transactions must therefore be systematically modelled, planned and controlled if one is to continue to operate power systems efficiently and reliably under the new emerging structures.

Bilateral contracts can be classified according to:

(a) the type of trading entity (generator, load, marketer, pool), and according to the service provided, that is,

(b) demand,

(c) transmission losses,

(d) load following and frequency control,

(e) reserve,

(f) reactive power,

(g) congestion management.

It is unlikely that bilateral contracts for services other than demand will be voluntary (at least initially), meaning that so-called ancillary services will be provided by an Independent System Operator and the costs of these services will be passed on to the users on a pro rata basis. However, if the execution of ancillary services by competing entities ever becomes practical and it can be shown to reduce costs, we anticipate that even they will also be open to competition.

In this chapter, we present a mathematical framework and a set of methods for the analysis of the various types of existing and potential future transactions. The framework is consistent with existing models and tools of power network analysis, namely load flow, optimal power flow, enonomic dispatch and unit commitment. The framework makes it becomes possible to systematically analyze all aspects of power system operation and planning in the world of transactions, including loss allocation, transmission network usage, congestion management and the calculation of available transfer capability or ATC.

We begin by proposing a general method for characterizing bilateral transactions between any type of trading entity, be it generator, load, pool or marketer. Similarly, the notion of a transactions network is presented as a means of ensuring that all possible transactions are consistent with the physical bus loads and generators and that, at each node, the power balance is satisfied.

It will be generally assumed throughout this section that bilateral transactions are independently determined by market forces and that they, therefore, represent the sytem in-puts. The exception is the generation-to-pool and pool-to-load transactions which are determined by a central optimization procedure that maximizes the global benefit to both producers and consumers after subtracting the market-defined independent bilateral transactions.

However, as will be shown in this chapter, all transactions, regardless of how they are obtained, including the optimal pool transactions, can be shown to have bilateral generator-to-load (or bus-to-bus) equivalents. The calculation of these equivalent bus-to-bus transactions is necessary to be able to allocate a component of the transmission network usage and losses to individual contracts.

We also present systematic methods for loss allocation to bilateral contracts according to the location of buyer and seller, the contract amount and the loss provider. In general, for bilateral transactions, it is assumed that losses can be purchased from any generator willing to provide this service. It is then possible to search for optimum locations of this service relative to the contract being considered, both in the sense of low incremental losses but also costs.

As with losses, tariffs for network usage can be calculated as a function of the selling and buying bus locations. Such pre-calculated rates can then be utilized by traders to establish the most economic bilateral trades accounting for both generation and transmission costs. This approach can also be extended to pool transactions by dispatching according to the total of social benefits plus transmission usage costs., in this chapter, the notion of Available Transfer Capability (ATC) is discussed as a measure of the ability of the network to operate within its limits. We argue that, because of the complex interaction among multiple transactions, the conventional deterministic ATC measures can be deceiving, possibly too optimistic, or alternatively, too pessimistic. We therefore present a probabilistic notion of ATC which provides more realistic measures of the ability of the network to transact without curtailment due to congestion. Such probabilistic measures are shown to be useful in examining the impact of transmission rights as well as transmission expansion.

FRAMEWORK FOR THE ANALYSIS OF POWER TRANSACTIONS

Figure 3.1 shows the most general virtual network of power transactions. In contrast with the physical network which is composed of lines, transformers and other real devices conducting real power flows from generators to loads, the virtual transactions network models power transactions among financial entities. The arrows in Figure 3.1, represent the possible types and directions of power exchanges.

Through the virtual network, the transactions, in turn, define the real power generations and loads. These, then, fix the real power flows within the physical network which, ultimately, define the security of the network. Note, however, that a set of generation levels and loads do not uniquely define a set of transactions.

The general transactions network shown in Figure 3.1 consists of three types of *financial entities*: (i) the *individual generator-serving entities* representing the selling interests of individual physical generators, (ii) the *individual load-serving entities* representing the buying interests of retail loads, and (iii) the *trading entities* which are of three types: (a) *group generator-serving entities*, serving the selling interests of groups of individual generators, (b) *group load-serving entities*, serving the buying interests of groups of in- dividual loads, (c)

112   POWER SYSTEMS RESTRUCTURING

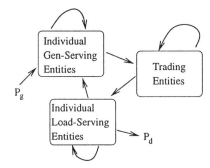

Figure 3.1 Virtual Network of Transactions among Financial Entities.

pure trading entities which trade for their own profit with individual or group entities of any kind.

Under these definitions, a utility that owns some generators will act as their group generator-serving entity in the open market. In this case, individual generators do not market their output, rather, the group entity markets the output of all its generators and dispatches each subordinate individual generator according to some economic criterion. An Independent Power Producer (IPP), on the other hand, can act as its own generator-serving entity and independently market its output to a trading entity or to a load-serving entity. Individual loads, also can join to form group load-serving entities, or each individual load could find its own suppliers in the market place. Load-serving entities should be independent of the generator-serving entities. A power pool is an example of a large load-serving entity that negotiates with all the generator-serving entities. Marketers are pure trading entities capable of both buying from or selling to anyone. A marketer could not, however, represent the interests of both generators and loads simultaneously without conflict of interest.

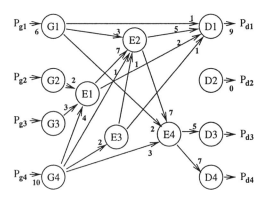

Figure 3.2 Example of virtual network of transactions.

An example of a virtual network of transactions is shown in Figure 3.2. The individual generator-serving entities are G1 through G4, the individual load-serving entities are D1 through D4, while E1 is a group generator-serving entity, E4 is a group load-serving entity and E2 and E3 are pure trading entities. Generator G1 represents an IPP trading directly with D1, E2 and E4. Generators G2 and G3 are uniquely represented by trading entity E1 which sells to E2 and D1. Finally, the load-serving entity E4 is supplied by G1 and G4 and by the pure trading entity E2. Note that load D2 does not trade with anyone as it has zero consumption.

Transactions matrices

The market defines the transactions network according to the type of bilateral trades that take place. The values of these trades can be represented by the following variables and relations,

(1) GD_{ik} = Contract for power sold by generator i to load k. DG_{ik} = Contract for power bought by load k from generator i. If, as will be shown later, transmission losses are a separate ancillary service, then,

$$GD_{ik} = DG_{ik} \tag{3.1}$$

or, in matrix form, defining the matrices of contracts, GD and DG,

$$GD = DG \tag{3.2}$$

Generally, $GD_{ik} \geq 0$, that is generators usually sell to loads and not vice-versa, however, if the interests of loads and generators are handled by serving entities, then these could also trade with each other and the above inequality need not necessarily apply.

(3) GE_{ik} = Contract for power sold by generator i to trading entity k. With no losses, $GE_{ik} = EG_{ik}$ = Contract for power bought by trading entity k from generator i. In matrix form, without losses, then,

$$GE = EG \tag{3.3}$$

(4) ED_{ik} = Contract for power sold by trading entity i to load k which must be equal to DE_{ik} = Contract for power received by load k from trading entity i. In matrix form,

$$ED = DE \tag{3.4}$$

(5) EE_{ik} = Contract for power sold by trading entity i to trading entity k. Since there are no losses in such transactions, it follows that this contract must be equal to the negative of the contract for power sold by entity k to entity i,

$$EE_{ik} = -EE_{ki} \tag{3.5}$$

or, in matrix form,

$$EE = -EE^T \tag{3.6}$$

Because a trading entity will not trade with itself, it is assumed that $EE_{ii} = 0$. The reciprocal nature of the EE-type transactions does not apply to GD or GE transactions, that is, for example, the power sold by generator i to load k, GD_{ik}, has no relation whatsover to the power sold by generator k to load i, GD_{ki}.

(6) It is also possible for generator-serving entities to trade with other generator-trading entities and likewise for load-serving entities. In this case, one can define the quantities GG_{ik} and DD_{ik} to characterize such trades. It is however possible to declare that the matrices GG and DD are zero without loss of generality if one considers the generator and load-serving entities as just another set of trading entities. In this case, the GG and DD matrices would be part of the EE matrix of trading entity contracts.

From the above definitions of bilateral matrices among generators, loads and trading entities, it is possible to express the physical generators and loads in terms of the contracts. We assume that the network has N_b buses at each of which, in general, there can exist both a generator and a load. Similarly, the number of trading entities is taken to be N_e. Then, assuming $GG = 0$ as argued above, the power generated at bus i, P_{gi}, can be expressed as,

$$P_{gi} = \sum_{k=1}^{N_b} GD_{ik} + \sum_{j=1}^{N_e} GE_{ij} \tag{3.7}$$

or, in matrix form,

$$P_g = GDu_b + GEu_e \tag{3.8}$$

where u_b is a vector of ones of dimension N_b while u_e is a vector of ones of dimension N_e. Next, the power balance for an arbitrary load at bus i, P_{di}, yields,

$$P_{di} = \sum_{k=1}^{N_b} DG_{ki} + \sum_{j=1}^{N_e} DE_{ji} \tag{3.9}$$

which in equation form becomes,

$$P_d = DG^T u_b + DE^T u_e \tag{3.10}$$

Finally, pure trading entities do not directly generate or consume power, so that any trading entity i must sell all of its purchases to either other trading entities or to loads,

$$\sum_{j=1}^{N_b} EG_{ij} = \sum_{k=1}^{N_e} EE_{ik} + \sum_{j=1}^{N_b} ED_{ij} \tag{3.11}$$

This can be put in matrix form as,

$$-EG^T u_b + EEu_e + EDu_b = 0 \tag{3.12}$$

FRAMEWORK & METHODS FOR BILATERAL TRANSACTIONS 115

All of the above bilateral matrices and relations can be systematically represented by the transactions matrix, T,

$$T = \begin{bmatrix} 0 & GD & GE \\ DG^T & 0 & DE^T \\ -EG^T & ED & EE \end{bmatrix} \quad (3.13)$$

such that,

$$\begin{bmatrix} P_g \\ P_d \\ 0 \end{bmatrix} = \begin{bmatrix} 0 & GD & GE \\ DG^T & 0 & DE^T \\ -EG^T & ED & EE \end{bmatrix} \begin{bmatrix} u_b \\ u_b \\ u_e \end{bmatrix} \quad (3.14)$$

Note that the diagonal blocks GG and DD have been assumed to be zero without loss of generality according to the arguments previously offered. Under lossless assumptions, equation (3.14) can be expressed in terms of the purchased quantities only, that is, DG, DE and EG by using equations (3.2), (3.3) and (3.4). Then,

$$T = \begin{bmatrix} 0 & DG & EG \\ DG^T & 0 & DE^T \\ -EG^T & DE & EE \end{bmatrix} \quad (3.15)$$

and,

$$\begin{bmatrix} P_g \\ P_d \\ 0 \end{bmatrix} = \begin{bmatrix} 0 & DG & EG \\ DG^T & 0 & DE^T \\ -EG^T & DE & EE \end{bmatrix} \begin{bmatrix} u_b \\ u_b \\ u_e \end{bmatrix} \quad (3.16)$$

As an example, for the transactions network of Figure 3.2 the transactions matrix takes the form,

$$T = \begin{bmatrix} 0 & 0 & 0 & 0 & 1 & 0 & 0 & 0 & 0 & 3 & 0 & 2 \\ 0 & 0 & 0 & 0 & 0 & 0 & 0 & 0 & 2 & 0 & 0 & 0 \\ 0 & 0 & 0 & 0 & 0 & 0 & 0 & 0 & 3 & 0 & 0 & 0 \\ 0 & 0 & 0 & 0 & 0 & 0 & 0 & 0 & 4 & 1 & 2 & 3 \\ 1 & 0 & 0 & 0 & 0 & 0 & 0 & 0 & 2 & 5 & 1 & 0 \\ 0 & 0 & 0 & 0 & 0 & 0 & 0 & 0 & 0 & 0 & 0 & 0 \\ 0 & 0 & 0 & 0 & 0 & 0 & 0 & 0 & 0 & 0 & 0 & 5 \\ 0 & 0 & 0 & 0 & 0 & 0 & 0 & 0 & 0 & 0 & 0 & 7 \\ 0 & -2 & -3 & -4 & 2 & 0 & 0 & 0 & 7 & 0 & 0 \\ -3 & 0 & 0 & -1 & 5 & 0 & 0 & 0 & -7 & 0 & -1 & 7 \\ 0 & 0 & 0 & -2 & 1 & 0 & 0 & 0 & 0 & 1 & 0 & 0 \\ -2 & 0 & 0 & -3 & 0 & 0 & 5 & 7 & 0 & -7 & 0 & 0 \end{bmatrix} \quad (3.17)$$

LOAD FLOW EQUATIONS IN TERMS OF TRANSACTIONS MATRICES

For simplicity, we assume that the load flow equations are governed by a relation between the net power injections, denoted by the N_b-dimensional vector, P, and the voltage phase angles, δ, of the same dimension[1], that is,

$$P = P_g - P_d = P(\delta) \quad (3.18)$$

116 POWER SYSTEMS RESTRUCTURING

where $P(.)$ is a function of the angles δ representing the load flow equations. Using (3.16), one can express the load flow equations in terms of the specified bilateral contracts. For the lossless case, this becomes,

$$P = (DG - DG^T)u_b + (EG - DE^T)u_e = P(\delta) \qquad (3.19)$$

while the lossy case is discussed in a subsequent section.

The importance of equation (3.19) lies in its ability to express network quantities such as transmission flows in terms of the market-specified bilateral contracts. From such a relation, as is shown next, one can derive how each individual contract affects the flows in the network and hence how usage-based tariffs can be systematically set-up. In more elaborate variations of the above load flow model, it is also possible to establish a relationship between the bilateral contracts and reactive power injections and voltage magnitudes.

Power flow distribution factors in terms of transactions matrices

One way to model the impact of a bilateral transaction on the network is to examine the power flow distribution factors in terms of the transaction matrices. Because of the special nature of the problem, it is useful first to go through the derivation of the flow distribution factors.

Under DC-load flow assumptions, the term $P(\delta)$ reduces to $B\delta$, where B is the symmetric load flow Jacobian matrix. From equation (3.19) then,

$$\delta = B^{-1}\left[(DG - DG^T)u_b + (EG - DE^T)u_e\right] \qquad (3.20)$$

If we denote the vector of network line power flows as, P_f, and its relation to the angles through,

$$P_f = M\delta \qquad (3.21)$$

where M is a matrix dependent on the network topology and the line susceptances, then,

$$P_f = MB^{-1}\left[(DG - DG^T)u_b + (EG - DE^T)u_e\right] \qquad (3.22)$$

The transaction participation factors are found by solving equation (3.22) as a linear combination of each individual contract. This can be readily done for the generator to load contracts DG_{ik} since the term $(DG - DG^T)u_b$ can be split into the sum of N_b^2 terms,

$$(DG - DG^T)u_b = \sum_{i=1}^{N_b}\sum_{k=1}^{N_b} u_{ik} DG_{ik} \qquad (3.23)$$

where u_{ik} is a vector of dimension N_b with all entries equal to zero with the exception of for entry i which is equal to 1 and entry k which equals -1. Since the term $B^{-1}u_{ik}$ exists[2], the vector of flow participation factors corresponding to contract DG_{ik} is then given by,

$$PF_{ik} = MB^{-1}u_{ik} \qquad (3.24)$$

so that component of the vector of line flows, P_f, due to the generation-to-load contracts becomes,

$$P_f = \sum_{i=1}^{N_b}\sum_{k=1}^{N_b} PF_{ik}DG_{ik} \qquad (3.25)$$

An illustrative example of the generation-to-load transaction participation factors is that of a 3-bus system with identical line susceptances and with a generator and load at each bus. The vector of line flows then becomes,

$$P_f = \begin{bmatrix} 2/3 \\ -1/3 \\ 1/3 \end{bmatrix}(DG_{12}-DG_{21}) + \begin{bmatrix} -1/3 \\ 2/3 \\ 1/3 \end{bmatrix}(DG_{23}-DG_{32}) \qquad (3.26)$$

$$+ \begin{bmatrix} 1/3 \\ 1/3 \\ 2/3 \end{bmatrix}(DG_{13}-DG_{31})$$

The flow participation factors corresponding to the contracts EG_{ik} and DE_{ik} are not as evident however. Whereas it can be shown that the term $(EG - DE^T)u_e$ is in the range space [3] of B, equation (3.22) cannot be solved for the effect on the flows of each individual EG_{ik} and DE_{ik} contract, given that the corresponding vector is not in the range space of B. Intuitively, this is correct, since a contract between a generator and a trading entity only specifies where the power is injected into the network but not the bus or buses where this power is consumed. Thus, unless one knows where the power bought by the entity is sold to, one cannot calculate its impact on the power flows. The difficulty is that the power bought by a trading entity may be sold in part to some loads and in part to numerous other trading entities which, in turn, may also trade this power in many ways. Eventually the power sold by a generator to a trading entity finds its way to the loads however the original power is split up into fractions which depend on the number and type of entity to entity and entity to load contracts.

Although the individual impact of generator-to-entity or entity-to-load transactions on the power flows cannot be evaluated, their combined impact can. This is done by showing that all transactions involving trading entities can be reduced to an equivalent parallel set of generation-to-load transactions whose impact can be analyzed through the generator-to-load transaction participation factors computed above. This approach essentially says that each generator-to-load transaction gets attributed a correction, the total of which accounts for the effect of all transactions with and among trading entities. Any charges for network usage due to transactions involving trading entities thus get passed on to the equivalent generator-to-load transactions which, it can be argued, are the ultimate payers for all transmission costs incurred.

118   POWER SYSTEMS RESTRUCTURING

Generality of bus-to-bus transactions

Pure trading entities, by buying and selling power from and to physical entities (generators and loads) as well as among each other, modify the flow of power through the network and, therefore, affect transmission flows and losses. Thus, such transactions, although carried out in the financial world, have an impact on the physical world and should be allocated some of the cost of this impact. By identifying equivalent bus-to-bus transactions, one essentially transfers the network usage responsibility from trading entities to the physical entities injecting and consuming power. The first step is to define the vectors,

$$de \triangleq DE^T u_e \tag{3.27}$$

and,

$$eg \triangleq EG u_e \tag{3.28}$$

Then eg becomes the vector of sending contracts between the generators and all the trading entities, while de becomes the vector of buying contracts between the loads and all trading entities.

Using the above definitions, the power balance relations can now be rewritten as,

$$P_g = DG u_b + eg \tag{3.29}$$

$$P_d = DG^T u_b + de \tag{3.30}$$

But since the total power bought by the loads from the trading entities must be equal to the total power sold by the generators to the trading entities, then,

$$eg^T u_b = de^T u_b \tag{3.31}$$

from which (3.29) can be rewritten as,

$$P_g = DG u_b + eg \frac{de^T u_b}{de^T u_b} \tag{3.32}$$

By a similar operation, the power balance for the bus loads, (3.30), becomes,

$$P_d = DG^T u_b + de \frac{eg^T u_b}{de^T u_b} \tag{3.33}$$

Thus, an equivalent generation-to-load bilateral matrix representing the impact of all trading entity transactions, DG^{eq} is,

$$DG^{eq} = \frac{eg\, de^T}{de^T u_b} \tag{3.34}$$

For the example of Figure 2 and equation (3.17), the specified generator-load bilateral matrix is,

$$DG = \begin{bmatrix} 1 & 0 & 0 & 0 \\ 0 & 0 & 0 & 0 \\ 0 & 0 & 0 & 0 \\ 0 & 0 & 0 & 0 \end{bmatrix} \tag{3.35}$$

and from (3.27), (3.28) and (3.34),

$$DG^{eq} = \frac{1}{20} \begin{bmatrix} 40 & 0 & 25 & 35 \\ 16 & 0 & 10 & 14 \\ 24 & 0 & 15 & 21 \\ 80 & 0 & 50 & 70 \end{bmatrix} \tag{3.36}$$

In essence, the equivalent generator-to-load transactions matrix takes the total power sold by each generator to the entities and distributes it among the loads in proportion to the fraction of the total power bought by each load. For example, from Figure 2, one can see that generator 1 sells a total of 5 units of power to the entities, that is, 5/20 of the total bought by the trading entities. Since load 1 buys a total of 8 from the trading entities, its equivalent purchase from generator 1 is therefore 8(5/20).

One of the obvious impacts of the equivalent transactions matrix is that equivalent contract links are created between generators and loads that do not have direct contracts. This can be clearly seen by comparing the original DG of equation (3.35) with the equivalent DG of equation (3.36).

The new DG is the sum of the two matrices.

$$DG^{new} = DG + DG^{eq} \tag{3.37}$$

USAGE-BASED TRANSMISSION FLOW PRICING

Under open access, the emergence of Independent Transmission Projects (ITP) is a distinct possibility. Each ITP could be permitted by the regulator to establish its own tariffs in \$/MWh through which the ITP would collect revenue on the basis of the level of usage of its facility by individual contracts.

Under a set of bilateral contracts, DG, a network tariff can be calculated for each contract based on the individual ITP tariffs and the transactions participation factors of equation (3.25).

Suppose that each line flow in the network, P_{fk}, has an assigned tariff, r_k, in \$/MWh, a charge which is applied to the absolute value of the contract component flowing through the ITP. Let r be the vector of flow tariffs. Then from (3.25), the total transmission charge for all contracts, TC, is,

$$TC = \sum_{i=1}^{N_b} \sum_{k=1}^{N_b} \left[r^T |PF_{ik}| \right] DG_{ik} \tag{3.38}$$

Thus, the network tariff for an arbitrary contract DG_{ik}, denoted by R_{ik}, is given by,

$$R_{ik} = r^T |PF_{ik}| \qquad (3.39)$$

where PF_{ik} is the vector of transaction participation factors derived in equation (3.24). The matrix of bilateral generator-to-load transmission rates, R, could be posted by the ISO for the use of potential traders.

Contracts from the equivalent transactions matrix would be charged in the same manner as direct bilateral transactions. However, one disadvantage remains when buying power from a trading entity, and that is that one cannot easily predict in advance what is the equivalent generation-to-load transactions imputed to the buyer seeing that equivalent trades depend on all contracts involving trading entities.

As an illustrative example of the notion of the matrix of contract transmission usage rates, R, consider the simple 3-bus example whose contract participation factors are given by equation (3.26). Let the individual flow tariffs be given by $rT = [1, 2, 3]$. Then, applying (3.39),

$$R = \frac{1}{3} \begin{bmatrix} 0 & 7 & 9 \\ 7 & 0 & 8 \\ 9 & 8 & 0 \end{bmatrix} \qquad (3.40)$$

so that contract 12 is the least costly in terms of total transmission tariffs.

One consequence of the availability of the matrix of bilateral transmission tariffs is that traders would have access to its contents and would therefore adjust their contracts accordingly. For the above example, the load at bus 1, all things being equal, would choose to buy its power from generator 2 with a transmission tariff of 7/3 rather than from generator 3 whose bilateral transaction has a tariff of 9/3. Note that the diagonal terms of R are zero since transactions between a generator and a load in the same bus do not load the network.

Combined bilateral and pool dispatch with and without independent transmission flow pricing

The existence of a matrix of bilateral transmission tariffs also influences the manner in which the generation under the control of a pool is dispatched. Since different flow patterns will result in different transmission charges to the pool, the minimum cost dispatch of the pool has to account for both generation and transmission costs. In the conventional economic dispatch problem, transmission usage rates are not a factor since it is assumed here that the network charges are the same regardless of the flow pattern.

To analyze the combined bilateral/pool dispatch problem, we first assume that the pool is voluntary, that is, unlike the England and Wales pool, for example, not all transactions need go through the pool. Next, the demand and generation vectors are split into two components each, one due to bilateral trades and another due to the pool dispatch. Thus, letting the superscripts p

and b denote pool and bilateral respectively, we have,

$$P_g = P_g^p + P_g^b \tag{3.41}$$

for the generation and,

$$P_d = P_d^p + P_d^b \tag{3.42}$$

for the load. The bilateral components are given by,

$$P_d^b = DG^T u_b \tag{3.43}$$

and, under lossless assumptions[4],

$$P_g^b = DG u_b \tag{3.44}$$

where the bilateral generation-to-load transaction DG is independently determined by market forces according to the generation bids and the transmission tariffs. It is also assumed that the matrix DG above includes both its original true generation-to-load transactions plus the equivalent values due to those contracts involving trading entities derived as indicated earlier.

We next assume that the pool generation component is dispatched according to the standard way, that is, by minimizing the total generation costs according to the bids submitted to the pool,

$$\min C(P_g^p) \tag{3.45}$$

where

$$C(P_g^p) \equiv \sum_{i=1}^{N_b} C_i(P_{gi}^p) \tag{3.46}$$

This minimization is subject to the load flow equations, expressed in the form,

$$P_g^p + DG u_b - P_d = B\delta \tag{3.47}$$

and to the usual limits on P_g and power flows. The solution of the optimization problem (3.45) yields the optimum levels of the pool generation dispatch, P_g^p.

In the presence of independent transmission usage tariffs, the pool will want to dispatch, P_g^p by minimizing both generation and transmission usage charges. The flow components due to the pool generation and load components can be shown to be (in the same manner as for bilateral contracts),

$$P_f^p = MB^{-1}(P_g^p - P_d^p) \tag{3.48}$$

so that the total transmission charge to the pool by the independent transmission projects is,

$$TC^p(P_g^p) = r^T |MB^{-1}(P_g^p - P_d^p)| \tag{3.49}$$

where r is the vector of individual transmission flow tariffs. The pool dispatch according to a minimum generation and transmission cost strategy then calls for the pool to solve,

$$\min \left[C(P_g^p) + TC^p(P_g^p)\right] \tag{3.50}$$

subject to the same constraints as above.

As an example of optimum pool dispatch accounting for transmission usage tariffs, consider a 3-bus network, each line having identical line susceptances of 10 per unit. Assume that each bus has both a load and a generator. Let the vector of loads, P_d, be given by $[2, 3, 4]^T$ per unit, while the vector of flow tariffs, r, is given by $[2, 0, 0]$ per unit. Let the non-zero tariff corresponding to flow 12 be 2 per unit, while flows 23 and 13 are assumed to be costless. Asume that the three generators submit identical price bids, namely, $C_k(P_{gk}) = 0.5 [P_{gk}]^2$ per unit. Without optimizing transmission costs, the minimum generation pool dispatch calls for each generator to produce equal amounts, that is, 3 units each. The corresponding flow in line 12 is then 1/3. The cost of generation is 13.5 while that of transmission is 0.67 for a combined total of 14.17. If the pool dispatches according to the minimum combined generation and transmission cost, the generation levels become $[5/2, 7/2, 3]$, that is, compared to the minimum cost dispatch, generator 1 decreases its output by 1/2 while generator 2 increases his by the same amount. The corresponding absolute flow in line 12 under this generation dispatch reduces to zero, resulting in zero transmission costs and generation costs of 13.75 for a combined total of 13.75. Naturally, this combined cost is less than 14.17, the amount the pool would have paid had it optimized only generation costs.

LOSS ALLOCATION CONTRACTS

Transmission losses form a significant component of the amount of power that has to be generated in order to meet the power demand. As an example, in a power network with a demand of 10,000 MW and 7% transmission losses, the implication is that the generation must be capable of supplying 10,700 MW, an extra 700 MW, fully a large power plant that must be built and operated, in essence to heat the atmosphere.

Clearly, someone must pay for both the capital investment and the fuel needed to generate the 700 MW of lost power. In the traditional utility, this cost is bundled into the rates together with other ancillary services and charged in some pro-rata fashion. With competition, this practice still persists but, more and more, there will be a need to allocate losses to transactions in a more systematic manner, particularly one that will account for the network location of the buyer and seller as well as the non-linear interaction among simultaneous transactions. For example, transactions where the seller and buyer are electrically close may not generate much in the way of losses. Similarly, some transactions may actually reduce overall system losses while others can have an opposite effect. Methods that can systematically identify such differences in behaviour are therefore required.

In this section, we assume that all transactions are bilateral between generations and loads. Generalizations to combined pool and bilateral transactions with losses, while also possible, is beyond the scope of this chapter. Similarly, the concept of equivalencing trading entity transactions to analogous generator-to-load contracts can also be ex- tended to the lossy case.

We begin by stating that because of the problem's non-linear nature, linear superposition cannot be used to analyze the impact of arbitrarily large independent transactions, DG_{ik}. Even if a non-linear load flow is run with one transaction at a time, the combined incremental transmission losses thus calculated will not generally resemble the calculated losses when all the transactions are treated simultaneously. To get around this difficulty, one must instead analyze the impact on losses of incremental contracts, dDG_{ik}, (since, for these, superposition does hold) and integrate over all increments.

Incremental power balance equation

The key relation in the subsequent loss allocation analysis is the incremental power balance equation which takes the form,

$$\sum_{i=1}^{N_b} \alpha_i dP_i = 0 \tag{3.51}$$

where dP_i is the net power injection at bus i given by,

$$dP_i = dP_{gi} - dP_{di} \tag{3.52}$$

The parameters α_i can be found in two ways, one is from the load flow equations and the other from the sensititivity coefficients of the loss with respect to the injections. These two equivalent methods are presented here.

From the power flow equations of equation (3.18), it follows that under incremental changes,

$$dP = dP_g - dP_d = \left[\frac{\partial P(\delta)}{\partial \delta}\right] d\delta \tag{3.53}$$

Since the load flow Jacobian matrix $\left[\frac{\partial P}{\partial \delta}\right]$ is singular (normally of full rank minus one), there exists a non-zero vector, α such that,

$$\left[\frac{\partial P(\delta)}{\partial \delta}\right]^T \alpha = 0 \tag{3.54}$$

Thus, from (3.53) and (3.54) this value of α, corresponding to a vector spanning the null space of the transpose Jacobian, satisfies the incremental power balance equation.

An alternative way of deriving the parameter vector α is to start with the standard power balance relation,

$$\sum_{i=1}^{N_b} dP_i = dP_L \tag{3.55}$$

where P_L stands for the system transmission losses. If we let bus k be the slack bus, from standard load flow sensitivity analysis, it is well known that dP_L can be expressed in terms of $N_b - 1$ of the incremental injections with the exclusion of the slack bus incremental injection, dP_k. Then,

$$dP_L = \sum_{i=1}^{N_b} \left[\frac{\partial P_L}{\partial P_i}\right]_k dP_i \qquad (3.56)$$

where $\left[\frac{\partial P_L}{\partial P_i}\right]_k$ is the sensitivity of the losses P_L with respect to P_i with bus k as the slack. Note that the kth sensitivity coefficient is zero. Thus, from (3.55) and (3.56),

$$\sum_{i=1}^{N_b}(1 - \left[\frac{\partial P_L}{\partial P_i}\right]_k)dP_i = 0 \qquad (3.57)$$

from which it follows that, within an arbitrary non-zero constant, a possible set of values for the α_i is,

$$\alpha_i = 1 - \left[\frac{\partial P_L}{\partial P_i}\right]_k \qquad (3.58)$$

We note in passing that, since the choice of the slack bus is always arbitrary, from (3.57) the following useful identity can be easily proven,

$$\frac{(1 - \left[\frac{\partial P_L}{\partial P_i}\right]_k)}{(1 - \left[\frac{\partial P_L}{\partial P_j}\right]_k)} = (1 - \left[\frac{\partial P_L}{\partial P_i}\right]_j) \qquad (3.59)$$

This identity allows us to calculate the loss sensitivities with bus k as the slack to the loss sensitivities with another arbitrary bus j as the slack.

Incremental loss allocation formulae

When a contract between generator j and load k varies by dDG_{jk}, by definition, the injection into load bus k must vary by,

$$dP_k = -dDG_{jk} \qquad (3.60)$$

while that into the generating bus j changes by,

$$dP_j = dDG_{jk} \qquad (3.61)$$

plus dL_{jk}, the incremental losses created by the transfer of power from bus j to bus k. It turns out, however, that the buyer at bus k need not necessarily buy its losses from the same bus from which it buys its load requirements, that is, from bus j. In fact, if the load supplier is far away, and the operating point is such that there is a net flow of power toward the buying bus, in such cases, it may be advantageous to purchase the required losses as close as possible to the consuming bus, as the value of the incremental losses, dL_{jk}, will then be lower. Alternatively, there may be situations where the incremental losses

FRAMEWORK & METHODS FOR BILATERAL TRANSACTIONS 125

will be lower if bought at a bus away from the load bus. In this theory of incremental loss allocation, it is supposed that the incremental losses associated with an incremental transaction, dDG_{jk}, can be purchased at an arbitrary bus s. Thus, since,

$$\sum_{i=1}^{N_b} \alpha_i dP_i = 0 \qquad (3.62)$$

then, by superposition, taking one contract increment at a time, if we consider an increment in contract jk, dDG_{jk}, then from (3.62) the following must hold,

$$\alpha_j dP_j + \alpha_k dP_k + \alpha_s dP_s = 0 \qquad (3.63)$$

where dP_k and dP_j are given by equations (3.60) and (3.61) while the associated incremental losses, dL_{jk}, are bought at bus s,

$$dP_s = dL_{jk} \qquad (3.64)$$

Then, from (3.60), (3.61), (3.63) and (3.64), the following *incremental loss allocation formula* is obtained,

$$dL_{jk} = \left(\frac{\alpha_k - \alpha_j}{\alpha_s} \right) dDG_{jk} \qquad (3.65)$$

Under the special case where the buyer and seller are at the same bus, the losses as defined above are zero, a result which is clearly reasonable. Similarly, there are special cases where the losses are bought either at the selling bus, j, or the at the buying bus, k.

Since the parameters α depend on the system operating point in a complex way, the incremental loss allocation formula (3.65) only gives a first order approximation of how an increment in a bilateral contract affects the transmission losses. One possible use of this result is for the ISO to periodically display the values of α as the operating point of the system varies. Independent traders can then decide based on (3.65), as well as on the prices for power and losses, whether a bilateral transaction is worthwhile including costs and losses. In such an approach, the buyer (bus k) would pay the supplier of its load at bus j and the supplier of its losses at bus s.

Exact solution of the loss allocation formulae

In order to obtain the exact values of the losses attributed to the specified absolute contracts, DG, it is necessary to add up (integrate) all the incremental contracts and their corresponding losses while tracking the operating point through the load flow equations. If the given bilateral contract matrix, DG, is assumed to vary linearly with a scalar parameter, t, then, $DG(t) = DG \cdot t$, and the process of finding the matrix of absolute loss allocations, L, involves solving a set of differential equations simultaneously with the load flow equations and the calculation of the vector α. These steps are:

(a) For all j and k, solve from $t = 0$ to $t = 1$,

$$\frac{dL_{jk}}{dt} = \left(\frac{\alpha_k - \alpha_j}{\alpha_s}\right)\frac{dDG_{jk}}{dt} \tag{3.66}$$

with $L_{jk}(0) = 0$,
(b) Solve for δ from the load flow equations,

$$P_g - P_d = (DG - DG^T)u_b + \tilde{L}u_b = P(\delta) \tag{3.67}$$

(c) Solve for α from,

$$\left[\frac{\partial P(\delta)}{\partial \delta}\right]^T \alpha = 0 \tag{3.68}$$

or from the loss sensitivity formula (3.58).

The matrix $\tilde{L}$ in (3.67) is found from the loss matrix, L, defined in (3.66), by reordering the L_{jk} elements according to the bus from which the losses are purchased. Thus, for example, if every bilateral generation-to-load contract wished to buy its losses from bus 1, then all of the L_{jk} elements would be added to generation 1 in (3.67).

An illustrative example of this theory is that of the two-bus system shown in Figure 3.3 consisting of a single lossy line with impedance $0.02 + j0.1$ per unit. Both buses have generation and load, and the bilateral contracts are such that the loads buy power only from the remote bus. Thus, the given DG matrix is,

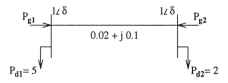

Figure 3.3 Illustrative two-bus example

$$DG = \begin{bmatrix} 0 & 2 \\ 5 & 0 \end{bmatrix} \tag{3.69}$$

We consider four cases and the results are shown in Table 3.1. Note that in all cases, the theory allocates a positive load component to contract DG_{21} and a negative one to contract DG_{12}. This is a reasonable phenomenon since in all cases the smaller 12 contract ($DG_{12} = 2$) tends to reduce system losses while the larger 21 contract ($DG_{21} = 5$) does the opposite. As such, on the whole, contract 12 gets rewarded while contract 21 is penalized. We note that Case A generates the highest system losses and the highest loss penalty to contract 21. This is apparent from the fact that, in this case, the 21 contract also buys its

Losses supplied by bus:	System losses	L_{21}	L_{12}
Case A: 1 for L_{12} and 2 for L_{21}	0.2252	0.3644	-0.1292
Case B: 2	0.2058	0.3429	-0.1372
Case C: 1	0.1803	0.3006	-0.1202
Case D: 2 for L_{12} and 1 for L_{21}	0.2004	0.3175	-0.1121

Table 3.1 Various cases of loss allocation. All quantities in per unit

losses from the distant bus 2 which happens to be a net generating bus. The lowest losses occur in Case C where both contracts, 12 and 21, buy their losses at bus 1, a net load bus. In this case, the loss penalty for contract 21 is also substantially reduced as this contract is now obtaining its losses from the local rather than the distant bus.[5]

Implementation Issues

The loss allocation scheme discussed here has the characteristic that it is able to distinguish between contracts which tend to decrease or increase losses. In contrast, other heuristic schemes based, for example, on a pro-rata distribution in proportion to the size of the contract, cannot identify such differences. In such schemes, every contract is assigned a positive part of the total losses, whether the contract tends to increase losses or not.

The difficulty with the above method is that the loss allocation terms are difficult to compute and to predict. Further research is presently under way to find quick methods which retain the penalty/reward feature while simplifying its calculation.

CONCLUDING REMARKS

The proposed framework for the analysis bilateral transactions has a number of additional potential applications beyond the ones discussed above. Among these one can mention the calculation of Available Transfer Capability (ATC) measures in the world of transactions, the study of the impact of firm and non-firm transactions on ATC, and the relief of transmission congestion through transaction based algorithms, both optimally, according to a specified criterion, or via simple schemes where certain transactions are modified by the ISO. Methods should also be developed for the analysis of pool/bilateral systems in the presence of losses. Finally, the fact that transactions and loads tend to be random and difficult to predict, plus the highly non-linear inter-relation among bilateral transactions, tend to make deterministic ATC values conservative and unrealistic measures of the ability of the network to transfer power. One alternative approach, is to simulate, through Monte Carlo methods, large numbers of random transactions from which one can extract probability distributions of those transactions which meet all the network security requirements. The result is a probabilistic measure of ATC for the transaction matrix DG. This is a more realistic measure than the deterministic ATC, as probabilistic measures allow all transactions to vary randomly and simultaneously. The resulting probabilistic ATC measures are specially useful as planning tools to analyze the impact of new network components or of long-term firm contracts.

Notes

1. That is, no specific angle is arbitrarily selected as the reference.

2. This is true since u_b belongs to the null space of B as well as being orthogonal to u_{ik}, thus implying that u_{ik} belongs to the range space of B.

3. This is true since it can be shown that u_b is orthogonal to $(EG - DE^T)u_e$ by the fact that the combined power bought from all generators by all the trading entities, $u_b^T [EG] u_b$, must equal the combined power sold by all trading entities to the loads, $u_b^T \left[DE^T \right] u_e$.

4. The lossy case is described in the next section.

5. A number of additional numerical tests have been carried out with this and other loss allocation schemes in M. Phelan, " Loss Allocation Under Open Access", M. Eng. Thesis, McGill University, January 1998

II INDUSTRY EXPERIENCES AND CHALLENGES

4 THE POLITICAL ECONOMY OF THE POOL

Richard Green[1]

University of Cambridge
Cambridge, England

Introduction

This chapter studies the Electricity Pool of England and Wales, better known simply as "the Pool". Practically all the electricity sold on the public supply system has to pass through the Pool, and the Pool sets an extremely visible marker price for electricity, even though most trading is actually based on contracts which hedge the Pool price. The chapter explains how the Pool was created, and why its rules and procedures were chosen. There are economic justifications for these rules, but there were often political reasons for choosing them, and the chapter title reflects this. The chapter then discusses some of the experience gained over the seven years for which the Pool has been operating. The Pool has succeeded in its task of coordinating the system's day-to-day operation, but it has attracted many criticisms for the level and volatility of its prices, and for the complexity of its rules. Those rules can be changed, but the Pool was deliberately designed to make such changes difficult, which has impeded some useful reforms. Outsiders, including the large customers who are among the Pool's fiercest critics, have also suggested potential changes to the Pool. Some of these seem unlikely to lead to greater efficiency, but the political need to "do something" about the Pool has ensured that any plausible suggestion can receive a hearing.

The Creation of the Pool.

Ten years ago, in the autumn of 1987, the British government decided that when it met its election pledge to privatize the electricity supply industry, it would also introduce competition. At the time, it did not know how to do this. The White Paper (Department of Energy, 1988) which set out the proposals for the industry in February 1988 accordingly concentrated on describing the new structure, which would split generation from transmission, and hinted at competition to supply large end-users of electricity, but said little on how it would be made to work. The White Paper contained an implicit assumption that competition would have to be organized around long-term contracts between generators and purchasers (either distribution companies or large customers), and when the industry started to design the details of the system, this was the model that was used.

Distributors would hold contracts with particular power stations, with a fixed availability component (to cover the station's fixed costs) and a unit rate related to the station's operating costs. The contracts were dispatchable, in the sense that when the distributor needed more power, it would call on one of the power stations which it had under contract, but it would not be obliged to take energy for which it had no need. In many ways, they were the standard contract offered to independent power producers, which leaves all the decision-making power, and the advantages of managing a portfolio of plant, with the buyers. In the longer term, companies planning to build power stations would be able to find buyers for their power (if the stations were needed) and sign new contracts which would give them the security to make a large, sunk, investment. Bilateral contracts like these work well in many industries, although the operator of a power system must retain the ability to over-rule proposed transactions which turn out to be infeasible.

In England and Wales, however, there was another problem. The White Paper had contained a pledge that the "merit order" system which the Central Electricity Generating Board (CEGB) had used to minimize its generating costs would be retained after the privatization. Under the merit order system, the dispatcher had always attempted to meet demand from the stations with the lowest variable costs, and the CEGB had actually used a suite of linear programming models, coupling generation with coal procurement and transport, in order to calculate and minimize these costs. After the privatization, however, there was no guarantee that a system based on bilateral contracts would even come close to the cost-minimizing solution.

To reconcile these two objectives, the industry started to design a "two Pool" system. Once they had forecast their demands, distributors would need to call on enough power stations to meet those demands, starting with the stations with which they held contracts. To reduce their costs, however, distributors who were about to call on expensive contracts would attempt to get the use of cheaper contracts which had been placed in a "distributors' pool" by companies which did not need to use them. If this pool worked efficiently, the portfolio of contracts which would be submitted to the generators would be the cheapest set of contracts capable of meeting the forecast level of demand.

The generators' contracts would inevitably be based on a particular set of power station availabilities, and the out-turn availabilities would almost always be different. This means that the least-cost combination of distributors' contracts would not be the same as the least-cost combination of available power stations. A second pool, the generators' pool, would allow lower-cost generators to offer to generate instead of higher-cost stations. If this pool worked efficiently, it would produce a generation schedule which would meet demand at the minimum cost. The industry spent nearly eighteen months trying to design this complex system, but the Department of Energy concluded that the proposed rules, which included a number of restrictions on the form of contracts and the generators' behavior, would be anti-competitive (Henney, 1994).

The industry's second attempt at a wholesale market conceded that station dispatch would have to be handled centrally, but required distributors to hold capacity tickets valid for the current period before they could buy from the market. The prospect of selling tickets should ensure that enough capacity was made available, and that new stations could obtain a contract before they were built. The problem with this system was that there will be occasional shortages of capacity (unless the industry has kept far too many stations open), and there would be a scramble to buy certificates at these times. Buyers were concerned that some companies might attempt to corner the market and drive the price up, while the generators were worried that "junk generators" with high running costs would be used to provide cheap capacity tickets giving access to the cheap energy from their capital-intensive plant. Since both sides of the market could see problems with it, the idea of capacity tickets was quickly abandoned.

The failure of separate markets for energy and capacity implied that both would have to be sold in a spot market, with, effectively, a single price. The theory of system planning implied that this price should be based upon the marginal cost of electricity, equal to the marginal operating cost of generation, plus the marginal value of capacity. When capacity is plentiful, its marginal value is zero, but when capacity is scarce, its price should rise. One way of measuring the value of capacity is by the marginal value (to the consumer) of the load which has to be displaced on the rare occasions when there is an actual shortage - this will produce a few very high values, but a value of zero whenever there is enough capacity. An alternative, which was adopted in England and Wales, is to use the expected value of capacity one day in advance: the value of the load which might be displaced, multiplied by the probability that there will not be enough capacity on the following day, and that load will have to be shed. This will produce lower payments, but many more of them, and the expected level of payments should be the same. Buyers also preferred the formula-driven approach, which appeared to leave less scope for the exercise of market power.

Although the "unified pool" was placed at the center of the new structure, it was still possible to hedge most transactions with bilateral contracts. The trick is that the contracts must be financial contracts for differences (CfDs), which specify a payment relative to the spot price. If a generator agrees to pay a distributor the amount (spot price - contract strike price) as part of a contract for differences, and the generator sells the same amount of electricity to the spot market, receiving the spot price, its net receipt is equal to the contract's strike price. If the strike price is high, and the contract allows the generator to receive payments from the contract as well as making them, the contract should be acceptable to both sides if the strike price is equal to the expected spot price. (In fact, the price for such a zero-premium two-way CfD might be used as a measure of expected spot prices, if we could be confident that the contract market is unbiased).

A different type of CfD can even replicate the bilateral contracts discussed above, at least in financial terms. Assume that the strike price is low, equal to the generator's marginal cost, and that the generator is unable to receive payments from the distributor - in the industry's jargon, it is a "one-way CfD".[2] To ensure that it is generating whenever the spot price is greater than the contract's strike price, so that it can cover the difference payments that it will then have to make, the generator should bid that strike price - its marginal cost - to the spot market. The contract will only be attractive to the generator if it contains a substantial fixed payment, equal (at least) to the expected difference payments. The combination of a payment which may cover the station's fixed costs, and a payment based on marginal cost for each unit generated, is similar to the bilateral option contract for physical delivery discussed above. The difference is that the task of dispatching the system is handled by the spot market, which can concentrate on finding an efficient dispatch. The spot market need not concern itself with the contract, which is purely a financial transaction, and involves no operational commitment. Indeed, some companies with no other involvement in the electricity industry have traded in the contract market.

In September 1989, the industry and the government decided that this could provide a workable wholesale market for a competitive industry. The remaining six months were spent in a frantic effort to put the idea into practice. The new wholesale market needed operating rules, and an organization to run it, and this organization needed a constitution. In practice, the Pool was created by the Pooling and Settlement Agreement, which is an agreement between the firms in the industry to create an unincorporated association for electricity trading. The Pool has a small central staff (the Chief Executive's Office), but most of the day-to-day work of running the market is sub-contracted to other companies.[3] The Pool Executive Committee, which had five generator and five supplier representatives, meets each month to oversee the Pool. Many issues are delegated to a network of sub-committees, while rule changes are subject to votes by all Pool Members.

The contracts putting the Pool into operation were signed (together with literally thousands of others) on the final day of the old system, March 30 1990, which was three months behind the original target date. Even so, the Pool's operating software was not yet ready, and a simplified version had to be used for the first few months. More importantly, many issues in the Pool Rules could not be decided in time, and, as we shall see, were postponed for later decisions. The need to create a workable system to a tight deadline meant that people had too little time to concern themselves with some of the longer-term strategic issues.

When the Pool began operating, it had 20 members: the twelve distributors (renamed Regional Electricity Companies, or RECs) and eight generators, most of whom also bought from the Pool as suppliers. The generators included three successor companies to the CEGB (National Power, PowerGen and Nuclear Electric) and three "external members" (ElectricitJ de France, Scottish Hydro-Electric and Scottish Power). By April 1997, the Pool had grown to have 55 members. The industry was overseen by the Director General of Electricity Supply, based at the Office of Electricity Regulation (Offer), who was entitled to attend Pool meetings. He was allowed to veto changes to the Pool Rules, and could hear appeals from companies which disagreed with potential changes, but had no day-to-day powers over the Pool.

The Pool's Pricing System

Between 1990 and 1997, Pool prices had three components: the System Marginal Price (SMP), the Capacity Payment, and Uplift. Since April 1997, Uplift has been divided into a Transmission Services Payment and a residual Uplift. SMP and the Capacity Payment together make up the Pool Purchase Price, while the Pool Selling Price also includes Uplift. SMP accounted for 85% of the demand-weighted Pool Selling Price between 1990/1 and 1995/6, while the remainder was evenly split between Uplift and Capacity Payments.[4]

To a large extent, the industry tried to continue to use its existing procedures for scheduling plant, which remains the responsibility of the National Grid Company (NGC). The CEGB used a computer

THE POLITICAL ECONOMY OF THE POOL 137

program to draw up the least-cost schedule which is capable of meeting the forecast demand, given cost information on each set, and the same program was used to draw up this unconstrained day-ahead schedule after the restructuring. (The term "unconstrained" means that transmission constraints, which will affect the actual dispatch, are ignored at this stage). Rather than their costs, generators now submit a set of prices for each of their generating sets, together with technical information on the sets' availability for the following day. In each half-hour, the bid of the most expensive set in normal operation is used to calculate the System Marginal Price (SMP), which is paid for every unit generated in the day-ahead unconstrained schedule. The computer also calculates the risk that the available plant will be unable to meet the forecast level of demand, the loss of load probability (LOLP). This probability is combined with the value of lost load (VOLL), set at £2/kWh in 1990, and uprated annually with inflation, to give the expected cost of power cuts. A capacity payment, equal to LOLP × (VOLL - SMP) is added to SMP to produce the Pool Purchase Price. Stations which are available, but not required to generate, are paid a capacity payment equal to LOLP × (VOLL - their bid price). The cost of these unscheduled availability payments is recovered as part of a charge called Uplift.

Hunt and Shuttleworth (1993) suggest that this part of the Pool is not a spot market, but a forward market in which generators agree plans to make deliveries on the following day, while the capacity payments represent an option market which gives the controllers the right to ask for additional generation if required. (In practice, these plans are not binding, and generators are not always penalized if they fail to produce their scheduled output.) The time at which the Pool comes closest to a true spot market is when the unconstrained schedule is reconciled with actual generation. The forecast level of demand will never be quite accurate, while some sets may have technical problems, and will have to be replaced by others. Some sets may be asked not to generate because of transmission constraints, when the grid system could not absorb their output safely, so that another station, closer to demand, will have to be used instead. If a station's generation is different from that in the forecast schedule, then it will sell the excess to the Pool (or buy back a shortfall), generally at its own bid price. This would result in the Pool making net payments, which are known as Operational Outturn, and were also recovered through Uplift, since the Pool has to balance on a daily basis. A third component of Uplift covered the cost of ancillary services bought by NGC, such as reserve, which are needed to keep the transmission system stable. The Pool Purchase Price plus Uplift make up the Pool Selling Price, paid for all electricity bought from the Pool.

Since April 1997, Uplift has effectively been divided into two parts, although the arrangements for paying generators have not changed. NGC now makes the payments which are part of "Transport Related Uplift", and recovers the cost through a regulated Transmission Services Use of System Charge. The remaining "Energy Uplift" payments are made by suppliers, as before. This follows on from an Uplift Management Incentive Scheme, and a Transmission Services Scheme, which exposed NGC to a share of the cost of those parts of Uplift (particularly Operational Outturn) which the company could influence, in order to improve its incentives to keep those costs down.

System Marginal Price

SMP is the price paid for each unit of energy in the unconstrained schedule, and might be seen as the marginal cost of energy. Given the haste with which the system was designed, the industry kept as close to its existing procedures as possible. The CEGB had used a computer program known as GOAL (Generation Ordering and Loading) to construct its merit order and its operating schedule, and this program was inherited by the Pool. GOAL had three main inputs: a demand forecast, cost information on each generating set, and operating parameters for each genset. The operating parameters include the genset's availability, its minimum on and off times, its ramp rates, and its minimum stable generation levels. These act as constraints on the schedule that GOAL draws up. The demand forecast was drawn up by the CEGB, using information supplied by the distributors, together with weather forecasts and television schedules. (There can be a "tv pick-up" of up to 1% of demand at the end of a particularly popular program, since there are few breaks in the middle of programs (two of the four main channels had no commercials) and live coverage of a major sporting event is practically never interrupted for commercials). GOAL scheduled plant to meet the peaks in demand, and then reduced output from the more expensive gensets at off-peak times. Under the CEGB, GOAL required five pieces of cost information for each genset: a start-up cost (per start), a no-load cost (per hour), and up to three incremental rates (per MWh) (corresponding to different ranges of output). Three rates were allowed in order to provide a linear approximation to the non-linear Willans line which gives the relationship between heat input and energy output from a genset. The cost components were based upon the station's heat rates, multiplied by its delivered cost of fuel. This was obtained by iterating between a model of coal purchase and transport, and a station output model, (with a time scale of several months) until the system's total costs were minimized. GOAL then attempted to minimize the cost of meeting the forecast level of demand.

After the restructuring, the Pool continued to use GOAL, but the generators submitted bid prices, rather than the cost information used by the CEGB. They were free to determine these prices, although their technical parameters are required to reflect the genset's actual operating characteristics. These bids have to be submitted by 10 a.m. one day ahead of operation. GOAL is then run to minimize the cost of generation (based on these prices) over a thirty-nine hour period (to ensure that decisions to turn plant on or off at the end of the first day take the needs of the following day into account). Once GOAL has identified its cost-minimizing schedule, a separate module calculates the System Marginal Price for each half-hour.

For most half-hours, SMP is equal to the average cost per MWh of the marginal genset. These Table A periods are those in which extra demand has to be met by turning on another genset, and so it may be appropriate to include start-up and no-load costs in the price of energy. The basic formula for this genset stack price adds a share of the genset's fixed costs (the cost of starting it, and running it unloaded for the duration of its period of operation) to its incremental price:

$$\text{Genset price} = \text{Incremental price} + \frac{\text{no-load} \times \text{duration} + \text{start-up}}{\text{total output}}$$

We could also write this formula as the genset's total cost, divided by its output:

$$\text{price} = \frac{\text{Incremental price} \times \text{output} + \text{no-load} \times \text{duration} + \text{start-up}}{\text{total output}}$$

If the genset has submitted more than one incremental price, and is generating above its elbow point (the level of output at which the second incremental price applies), an adjusted no-load price is used, equal to

$$\text{adjusted no-load} = \text{no-load} + (\text{incremental}_1 - \text{incremental}_2) \times \text{elbow point}$$

If this adjusted no-load price was applied to each individual half-hour for which the station is generating above the elbow point, we would have a genset price equal to:

$$\text{price} = \frac{\text{Inc}_1 \times (\text{output} - \text{elbow}) + \text{Inc}_2 \times \text{elbow} + \text{no-load}}{\text{output in that half-hour}} + \frac{\text{start-up}}{\text{total output}}$$

The left-hand fraction is equal to the total cost of running the genset in a half-hour, divided by the output in that half-hour, while the right-hand fraction adds a share of the cost of starting the set, divided by the total output from the period of operation. In practice, the no-load prices are averaged over the day as a whole before the stack prices are derived, and there are circumstances in which this averaging will mean that the stack price is not close to the genset's average cost.[5] There are also circumstances in which a genset which is operating for a very short period (and has high start-up costs per unit generated), or one which has been declared to be inflexible (and is scheduled on the basis of its operating parameters, not its price bid) will not be used to determine SMP. On the whole, however, SMP in table A periods, equal to the stack price of the marginal genset, can be seen as an approximation to the cost of the marginal genset.

There are some times when it is cheaper to operate several gensets at reduced loads than to turn some of them off, and on again a few hours later. The Pool rules set out the criteria by which these times, known as Table B periods, are identified and specify a maximum number in any day. At these times, SMP is equal to the marginal genset's incremental price, ignoring start-up and no-load prices, since no stations are being started, or kept on load, specifically to meet demand at those times. While this is a plausible economic argument for the distinction, its main result is to reduce night-time prices. This helps electricity to compete with gas in the domestic heating market, using the "Economy 7" tariff, which offers seven hours of lower prices at night. The RECs would have been keen to keep this tariff competitive against a falling gas price.

Lucas and Taylor (1994) use another cost-based argument to suggest that start-up prices should only be included in SMP at peaks in demand, since it is only at these times that plant has to be started specifically to meet demand in that half-hour: the plant which is started to meet demand at other times would be needed for the peak anyway. They argue that Pool prices are likely to be excessive because of this. In practice, however, adjustments of this type are likely to change the pattern of Pool prices in equilibrium, but may not change their average level. In equilibrium, Pool prices should be just sufficient to remunerate all the plant on the system: higher prices would tempt entrants into the industry, while lower prices would lead to plant closures. If entrants are likely to build an efficient modern station for baseload operation, then they will receive the time-weighted average price, and the threat of entry might keep this average close to the cost of such a station.

Capacity Payments

Similar considerations affect payments for capacity. If marginal stations cannot cover their costs, then they will tend to shut down, reducing the margin of spare capacity. A well-designed system will ensure that when the industry has an appropriate level of capacity, the marginal station just covers its costs. Note that the appropriate level of capacity is in fact endogenous - it should be that level where the cost of the marginal station is just equal to the benefits which it brings, in terms of a reduction in the expected cost of power cuts. System planners frequently use fixed margins (the CEGB once used a 28% planning margin in its seven year projections!), but in the Pool, the level of capacity is decided by the generating companies, based upon price signals.

Every MW of capacity which is declared available in a half-hour receives a capacity payment for that half-hour, whether or not it is scheduled to generate. The payment is intended to reflect the value of that capacity, in terms of the cost of the outages which it is expected to prevent. This depends on the Loss of Load Probability (LOLP) and the Value of Lost Load (VOLL). LOLP is calculated by the Pool to measure the risk of a power cut (and hence the level of lost load which an extra MW of capacity would prevent), while VOLL was set by the government to equal £2/kWh in 1990, and is uprated annually in line with the retail price index. Some earlier studies had suggested that £2/kWh reflected the value that consumers placed on avoiding power cuts (at a time when the general price level was rather lower!), but it was also presented as the number which was consistent with the industry's existing operating margin for capacity, implying that security standards would be maintained. The net benefit of avoiding the loss of a MWh is equal to VOLL, less the cost of generating that MWh. For a station which is scheduled to generate, that cost is taken to be SMP, while for a station which was not scheduled, its own bid is used. The capacity payment for a station which is scheduled to generate is therefore equal to:

Loss of Load Probability × (Value of Lost Load - SMP)

This is added to SMP to give the Pool Purchase Price, paid for every unit generated in the unconstrained day-ahead schedule. The capacity payment for other stations is equal to:

Loss of Load Probability × (Value of Lost Load - bid)

The cost of these Unscheduled Availability payments made in each half-hour is recovered through Uplift. LOLP is calculated by a commercial program, which takes a forecast of demand (and its variance), together with a disappearance ratio and the capacity for each genset which has been declared available in the preceding eight days. A period of eight days is used in order to reduce the opportunities for increasing the value of LOLP, and hence capacity payments, by withholding capacity for short periods. A generator could still raise LOLP by delaying the return to service of a station which had already been out of action for eight days, but could not affect LOLP by taking a station out of service for less than a week. The disappearance ratio measures the probability that a genset which has been available at some point in an eight-day period will become unavailable by the ninth day, and is therefore consistent with the eight day availability record. The disappearance ratios for older stations are based on their performance before 1990 (which is often much less reliable than their more recent record), while newer stations use their performance over the previous year (or shorter periods while the stations are still very new). Given this data, the program calculates the probability that the out-turn level of demand will be greater than the available capacity. High values of this probability, known as JOLP, are then reduced to take account of the assumed elasticity of demand and supply (load management and increased autogeneration in response to high prices). The resulting value, KOLP, is then averaged over three adjacent half-hours to produce the smoothed variable used by the Pool, LOLP.

LOLP is extremely sensitive to the level of capacity relative to demand. The table below shows figures for KOLP, taken from January 23, 1995, for five levels of demand that occurred during the early evening. The peak demand level at 5.30 p.m. raised KOLP to 0.368, implying a charge of around £850/MWh, although the averaging process meant that this value of KOLP actually contributed 0.123 to the value of LOLP for 5.00, 5.30 and 6.00 p.m. that day, and the peak capacity payment, at 5.30, was £570/MWh.

Demand (GW)	40.3	42.6	43.0	44.3	46.5
KOLP	0.001	0.019	0.033	0.093	0.368

Capacity payments have been heavily criticized. For much of the time, they are negligible, but when capacity is scarce, they can rise to several hundred pounds per MWh. At these levels, some consumers who have chosen to face Pool prices feel forced to reschedule their production. As long as LOLP truly measures the risk of a power cut, this is an appropriate response, for these consumers should respond to the price signal in relation to the value they derive from being able to consume electricity, with those who gain least from their consumption reducing demand the most. One concern is that LOLP may overstate the risk of an outage, and hence the need to reduce demand.[6] A second problem is that some customers facing Pool prices may not have been warned about the true nature of the risky contracts which they were signing. On average, Pool-related supply contracts have often turned out to be cheaper than the fixed price alternatives (which include a hedging premium), but this is quite consistent with the customer on Pool prices having to pay more than the fixed price in one month, or even one year.

Despite this, capacity payments are an appropriate way to pay for capacity, and to allow the market to determine the correct level. The decision to build a power station depends upon long-term forecasts, and once a decision is made, there is an inevitable delay for construction before the station can affect the level of capacity. This means that from year to year, the out-turn level of capacity usually depends on decisions on whether or not to close power stations. For the market signal to work correctly, the owner of a single station should have an incentive to keep it open if the value of having that station on the system is greater than the cost of doing so, and to close it if not.[7] Capacity payments are intended to equal the value of having a station available to generate, so that if they are expected to be greater than the cost of keeping the station open, then it should be kept open. The station owner might consider it too risky to keep the station open for the prospect of a series of half-hourly payments, determined day by day, but this risk can be hedged. As long as there is a liquid market for contracts for differences which hedge capacity payments, then the station owner should be able to sell a contract which fixes its revenues in advance.[8] The contract's buyer could use the contract to hedge its payments under the unscheduled availability component of uplift.

If the price of a one year "LOLP hedge" is greater than the cost of keeping the station open, this implies that the expected cost of power cuts is greater than the cost of preventing them, and the station could be kept open. If the price of the contract is less than the cost of keeping the station open, then it is not worth keeping it on the system, and the owner should be encouraged to close it. In the absence of a contract which would cover its costs, this is likely to happen. The contract therefore leads to an appropriate level of capacity, and allows both parties to know their payments in advance (the buyer to a lesser extent), but they still both face appropriate price signals at the margin. The station owner is committed to passing on the value of the capacity payments it could receive, whether or not the station is open and earns them, but if these are low, it may be cheaper to keep the station closed for a day than to incur the cost of manning it, in return for very little revenue. If capacity becomes scarce, and payments rise, then the gain from keeping the station open will be higher, and the owner will make every effort to do so. Similarly, the contract's buyer knows how much they would have to pay for a steady demand, since their receipts from the contract should cover their payments to the Pool, but if the price in a period rises above the value of their consumption, then

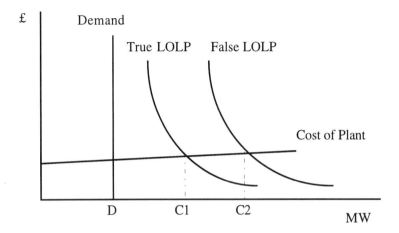

Figure 1: LOLP and Capacity

they can make themselves better off by reducing demand and keeping the payments from the contract.

We can also use this way of looking at the relationship between the out-turn level of capacity and capacity payments to consider the impact of mis-calculating LOLP, which was mentioned earlier as a possible problem. If LOLP is over-estimated, then at a given level of capacity, capacity payments will be higher than with the true value. This will encourage generators to keep more plant on the system than with the true value (since the system appears to be less reliable). The equilibrium level of capacity occurs when the value of capacity payments is just equal to the cost of keeping capacity open. In figure 1, the equilibrium level of capacity should be C1 (based on the "True LOLP" curve), but the actual level would be C2 (based on the "False LOLP" curve). If all the marginal plants have a similar cost, then the error in LOLP will increase the number of plants kept open, but the average out-turn level of LOLP should not be significantly affected by the error in its calculation. The actual risk of a power cut will be "too low", giving consumers excessive quality. They will have to pay for this through increased Unscheduled Availability payments (since the level of spare plant on the system is C2-D rather than C1-D), but they should not expect the Pool Purchase Price to be higher because of the error.

Uplift

Uplift was the least sophisticated part of the Pool's pricing arrangements. It was created to recover all the costs of generation which were not met by SMP or capacity payments for generation in the unconstrained schedule. These included capacity payments for stations which were not scheduled to generate (unscheduled availability payments) and payments for ancillary services bought by NGC, such as short notice reserve, and reactive power. In most years, however, the largest component of Uplift has been the Operational Outturn payments which reflect the cost of differences between the unconstrained schedule, and actual generation. There are three main causes of these differences. First, the demand forecast will generally be slightly inaccurate, so that more or less generation is needed than was predicted. Second, some generators will deviate from their operating schedule (by accident or otherwise), and some stations may become unavailable for technical reasons. Other stations will have to generate more or less to compensate for this. Third, the grid controllers will have to instruct some stations not to follow the unconstrained schedule, in order to take account of transmission constraints. Stations in an 'export constrained' region will have to produce less, while those in an 'import constrained' region will have to produce more.

In general, stations which generate more than in the unconstrained schedule sell the excess to the Pool at their own bid price, while those which generate less buy back the difference, also at their bid price. Since stations which had been scheduled to generate, and are now buying back from the Pool, will have a bid price which was lower than SMP, any reduction in their output will (weakly) involve a cost to the Pool. Similarly, stations which had not been part of the unconstrained schedule will have a bid price which is greater than SMP, and so the increase in their output will also raise the Pool's costs. When one station replaces another, the difference in their bids is added to Operational Outturn. If demand is greater (less) than the level in the unconstrained schedule, the Pool receives (loses) SMP for the divergence, and so the cost for Operational Outturn is the difference between SMP and the station's bid.

Unscheduled Availability payments are mainly recovered in the half-hour to which they relate. The other components of Uplift are spread across every Table A period as an equal charge per MWh, with no differentiation between regions. This meets two objectives - that the payments into and out of the Pool should balance on a daily basis, and that the system should be simple, but may not reflect costs very well.

Transmission Services

Uplift rose rapidly in the first years of the Pool's operation, and because most CfDs are based on the Pool Purchase Price (SMP and Capacity Payment alone), many suppliers were exposed to this rise. This meant that the regulator came under pressure to "do something", but it was also clear that the incentives in the system could be improved. Many of the costs recovered through Uplift depend upon decisions taken by NGC, but the company had no incentive to keep these costs down. In particular, investment in the grid could sometimes relax a constraint which would otherwise lead to large payments through Operational Outturn. Similarly, reactive power could sometimes be provided either by generators (in which case payments would be made through Uplift) or by static compensation equipment provided by NGC (in which case no payments would be made). Following pressure from the regulator, the industry negotiated the Uplift Management Incentive Scheme, implemented in 1994/95.

This defined the elements of "Incentivised Uplift" (basically Ancillary Services and Operational Outturn) and set a target level of £570 million. If Incentivised Uplift fell below this level, then NGC would receive 30 per cent of the saving, subject to a maximum of £25 million. If it rose above £587 million, NGC would pay 20% of the excess, up to a maximum of £15 million. Interim targets were set for each month of the year, and NGC made some profit from the scheme in the first half of the year, although it was possible that this would have been "clawed back" during the winter. As discussed below, however, almost every station's bid was very low in February and March 1995, leading to very low Uplift costs, and NGC ended the year with a profit of the full £25 million.

A second scheme, the Transmission Services Scheme, attempted to divide Uplift more finely, and to give NGC some more targeted incentives. To do so, GOAL is run a second time, after the end of each day, to calculate an ex post unconstrained schedule. This uses the actual levels of demand, rather than the day-ahead forecast, and also takes account of under- and over-generation, compared to the instructions given to stations by NGC. It is an attempt to find the unconstrained schedule that NGC would have calculated, had it had perfect knowledge about what was going to happen. The difference between this ex post unconstrained schedule and actual generation is deemed to be due to transmission constraints. Since the cost of these constraints can be influenced by NGC, it can be given a stronger incentive than with the cost of demand forecasting errors (although it has been suggested that NGC should bear some exposure to the cost of these), or generators' errors.[9]

When NGC's price control was revised for the second time, in 1996, (for a control to run from 1997/8 to 2000/1), the regulator took over the task of setting the target level for Uplift. There had been a tendency for each year's target to be based upon the previous year's out-turn level, so that any savings made by NGC were immediately clawed back. If the regulator can agree targets for several years at a time, then NGC may have more incentive to undertake investments that can reduce Uplift costs. This change was not popular with all participants, for some felt that they could strike a better deal themselves, and wished to keep control of the process. The negotiations between NGC and the suppliers have been somewhat artificial at times, however, and so the change to a regulated system may well produce an improvement.

The Pool in Practice

The Pool has now been operating for seven and a half years: what have we learned from that experience? The first lesson has to be that it has a workable system: the lights have stayed on. This is perhaps hardly surprising, given that the system was designed to retain as many as possible of the nationalized industry's procedures, but the Pool's ability to coordinate the industry was questioned before the privatization. That is not to say that there have not been problems, however. The Pool has suffered from the exercise of market power (which is not its fault), and its rules have also allowed gaming, and occasionally produced price spikes which seemed hard to justify. Both topics are discussed below.

The Pool also suffered a public relations disaster with the opening of the 100 kW market in April 1994, when the number of customers allowed to take a competitive supply rose from 5,000 to 50,000. To take a competitive supply, a customer of this size must have a meter which can record the level of demand in each half hour, and send the information to the Pool regularly, so that the customer's supplier can be billed for the electricity taken. Several changes to systems and procedures were needed to cope with the larger number of customers, many but not all of which fell within the Pool's remit. This meant that there was no single forum where all the changes could be debated and resolved, and no person or organization with overall responsibility for implementing the changes. Many important decisions were not made early enough to allow systems to be developed and tested. Customers who wished to change supplier would need a new meter, and a communications link to transfer data to the Pool, but many of these were not installed, or installed but not properly registered. It was decided that these consumers should be allowed to change supplier anyway. In the event, some customers received two bills, some received none, and many of those bills were based on estimated data, sometimes wildly inaccurate. It took more than a year to sort out some of the problems, amid acrimonious debates over who had caused the chaos.[10]

Preparations for full competition in 1998, when even domestic customers will be allowed to choose their supplier, started soon after the 100 kW market was opened. It would be impossibly expensive to require every household to fit a half-hourly meter, and so a system of load profiles will be used to allocate each customer's demand, measured at intervals of several months, into each of the intervening half-hours. Writing the rules for this new system, and designing the software to transfer data between companies and perform the calculations has taken so long that many companies may not be ready for competition by 1 April, the scheduled date. Learning from the failures in 1994, companies will not have to open their markets until they are ready, although the laggards will probably be penalized. Another lesson concerned the need for a central authority to control the process, and the regulator has taken on the role of coordinator, employing a project management consultancy to check that participants are making progress. Even so, the regulator has argued that he does not have the power to force decisions on a reluctant industry, and this has undoubtedly slowed the process.

Market power

The Pool has been plagued by problems of market power. In 1988, the government hoped to privatize the entire electricity industry, including fourteen nuclear power stations, even though five of these had not yet been commissioned. (Work on one of them had only just started, but three had been under construction since before 1970!) To offset the risks of nuclear power, these stations were originally to be placed in a company with three-fifths of the CEGB's conventional power stations, giving 70% of the total capacity. A second company, with the remaining conventional stations, would provide some competition to "Big G". In 1989, it seemed impossible to privatize the nuclear stations without financial guarantees concerning decommissioning and reprocessing costs which the government was not willing to give. They were withdrawn from the sale, and placed in a separate company, Nuclear Electric, to remain in state ownership. Given the government's timetable, which aimed at completing the sales by the summer of 1991, to allow the option of an election in that year, it did not seem possible to change the structure of the conventional generators, and so National Power was privatized with 50% of the industry's total capacity, and PowerGen with 30%.

These companies had the ability to determine prices in the Pool, because they owned practically all the price-setting plant. (NGC owned two pumped-storage stations which set the price approximately ten per cent of the time). Furthermore, they could have an incentive to bid prices well above their costs, for while this might reduce the output of their marginal stations, the intra-marginal stations would gain from any resulting increase in SMP. Green and Newbery (1992) showed how a supply function equilibrium, in which the generators were not restrained by the threat of entry or of regulation, would imply prices well above costs, and significant deadweight losses. The possibility of entry, while it could lead to lower prices, might create excess capacity which would be almost as costly.

In practice, the companies have hedged most of their output with contracts for differences, which reduces their reliance on the spot price, and hence their short-term incentive to increase it. In the longer term, of course, the generators would like to see a higher spot price, because this will influence the future level of contract prices. In 1990/1, Pool prices were well below the expected levels, largely because of the contracts held by the generators. The recession had reduced demand below the expected level, leaving the generators over-contracted for part of the time, which actually had the effect of making them net buyers in the Pool, with an incentive to seek lower prices. The generators were also concerned to maintain their market shares and to burn as much as possible of the large quantities of coal which they were contracted to take from British Coal, leading them to favor increases in output at the expense of price.[11]

In April 1991, however, some of these contracts expired, and the generators have gradually had more and more of an incentive to raise prices. Gray, Helm and Powell (1996) show how prices have risen over time, responding to the successive expiry dates of the generators' contracts. The regulator reacted to these price increases (and customers' complaints about them) with a series of inquiries into Pool Prices. The first inquiries found that Pool Prices were below the generators' avoidable costs (contract prices were higher), and concluded that it was difficult to object to an increase. By 1993, however, the regulator was concluding that "the need to cover avoidable costs [did] not justify any further price increases - nor did it justify a price increase as high as the recent one" (Offer, 1993, p.iii). He was also clear that "both companies wanted a price increase and they were able to achieve it" (p.ii).

The regulator's problem was that he had few powers in relation to the generators, because the government had anticipated that the threat of entry would keep this part of the market adequately competitive.[12] The companies' generation licenses contain few clauses restraining their behavior, and there are few restraints which would be appropriate for companies in a potentially competitive market like the Pool. The one restraint which the regulator has imposed (Condition 9A) requires the companies to give him information on their closure decisions, and their station availability plans. The companies can choose these plans, but they must be able to justify them. The regulator could either negotiate a solution with the generators, or refer them to the Monopolies and Mergers Commission (MMC), the UK's competition authority. The MMC could force a change in the companies' licenses, or require them to divest some of their capacity, but there was also a risk that the MMC would decide to do nothing: the generators had never fully exploited their market power, and the MMC might decide that it was a potential problem which could be left alone until they did so.

In the event, the regulator negotiated two undertakings with the companies, in return for not referring them to the MMC. One covered the sale or disposal of plant: National Power agreed to use all reasonable endeavors to sell or otherwise dispose of 4,000 MW of coal- or oil-fired capacity by December 1995, and PowerGen 2,000 MW. This represented around 17% of the larger company's capacity, and 12% of PowerGen's. The aim of this undertaking was to increase the number of competitors able to set the Pool price. The significant amount of entry by combined cycle gas turbine stations had affected the level of the price, but those stations almost never set the price, and the regulator wanted to see more competition among mid-merit and peaking plant. Both companies eventually disposed of some of their stations on long leases to Eastern Group, one of the regional electricity companies (RECs), the privatized successors to the Area Distribution Boards.

The second undertaking referred to the level of prices in 1994/5 and 1995/6. Each company undertook to bid in such a way, that, under reasonable assumptions of other generators' bids and taking seasonal fluctuations into account, the annual average Pool Purchase Price would reasonably be expected not to exceed specified levels. The aim of this undertaking was to offer some relief on prices to larger customers while the plant disposals were being negotiated. It had little to offer small customers, whose prices were largely determined by five-year contracts which would not be re-opened, and (for the same reason) would not have a significant impact on the generators' profits.

The experience with the undertaking showed some of the difficulty in trying to cap Pool prices. For the first six months of the financial year, prices were well below the annual average of the undertaking. Below average prices might well be expected during the summer months, of course, although the Pool had not shown much of a seasonal pattern previously. Prices were higher during the autumn, and had reached the level of the undertaking by the end of December 1994. At that point, Nuclear Electric announced that two of its nuclear stations would be out of service during January (one had been undergoing inspections for several months). The reduced level of spare capacity led to a further increase in Pool prices (via high capacity payments), exacerbated by a further breakdown (at a National Power station) in mid-January.

The regulator issued a statement reminding the companies of their undertaking, and saying that if prices were above the undertaking levels by the end of the year (in two months' time) he might take action. Both companies responded with sharp reductions in their bids, which were maintained for the rest of that financial year. The time-weighted average price for the year was 2% below the level of the undertaking, but the demand-weighted price was 1% above. In a further statement (Offer, 1995), the regulator decided that the plant break-downs had been abnormal, and that the generators had not been in breach of their undertaking, but he also took the opportunity to remind them of his concerns over their ability to manipulate prices. The following year, Pool prices remained (slightly) below the level of the cap, despite even higher capacity payments.

Spikes and Gaming.

The problems with market power were not of the Pool's making, but there have also been concerns that the Pool rules have sometimes created unjustified price spikes, and have created opportunities for gaming by generators. The System Marginal Price is mechanically derived from the bid of the marginal plant, and is therefore subject to the risk that when a plant which has submitted a very high bid is required to operate, that bid will set the price. There are some exceptions to the rule that the most expensive station in operation sets the price (for example, stations which are only operating for very short periods are ignored), but they will not prevent every spike.

The first significant spike occurred in September 1991, and helped to trigger Offer's first Pool Price Inquiry (Offer, 1991). GOAL had been unable to calculate an operating schedule which would meet the forecast level of demand (plus reserve), and a secondary program, GOALPOST, was called in. This post-processor had fewer constraints to obey, and was therefore able to find a solution when GOAL was not. On this occasion, however, GOALPOST scheduled an open-cycle gas turbine station, and sent SMP to more than £150/MWh, even though other stations were available. In the event, that station was not needed, as some customers reduced their demand in response to the high price, but it had already been set. GOALPOST survived that episode, but the spikes which it provoked remained a cause for concern, and the program's use was suspended in August 1993, following a further series of spikes.

There have also been times when GOAL has been able to produce significant price spikes, without any help from GOALPOST. For example, a set may submit a low bid for most of its capacity, ensuring that it runs at that level of output, but a high bid for the remainder. Each of the last remaining megawatts of output will be very expensive, but it may genuinely be cheaper for GOAL to meet a small increase in demand by calling on them, than to incur the cost of starting another station (with a lower cost per unit). Unfortunately, the price-setting algorithms may produce a value of SMP which is based on the bid for those few MWh, and is far above the average cost for the set. It is extremely difficult to explain incidents of this sort to customers, and they have contributed to the Pool's poor reputation among industrial customers.

A significant opportunity for gaming in the Pool has concerned the bidding of stations which might be constrained on or off the system because of transmission weaknesses. Stations which are constrained on will receive their own bid, whatever the level of SMP, and have an incentive to bid high, while those which are constrained off will receive a lost profit payment equal to the difference between SMP and their (lower) bid, and therefore have an incentive to bid low. If there are several stations owned by different companies inside a constraint, then competition between them may limit the use of these tactics, but there will frequently be constraints which only one station can relieve, and that station will then have significant local market power.

The most extreme example of this concerned two small stations owned by PowerGen, Ferrybridge B and Hams Hall. Each was connected to a local 132 kV distribution network, rather than the main transmission grid, and was occasionally constrained on when the local demand was greater than the guaranteed capacity of the transformers connecting that network to the grid (that is, the capacity after assuming that one transformer had failed). When PowerGen announced that it planned to close the stations, NGC was forced to upgrade its transformers, and had to reduce their capacity while the work was in progress. This increased the amount of constrained-on output which the stations could sell, and they raised their prices to more than £120/MWh, when similar stations elsewhere were bidding between £20 and £30/MWh. The two stations earned constraint payments of more than £100 million, and made a profit (over station costs) of more than £60 million during 1991/2. The regulator found it "difficult to avoid the conclusion that PowerGen took advantage of the monopoly position which it enjoyed at these two locations to withdraw its plant on exceptionally profitable terms." (Offer, 1992, page iv).

In contrast, National Power had adopted a policy of setting bids which it calculated would just recover the constrained station's "costs plus a reasonable contribution to company overheads and profits" (*ibid.* p.iii). The larger company may have felt that it was more vulnerable to accusations of misusing its market power, and deliberately adopted a more restrained policy than PowerGen. In any case, the regulator avoided criticizing the company's policy, but concluded that the rules governing constrained stations were unsatisfactory. He suggested several ways in which constraint costs might be reduced, and most have been adopted.

NGC has increasingly adopted contracts with both generators and customers as an alternative to paying for constrained running through the Pool. A station which is likely to be constrained on might sign a contract with NGC under which it got a fixed payment in return for any constrained-on revenues it might have received from the Pool. These contracts were particularly useful for the last remaining stations from the 1950s, most of which have now been closed. Most of these were directly connected to the distribution system and, like Ferrybridge B and Hams Hall, they were frequently constrained on while NGC prepared for their closure, while their revenues from the Pool were uncertain. The contracts gave their owners an incentive not to close the stations until NGC was ready, and prevented them from earning excessive rents from their position.

Agreements with customers who are able to reduce demand on request (with varying notice periods) are an alternative to constrained-on generation in the same area, when the constraint is due to security requirements rather than a binding limit on actual flows. NGC now holds an annual tender in which customers (and small, non-Pooled, generators) can bid to provide reserve. The Uplift Management Incentive Scheme and its successors have also given NGC an incentive to consider investments which will reduce constraints.

Changing the Rules

The Pool was designed to make changes to its rules difficult to achieve. There was perceived to be a significant danger that, for example, the incumbents might try to change the rules to the disadvantage of entrants, or that generators would attempt to harm suppliers (or vice versa). The response was to entrench the rights of minorities, and give them the right of appeal to the regulator, if they felt that a proposed rule change would unfairly prejudice their interests.

The Pool's main decision-making body is the Pool Executive Committee, which meets once a month. At first, this was set up to have five generator representatives, and five suppliers. Four of the suppliers were RECs (each seat was rotated among a group of three companies), and the fifth, independent suppliers' representative, actually worked for a large consumer buying on its own account. The three major generators each had a seat, and two seats were held by other groups of generators. For several years, the four REC representatives voted as a block, after a 'pre-meeting' at which the majority view among the twelve RECs was determined. The distinction between generators and suppliers has become blurred as many companies are now involved in both activities, and elections to the executive committee no longer distinguish between them. The Pool has also invited two consumer representatives to join the executive, in an attempt to deflect the criticism that it is an "industry club".

When a rule change is proposed, the executive will generally set up a working party (or more than one) to consider the case for it. After a suitable delay (for the relatively small number of "Pool professors" in the industry may have to serve on several such groups at the same time), a rule change can be formally proposed to the executive. If it passes, Pool Members then have an opportunity to object and call a meeting, at which further votes can be taken - first on a show of hands, and second on the basis of a weighted vote (with weights reflecting each company's sales). After losing a weighted vote, the minority may appeal to the regulator, who will ask for evidence, hold an oral hearing, and then issue his decision on whether to uphold the change. The regulator is also given the power to disallow a change to the Pool rules if he feels that it conflicts with the Pool's objectives.

The motive for these procedures, to protect minorities, was a good one, but they have had the effect of making the Pool very resistant to change. The network of committees provides ample opportunities for a company to delay consideration of a proposal which it dislikes, while an appeal, even one which is almost certain to fail, could well add three months to the process. If the rule change could significantly reduce the company's revenues once it is implemented, achieving such a delay with an appeal which costs practically nothing must feel like good value.

The possibility of an appeal also affects the regulator's staff, who attend every Pool committee as observers. It would be wrong for the regulator to hear an appeal on which he had already expressed a strong opinion, and this caution extends to his staff. There may be times when a proposed rule change, which will harm some participants, will clearly lead to greater efficiency, but the regulator's staff will be unable to give it much support. The Gas Act 1995, which extended competition in the gas industry, allowed the Director General of Gas Supply to be much more proactive in designing the spot market. She was even allowed to impose changes to the rules at short notice, subject to consultation and later ratification by the industry. Several members of her staff had previously worked for Offer, and so this difference probably reflected the problems experienced in electricity.

Transmission losses

The treatment of transmission losses is a good example of the appeal process in action. By the late 1980s, the CEGB had included transmission loss factors as an input to GOAL, so that the effective cost of a station in the north would be increased (reflecting the marginal transmission losses which would be incurred), and the effective cost of a station in the south would be reduced (reflecting the losses that could be avoided by generating close to the load). The Pool could have adopted a similar system when it was set up, but it proved impossible to reach agreement in time. There was certainly a political reason for not disturbing prices to customers at the time of the privatization, but this need not have extended to the prices paid to generators. In any case, the issue was placed on a list of items to be considered after the restructuring. The Pool was required to consider the possibility of introducing cost-reflective charges for losses, with a view to introducing a new system by December 1995. In the meantime, every company's metered demand was evenly scaled up, so that the adjusted demand was equal to metered generation in that half-hour. This scaling is generally equal to between 1% and 2% of demand, while marginal losses actually vary between 5% in the north, and -7% in the south-west, where additional generation can lead to a big reduction in power flows and losses.

After prompting by the regulator, the Pool began to consider the issue of transmission losses, and a scheme reached the Pool Executive Committee in March 1996. Metered demand and generation would be scaled in proportion to preset factors based on marginal losses, but only to the point where adjusted demand and generation were equal, recovering average losses rather than signaling the true cost of marginal losses. This scheme would have reduced the price in the north of England (by about 6%), and increased it in the south (by between 1 and 2%), and was therefore opposed by northern generators and southern suppliers. The regulator heard one appeal on the subject in June 1996, and rejected it, allowing the Pool to work on the details of the scheme. These came to the Pool Executive Committee in January 1997, and were again appealed to the regulator. In his judgment, (Offer, 1997) the regulator signaled his impatience with the companies who persisted in appealing against the principles of a scheme which had been foreshadowed several years earlier, but this has not stopped two northern generators from taking him to court ("seeking leave for a judicial review") over the changes. The regulator could well have to hear yet another appeal when the Pool considers the rule changes needed to implement the scheme. Introducing cost-reflective prices on a regional basis will bring some benefits overall, but the transfers between companies (positive and negative) are very large relative to these benefits, making the change very controversial.

Major Energy Users and the Pool

The large industrial customers have been among the Pool's fiercest critics. Before the privatization, many of the largest industrial customers enjoyed low prices, either in return for offering load management (which was not often required in practice), or through the "Large Industrial Customers Scheme". Under this scheme, British Coal sold some of its output to the CEGB "at world market prices", and the saving (compared to British Coal's normal prices) was passed on to selected large customers. In the first year of the new system, special contracts guaranteed that none of these customers faced a price increase above the expected rate of inflation, but some of them subsequently saw significant increases. Even though electricity prices have since fallen, so that the average for every class of customer is below what it was in 1989/90, the largest customers have seen a relative worsening of their position compared to other groups, and have lobbied for changes which might reduce their prices further. Two of these proposals have received support from Parliamentary Select Committees: demand-side bidding, and "trading outside the Pool" (Energy Committee, 1992; Trade and Industry Committee, 1993).

Demand-side Bidding

The demand side of the Pool is rather less sophisticated than the supply side. NGC needs a demand forecast in order to construct the day-ahead schedule, and this is based on information provided by the RECs and large customers, together with the company's own knowledge. Demand predictions should take account of the weather forecast, for example, but the Pool Rules explicitly prohibit NGC from taking the response of demand to price into account when drawing up its prediction.

This prohibition has been called into question on a number of occasions, since many customers now face Pool Prices in real time, and some have shown their ability to respond to price signals. The

Pool eventually introduced a demand side bidding scheme on a trial basis in December 1993. A scheme which had been designed to meet the needs of customers would require changes to the Pool which were never seriously contemplated by the industry, and so customers were required to bid blocks of "demand reduction" as if it was a kind of generation. The bids had to be submitted manually, which limited the number of participants (although there was never a huge take-up for the scheme), and the customer then had to "self-dispatch", reducing demand if SMP rose above their bid. There seems to have been little monitoring of whether the customers were actually doing so in practice. Indeed, there is even a story (probably not apocryphal) about a customer which only discovered that its supplier had entered it into the scheme when it asked why it was getting a payment from the Pool!

The payment for participating in the scheme - the Pool's capacity payment, multiplied by the amount of demand reduction offered - was not seen as sufficiently generous to interest most industrial customers. At the same time, the Pool has been reluctant to offer more than it feels is the value of the potential demand reduction, as measured by the capacity payment. Although few people are satisfied by the scheme, it has proved impossible to create a consensus on anything better, and because its existence meets a political need, the present "experimental" scheme has now effectively become permanent.

Trading Outside the Pool

Large industrial customers with unvarying demands have argued that they are not responsible for the peak levels of demand which lead to high Pool prices, and should not have to pay them. They suggest that they should be allowed to strike bilateral deals with generators, and that their demand, and the generators' output, should be "netted off" from the Pool's calculations. There have been suggestions that this would also give the generator more certainty about its output, and help to reduce its own costs. The regulator was asked to hold an inquiry into these proposals (Offer, 1994), and decided that the case for them was not convincing at the present time, although he would not rule them out in future.

In practice, the generator's output would rarely be exactly matched to the customer's demand, and so a "residual pool" would be required to settle the imbalances. There were fears that since this residual Pool would be a thinner market, it might be more susceptible to manipulation. Trading outside the Pool would also require a mechanism to recover an appropriate proportion of Uplift costs from the participants. Much of Uplift concerns payments for system security, which benefits all consumers, and so it is entirely fair that they should pay their share of the costs, although it should be admitted that the present Uplift charges do not fully reflect cost differences due to location, for example. (It might be added that some customers may have seen trading outside the Pool as a way of avoiding Uplift).

Some generators supported trading outside the Pool, suggesting that the scheduling procedures could easily allow it if generators were allowed to declare themselves "commercially inflexible". NGC is already required to schedule "operationally inflexible" plant in the way which its bid asks, and the technique would be the same. This would allow some stations to become "must-run plant", although in practice, this is already an option under the existing Pool Rules: all the generator has to do is to submit a bid which is low enough. Many of the new entrants bid close to zero, ensuring that they run, since the price they receive is fixed by long-term contracts. For a multi-station generator, however, this has the disadvantage of tending to depress the Pool price received by the company's other stations, and it may be that the companies advocating trading outside the Pool thought that this would allow an increase in their output without having to reduce the Pool price. It is not certain that this would be the case, since while a station with a bid of zero effectively pushes out the industry's supply curve, taking some demand outside the Pool should push the demand curve inwards by the same amount. The possibility of helping a customer to avoid Uplift payments, and sharing the rent created, may also have been a significant motive.

Trading outside the Pool is unlikely to improve the efficiency of the electricity industry. The more constraints which NGC faces when dispatching plant, the less likely it is to be able to minimize the industry's costs. The argument that a station which has a more predictable schedule (through trading outside the Pool) will have lower costs ignores the fact that other stations are likely to have less

162   POWER SYSTEMS RESTRUCTURING

predictable schedules as a result, increasing their costs. And in practice, most electricity trading takes place outside the Pool in any case, in the sense that prices for generators are determined by the contracts for differences which cover around 85% of the output sold through the Pool. Buying a contract for, say, 10 MW allows the customer to cap the cost of that level of consumption, while the marginal cost of each MWh is equal to the Pool price. (This is straightforward for an additional MWh, but also holds for a reduction in demand - the customer's bill for 9 MWh would be less than its total bill for 10 MWh by the amount of the Pool price for 1 MWh). Unfortunately, although a strong economic case can be made in support of the present system, the political need to "do something" may continue to create pressure to make changes.

Similar pressures help to generate support for a proposal known as "wybiwyg" (or "what you bid is what you get"), under which generators would receive their own bid, and not SMP. The price to customers would be based on the average payment to generators. This seems attractive to those customers who have noticed that nuclear stations (amongst others) bid zero, and have not realized that those bids would change in response to such a change in the market rules. Generators could not be dispatched in merit order unless they were allowed to change their bids frequently, so that the marginal plant would set the market price with a bid close to its cost, and other plants "in merit" would attempt to bid just below this level, which would change from hour to hour. If generators succeeded in following such a strategy, the change in market rules might have very little impact on prices, while if the generators failed to follow the strategy, the market might become significantly less efficient. Incumbent generators with a portfolio of stations would almost certainly be at an advantage compared to entrants with a single station. The regulator examined the proposal during his inquiry into trading outside the Pool (Offer, 1994), and dismissed it for these reasons. Unfortunately, the apparent fairness of the proposal means that it still gets "political" support, even though the economic case for it is extremely weak.

Conclusions

The Pool has fulfilled its most important task, that of ensuring that generation met demand after the electricity industry was restructured. Contracts for differences written around the Pool price have allowed generators and suppliers to hedge price risks, and have facilitated entry into the industry. Almost all of these contracts have been bilateral instruments between industry participants, however, and the industry lacks the liquid futures market, involving outsiders, which many other commodities have. Outsiders view the Pool as too risky at present: the major generators have too much influence over the prices, the regulator can affect the market in unpredictable ways (as with the price cap of 1994-6), and they feel that the rules are too complicated, occasionally producing unjustifiable prices.

The electricity industry is aware of these criticisms, and its behavior has been affected by them. National Power and PowerGen have deliberately avoided using the full extent of their market power, fearing the regulatory consequences if they pushed prices to excessive levels. Despite this, Pool prices have been high enough to attract a lot of criticism. Many aspects of the market have been seen as over-complicated and liable to favor generators at the expense of consumers. Capacity payments have come under repeated criticism on these grounds. Just as this chapter was being completed, the government announced a "fundamental review" of the Pool, to ask whether changes which the electricity industry was unwilling to make, or incapable of making, should be imposed upon it.

Like most of the Pool's rules, however, the capacity payment system was designed to reflect economic principles. Some rules have turned out to cause problems, and many of these have been changed. Rule changes which promise benefits for most parties have been much easier to achieve than those which are likely to create significant losers. This particularly applies to issues which were considered too difficult or controversial to resolve before the restructuring was implemented. If there had been more time, it might have been possible to do deals which covered several aspects of the proposals, creating majorities for controversial issues through a process of log-rolling, while the date for implementing the restructuring might impose the helpful pressure of a deadline. Once the restructuring has taken place, the deadline has disappeared, and the opportunities for log-rolling decline with the number of decisions to be made. Countries which are considering reforms to their electricity industries might be well-advised not to act until

164 POWER SYSTEMS RESTRUCTURING

they are ready to agree the complete package. Whether that package should be based upon the Pool, or whether one of the other wholesale markets which has been set up around the world would provide a better model, is beyond the scope of this chapter.

---

[1] Support from the Economic and Social Research Council under project R000236828, *Developing Competition in the British Energy Markets*, is gratefully acknowledged. I would like to thank Alex Henney, David Newbery, and Julian Sondheimer for helpful comments. The views expressed are mine alone.

[2] One-way CfDs were attractive to the RECs when the system was set up, perhaps because they implied that the Pool price would be capped at the level of the contract's strike price, while still offering the possibility of paying less than the contract price if the out-turn Pool price was lower.

[3] Energy Settlements and Information Services, a subsidiary of the National Grid Company, calculates Pool prices while NGC itself is responsible for the actual dispatch of plant. Another subsidiary, Energy Pool Funds Administration Limited, is responsible for administering the payments through the Pool.

[4] Those figures applied to the period as a whole. In individual years, the proportion of the Pool Selling Price (PSP) due to SMP varied between 95% and 71%, while Uplift made up between 5% and 9%. In two years, capacity payments were less than 1% of PSP, but in 1995/6, they accounted for 21% of it.

[5] If the second incremental price is much higher than the first, then the second no-load price will be much lower than the unadjusted price. If the station spent all the time generating above its elbow point, then the high second incremental price would be combined with a low (or negative) no-load price, giving a reasonable stack price. If the station only generates above its elbow point for a short period, the averaging means that the high incremental price is combined with an average of the no-load prices which is weighted heavily towards the high unadjusted value. The effect of this is to inflate the stack price during short periods of peak generation (when it might set SMP) and to reduce it at other times (when it is unlikely to do so, not least because of the reduction). A small adjustment to the Pool's equations could avoid this problem.

[6] England and Wales have not experienced any outages due to capacity shortages since the restructuring, despite values of LOLP which imply that we should have expected several such outages by now. The disappearance ratios for older plant, which has become more reliable since privatisation, may now be unrealistically high, raising LOLP.

[7] A generator with a portfolio of plant may have other considerations, such as keeping a balanced mix of fuels for the future. A portfolio generator will also be aware that closing a station will increase the value of LOLP, and the revenues received by its remaining stations, but the large generators in England and Wales are now required to justify their closure decisions to the regulator. It would be hard to justify closing a station which appeared to be individually profitable on the basis of projected capacity payments.

[8] The contract would give the station a fixed sum in return for the out-turn capacity payments on a given amount of capacity. As long as the station is actually

available at the times when capacity payments are made, to cover its payments under the contract, then this system fixes its revenues.

[9] It has been suggested that generators should bear the costs of their errors, but they have not yet shown much interest in developing such a scheme.

[10] Candidates included the electricity companies, for delaying decisions; the regulator, for insisting on several simultaneous changes (which included requiring customers to choose between competing meter operators); and the meter operators, for failing to install meters when promised. None of the parties faced financial penalties for failure to meet their obligations, and the industry recovered the additional cost of sorting out the chaos from a levy on consumers!

[11] In subsequent years, prices have often fallen near the end of a financial year as the generators sought to increase their output and reduce coal stocks.

[12] The threat of entry might be expected to cap Pool prices, and they may not have significantly exceeded this level. The regulator's complaint was that, given the significant excess capacity in the industry, a fully competitive market could well have produced much lower prices than actually occurred.

References

Department of Energy, (1988) *Privatising Electricity*, Cm 322, London, HMSO

Energy Committee (1992) *Second Report: The Consequences of Electricity Privatisation* HC 113 of Session 1991/2, London, HMSO

Gray, P., D.Helm and A.Powell (1996) "Competition versus Regulation in British Electricity Generation" in (ed.) G.McKerron and P.Pearson *The British Energy Experience: A Lesson or a Warning?* London, Imperial College Press

Green R.J., and D.M.Newbery (1992) "Competition in the British Electricity Spot Market"' *Journal of Political Economy* 100 (5), October, pp. 929-53.

Henney, A. (1994) *A Study of the Privatisation of the Electricity Supply Industry in England and Wales*, London, Energy Economic Engineering

Hunt, S. and G. Shuttleworth (1993) "Forward, Option and Spot Markets in the UK Power Pool", Utilities Policy, Vol.3, No.1, January, pp. 2-8

Lucas, N and P.Taylor (1994) "Structural Deficiencies in Electricity Pricing in the Pool" *Utilities Policy* Vol. 4, No 1, January pp. 72-6

Offer (1991) *Pool Price Enquiry* Birmingham, Office of Electricity Regulation

Offer (1992b) *Report on Constrained-On Plant*, Birmingham, Office of Electricity Regulation.

Offer (1993) *Pool Price Statement, July 1993* Birmingham, Office of Electricity Regulation.

Offer (1994) *Report on Trading Outside the Pool* Birmingham, Office of Electricity Regulation.

Offer (1995) *Generators' Pool Price Undertaking 1994/95* Birmingham, Office of Electricity Regulation.

Offer (1997) *Decision on the appeal regarding the works programme for the zonal allocation of the cost of transmission losses* Birmingham, Office of Electricity Regulation.

Trade and Industry Committee (1993) *First Report: British Energy Policy and the Market for Coal*, HC 237 of 1992/3, London, HMSO

5 PRACTICAL REQUIREMENTS FOR ISO SYSTEMS
Ralph Masiello

ABB, Inc.
Santa Clara, CA

Introduction

Much of the literature and debate about the development of ISOs focuses on two or three problems: market rules, congestion management and pricing, and codes of conduct. This is appropriate since only a few ISOs are formed as of this writing and only 2 or 3 actually have detailed protocols, systems to implement their business operations, and certain operational dates for a market that includes retail access.

The development of the California ISO has made crystal clear that mandated retail choice and retail access has tremendous implications for ISO operations and systems. An ISO in an environment without retail access, such as ERTCOT, is providing open and comparable transmission access, per FERC 888. However, it does not have to deal with the increased numbers of suppliers, marketers, and customer delivery points that exist in a retail choice environment. Retail choice implies more bilateral transactions directly to large suppliers or multiple aggregators of small customers; it attracts more Energy Service Providers to the market; and it as a result emphasizes the importance of business systems to support scheduling and settlements over the technical systems that monitor transmission conditions and deal with congestion. If we compare the map of prospective ISOs in North America and their operational dates (Fig 1) with a map of states committed to retail choice (Fig. 2) we can see that New England and California are unique in this regard. (New York actually has the same complexities, but is so far absent a legislated date certain for retail choice).

168 POWER SYSTEMS RESTRUCTURING

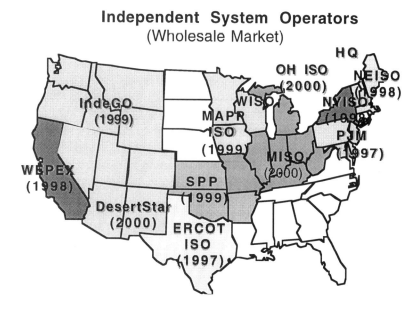

Figure 1 - Planned ISOs and Operational Dates

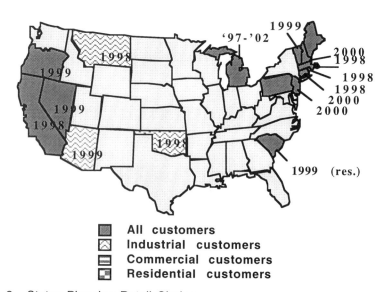

Figure 2 - States Planning Retail Choice

The devil is in the details indeed, and the systems development efforts for the early ISOs, notably California and New York, afford many lessons and principles for those that will follow, This article covers the practical details as they are evolving this winter. It will cover the following key points: issues around implementing congestion management solutions; practical problems in the network analysis solutions needed for congestion management, especially in "greenfield" situations as opposed to a migration of existing power pool systems; business systems needed to make the markets operate and support infrastructure for market participants; the settlements problem and implications for market participants and the member wire companies; and the realities of complex systems development under politically established protocols and "dates certain."

Most of the theoretical discussions and methods/algorithms in the literature, including this compendium, have to do with the congestion management and/or transmission pricing problem. This is understandable, as fairly establishing the price and access to limited transmission capacities is at the heart of the comparable access and open market problem. Also, having a scheme that results in an "orderly" market, meaning one that doesn't dissatisfy too many participants too much of the time and too badly, is very important. This article will not elaborate on any of the schemes being put forth, much less on the relative merits of any of them - instead it will take the position of a morally uninvolved "programmer for hire" that only wants to know what it is supposed to do and how to know that the end system is fulfilling its objectives.

One very key point I would like to make is that the market, like any other market, can only tolerate so much complexity. It is true that electric energy today is a commodity with geographic costs of delivery and pricing, and that any rigorous theoretical approach to the problem will arrive at something that looks like nodal prices derived from AC network models used for loss analyses and shadow pricing of transmission and perhaps contingency constraints. Taking our ability to perform these calculations as a given, consider the problem of an Energy Service Provider formulating contracts and offers in this environment. In effect, the cost of product transport has a variable component (losses) and a potentially very variable component and even uncertain availability in congestion pricing. The ESP, except for the largest customer, has to take the market risks of these components. Assuming a fixed transmission tariff on either a postage stamp or MW-mile basis for the basic service, the costs of ancilliary services, losses, and congestion pricing are not firmly known days or weeks ahead of the delivery time. Yet few customers are sophisticated enough to want these unbundled and to manage the risks themselves. So the ESP is faced with the need to develop or acquire the tools, systems, and people to forecast these costs with sufficient accuracy to offer firm prices or prices linked to transparent commodity (gas, futures indexes, exchange prices) costs for the delivered energy.

The net of this is that the ESP can deal with zonal bases for transmission pricing if the zones are large enough and relate to well-organized sub-markets. The ESP cannot easily deal with potentially large numbers of individual nodal prices. It makes pricing, settlements, and the conduct of business too difficult. Few other commodities are priced with such fine granularity in delivery costs - package delivery services, for instance, are on a zonal basis. Wholesale markets for bulk commodities may have point to point rail and trucking tariffs, but these are not characteristic of retail markets. The consumer rarely sees delivery prices for anything that are specific to a street address. Most importantly, the consumer does not see such pricing for other analogous services - phone, cable, gas, water, sewer, garbage. It is true that the consumer sees individually variable capital recovery pricing for some municipal services such as water/sewer but not for the actual service delivery. Thus both sides of the market will find detailed zonal pricing on a time-variable basis to be difficult to accept. This has led California, for instance, to move towards a reduced set of zones for congestion management. (Fig. 3)

Applying a zonal rather than a nodal analysis to congestion pricing actually makes the analytic problem more complex, not less so. The nodal problem is straightforwardly derived from an optimal power flow (well, straightforward from an academic point of view and practically do-able for large, real networks for one of a very few organizations skilled in the art). The zonal problem immediately leads to another round of philosophical issues - what do we mean by zonal prices - the average in a zone, the maximum, and so on. And - for whatever definition we select will we get an orderly market outcome and a physically acceptable (absence of constraint violation) answer? There is no theoretical answer to this question, and even practical answers are hard to come by since "no one knows" - meaning no one has enough experience with the problem and the ramifications of different ways of solving it.

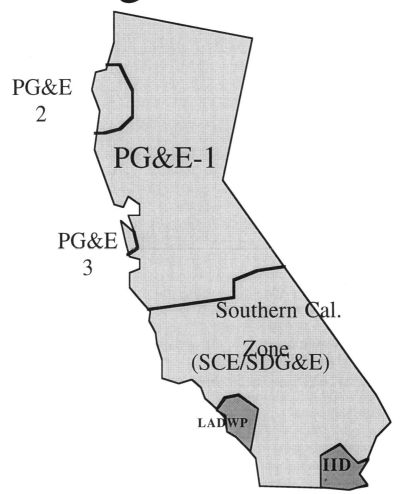

Figure 3 - California Transmission Congestion Zones

This leads to another pragmatic difficulty - the detailed protocols for how to handle zonal prices are necessarily buried in FERC filings.

Once approved, they become somewhat inflexible. Unfortunately, the mathematical "solvability" of the problem posed may not be known until much later when the need for different approximations or algorithmic compromise become clear; or worse, when market outcomes are experienced in the real world and flaws in the protocols exposed. In both cases, the ISO needs to make rapid adjustments in the details. Even though the adjustments may not be significant to the market in a broad sense, or even may be desirable, the sad reality is that these changes can then become bones of contention between rival market participants armed with their own hired economists, engineers, programmers, and attorneys. Considerable good will among the participants and a desire to see the whole thing work is required in order to work through these issues as they come up. How to assure that good will and spirit of accommodation is a subject beyond the scope of this treatise, that is for sure.

The other broad category of approaches to handling transmission constraints can be classed as market based - either allowing for a market in transmission rights or an iterative market (for example, as proposed by Ilic in Electricity Journal) with varying degrees of price and limit information available to the participants. So far these approaches have not been embraced by any formative ISOs as compared with the "analytical" approach based on nodal pricing or iterative market schemes using the clearing price of energy in a congested zone as a proxy price for the use of the congested transmission. (This latter is actually the case in the California model) We can only speculate why this is the case - perhaps the utility and other engineers driving the details in the protocols are naturally more comfortable with an analytical approach guaranteed to provide an answer as opposed to the uncertainties of an untried market approach; perhaps the market approaches are too complicated in some cases; and perhaps we are all uncomfortable with a "rights" based approach that either grants grandfathered preferential access or affords the potential of a large player "gaming" the transmission market.

Another drawback to the "rights" based approach can be put forth: under today's normal transmission operating conditions the bottlenecks and interfaces are fairly well known and it is sensible to talk about rights. However, under operating conditions that could develop as the markets evolve, new merchant plants come on line, and as outage and contingency conditions occur, the constraints and effects may be different than we know today. How quickly can a market react to such situations - there is no certainty that the hour ahead market may be able to react in an orderly way, and less so that the spot market can react to a contingency event in a timely and rational way. Knowing that the ISO is first and foremost responsible for system reliability makes it improbable that market schemes will be implemented, especially given the tight time frames established for protocol and software development.

Finally, a point not well understood by most of the debaters of these issues is that the market and rights schemes all give the larger players an advantage. Only they can afford the people and tools needed to manage risk and even game in these environments. The more autocratic analytic approach may appear to be less of an open market, but it is probably friendlier to the smaller market participant and less vulnerable to gaming.

Getting the congestion management software to work in the market raises several non-theoretical issues. First, a single, accepted network model has to be developed that can be used in the programs. Individual utilities today can contribute their own models used in operations - these will be accurate for their internal systems and considerably less so for the external systems; getting the different utility models "harmonized" with each other in a way that all agree with can be tedious and difficult. Regional planing models exist, of course, but it is always, always the case in practice that these models are not sufficiently representative or the real world or sufficiently current to be used as is without considerable rework. The planning and NERC regional staffs would not agree, of course, but every control center that has gone through the effort of getting a state estimator and security suite working will attest to this point. Even once this effort is complete, the same problem exists with the network model external to the ISO - where it comes from, how it is validated, and so on will be issues. And, since the utilities external to the ISO will probably be market participants whose business is affected by ISO operational and congestion management decisions, they in time will take a stronger interest in and want approval of the models used of their systems. This is a problem the industry has not faced before.

Getting the participants to buy into the models and the algorithms/codes used leads to another problem: these codes have to be tested with a degree of rigor not normal for planing or operational analytic codes. Specifically, test suites have to be developed that cover not only the expected conditions but a broad range of "pathological" cases and the codes tested and verified against them. And, since the codes will in many cases be solving problems never attempted before, defining the correct result may be difficult indeed.

This point cannot be overstressed, and it goes to the heart of the differences between a *planning* program and an *operational* program. Planning programs are used by engineers looking at potential transmission expansion or studying the system to determine seasonal operating conditions, among other uses. These uses often lead to cases which may not solve readily or well, and the engineering user understands and accepts that. The engineer is often interested in the intermediate results of, for instance, a load flow or optimal load flow, to see where the convergence or feasibility problems are in order to gain insight into how to pose an altered case as a step in the planning process.

The operational program, on the other hand, is used by operators who may or may not have an engineering background. They want clear-cut answers in a short time frame to a problem based on real world conditions. Ambiguous or non-convergent answers are not of any use and undermine faith in the program. Thus the operational program must in some ways be more robust than the engineering version. It is a little like the difference between a basic pick-up truck with four wheel drive for snowplow use but no other consumer options and the fully loaded sport utility that goes skiing once a year.

The pathological cases that the congestion codes have to be tested against are not ones of far-fetched planning scenarios but rather the difficult cases of over-constrained/infeasable problems where protocols for how to relieve the constraints have to be followed. In other words, some business/market rules may dictate how the program is to behave in infeasable conditions- and the program has to be methodically tested and verified against all the infeasable conditions that might conceivable arise. It is more important to get the "correctly derived" answer than the "optimal" answer - a point often lost on the people responsible for developing the algorithms and codes.

Thus the ISO needs an up-front test protocol that is somehow accepted by the market participants' representatives; it needs a way of documenting and validating the tests, it needs extraordinary configuration management and regression testing against code changes, and it needs a way to quickly resolve new technical problems as they come up. This is all very new to the utility industry - OPF is still a new tool not widely used in operations, and when used only used by engineers who can cope with the propensity for the code to diverge, fail to solve, get strange answers, and so on. Now OPF will used routinely to derive pricing - it has to be right, as defined by an agreed test suite. A new and large challenge for all involved.

A related problem will come up in defining the contingency constraints - thermal, voltage, and stability constraints that the ISO has to respect. The West Coast outage of 1996 showed that the economic cost of observing the stability constraints is real and significant. We can guess that the California ISO, for instance, is going to have visibility of transmission conditions never before available and a charter to maintain reliability that translates into conscientious observation of the constraints beyond what the region is accustomed to. Thus in time the constraints themselves may become the subject of market focus and debate, especially the more esoteric ones such as stability and voltage stability. The ISO does not have the tools today to make a rigorous probabilistic reliability-economics trade off.. A next evolution in transmission pricing may well be to develop a methodology for formulating this problem and enabling the ISO to balance economics against a probabilistic analysis of system reliability. We can expect that if a more rigorous adherence to second contingency limits results in increased congestion pricing then there will be an increased interest in developing a way to deal with the probabilities analytically.

We are still grappling with the problem of how to create economic rewards and incentives for investment in transmission capacity that relieves bottlenecks - this line of thinking says that we also should develop incentives for investments that improve reliability, such as new control and protection equipment. Today the FERC identified process is that RTGs will determine what expansion is needed and then somehow whoever makes it will earn a regulated rate of return. This is, in effect, a vestige of the old way of doing things and not market based at all. A market based approach would reward the transmission investor with a share of the congestion savings for a period of time. This is not a new idea, but is one that is needed in order to attract investment to the transmission side. Such an investment will by its nature destroy the value of existing transmission rights and/or the profitability of higher cost generation on the demand side of a congestion bottleneck. The only source of funding in a market based world for the payoff is, in the end, the increased profits of lower cost generation able to sell more power into the downstream zone. There are today no mechanisms for transferring some of those profits from the generator who benefits to the transmission investor, and there are plenty of disincentives for the vertically integrated utility that owns the transmission bottleneck and the downstream high cost generation not to make the investment. A simple market based mechanism would be available when the investment takes the form of a controllable transmission resource such as a DC line - then the owner can charge a fee for the exercising of the control. It is not clear how the other market participants and the consumers, especially, would feel about paying for the use of an incrementally "free" resource. The transmission investment which relieves a contingency constraint even when it itself is not heavily loaded poses an even more difficult question - no straightforward transmission tariff will reward that investor.

One can speculate that the likely compromise outcome is that the ISO/RTG will identify the need for transmission capacity in order to relieve congestion, and the projected economic benefit. This can then be used by regulators to put out for bid the construction of the needed facility - but on a priced basis not necessarily equal to rate of return compensation. Once the asset is constructed and transferred to a regulated entity (logically the surrounding wire company for operational reasons) that entity can get a regulated return on capital. While not truly market based, this approach is likely to be one that the utility industry and regulators will find more acceptable. It leads to the need for the ISO to have some sophisticated and defendable tools for assessing not only the need but the projected value of the investment.

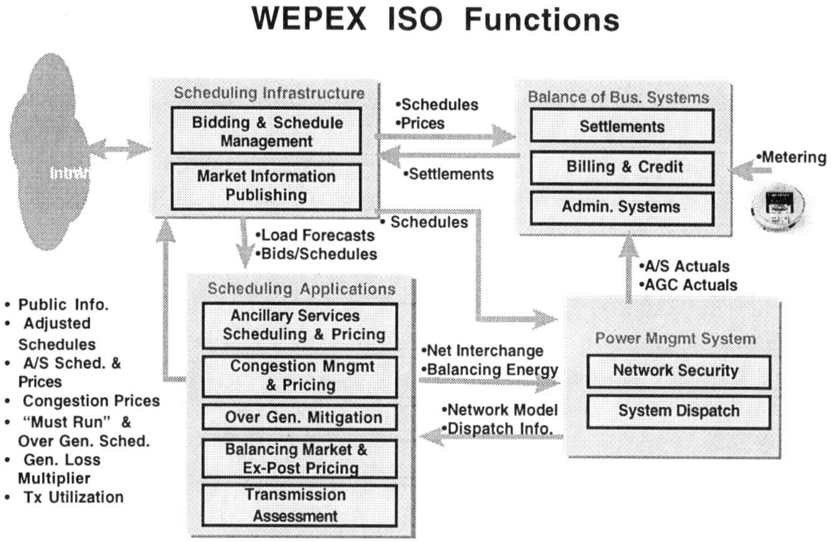

WEPEX ISO Functions

Figure 4 - The overall ISO system requirements in the California case.

The Scheduling Infrastructure system, or SI, is the heart of the transmission reservations and scheduling function. It runs the day ahead and hour ahead markets for anciallary services, balancing energy, and transmission congestion, and enables energy market participants to schedule their transmission services with the ISO. By requiring all transmission users to schedule their energy deliveries through the ISO, including the schedules coming from the Power Exchange, the ISO is able to build a comprehensive picture of the day ahead transmission conditions for the purposes of managing transmission congestion. In order to run the markets, the ISO publishes via the Internet based market interface, the load forecast, the forecast requirements for reserves, balancing energy, and other services. It also publishes the outcome of the daily and hourly auctions in terms of final clearing prices and, to the individual market participants, their accepted schedules.

The settlements system is responsible for reconciling the before the fact (ex ante) schedules for energy and services with the after the fact (ex post) meter reads and for distributing the actual financial responsibilities for balancing energy. This requires a state wide revenue metering system for wholesale energy that is capable of providing "settlement ready" meter data to the ISO. "Settlement ready" implies that all reads have been validated and all missing reads have been estimated.

The "Scheduling Applications" are the technical applications responsible for determing the pricing and schedules in the markets the ISO is responsible for, and for the analysis and management of transmission congestion. The names and purposes of the functions are what one would expect for any ISO with similar responsibilities; the details of the congestion pricing would vary from one philosophy to another. The one function peculiar to California is the over generation mitigation - special market rules needed at low load periods when the regulatory must run plants (nuclear and QF) may exceed forecast load conditions and generation must be curtailed.

Finally, the "Power Management" system provides business as usual EMS functionality for the entire state - a state of the art EMS but on a scale associated with large pools and national centers.

The business systems that allow market participants (ESPs) to schedule transmission services are in themselves non-trivial undertakings. The California ISO has defined day ahead and hour ahead markets for services, and schedules transmission services on an hourly basis with defined market windows and time sequences for both. This is at variance with OASIS both in the more fixed time structure for the market and in the lack of a longer range transmission reservations capability. The ISO has heavy-duty, industrial strength scheduling transaction systems sized for a large number of anticipated transmission schedules each hour. (10,000). (See the table, below, for overall performance and dimensionality) By comparison, most OASIS implementations are desk-top lowest common denominator solutions derived to satisfy FERC at low cost. The widespread and fervent distaste for OASIS I by the non-utility market participants should be a signal that the more robust and structured ISO solution will serve the market better. The ISO presents a "one-stop shopping" approach to transmission services - the basic service, congestion information, auxilliary services markets, auctions, and pricing, all accomplished in one user-friendly place. The ISO systems and support tools are developed with the goal of making the problem easy for market participants. (more about this later) OASIS has been developed with the goal of minimizing the hassle for the supporting utility operations organization.

Detailed Set of Performance Requirements

- SI Highlights:
  - 10,000 SC's (Stage 1), 4000 SC's (Stage
  - 10,000 Schedules/Hour (x 24 Hours) (Stage I) Increasing to 50,000 (Stage
  - 5 Sec average Schedule Retrieval Time (10 sec peak)
- PMS Highlights:
  - 80,000 Status and 60,000 Analog Points
  - 1,020 Generators
  - 2 Sec Screen Callup (4 Sec peak)
- SA Highlights:
  - 12,000 Buses, 4 Congestion Zones (10 Stage II)
  - 1,000 Contingency Cases Screened, 100 analyzed
  - 25 Sequence Executions NAC+NMB+DPF+ATC+CONG per day in a 7 window/hour
- BBS Highlights:
  - 1.3 million records/day
  - 95% Compound Business Object Retrievals <5 secs

The ISO scheduling system is analogous to a one stop travel service - one agent, one screen, one set of travel schedules. By contrast, OASIS today is like physically accomplishing your travel arrangements at the airport - going from airline to airline to find out about flights, availability, and fares, and perhaps making several false starts before getting schedules that work. And then, of course, one has to go down to the rental car counter and the hotel phone board for ancilliary services. Second, OASIS by the nature of the low cost PC solutions implemented is subject to poor performance and frequent down time. None of this makes for happy market participants.

Beyond this, there is the problem that in the OASIS world the power marketers have to "tag" their transactions thus identifying proprietary business information to a variety of utility operations offices. Code of Conduct notwithstanding, OASIS systems do not provide the security and guarantee of privacy that market participants desire. The ISO organization inherently provides a neutral ground for the market; the ISO scheduling input inherently performs the needed tagging function; and ISO software and systems are robust enough in their implementation to provide commercial grade firewalls and user security. All this argues that the market will drive towards the ISO solution and that the dream of a national OASIS as the basis for all transmission scheduling will have to be adapted to the reality of regional ISOs and defined markets.

The needs of low-cost and easy Internet access for transmission scheduling, absolute isolation of one scheduler from another and protection of confidential information, and high volume/high performance systems to serve a large real time market lead to technical solutions typical of the largest electronic commerce installations. The short development time frame of the California ISO also drove, in its case, a system development based entirely on standard tool sets with minimal unique code development. Figures 5 and 6 show the architecture of the California ISO market interface.

ISO Scheduling Infrastructure

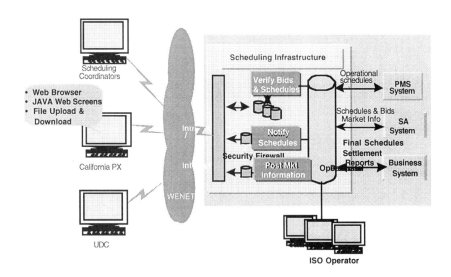

Figure 5 Scheduling Infrastructure Functional Architecture

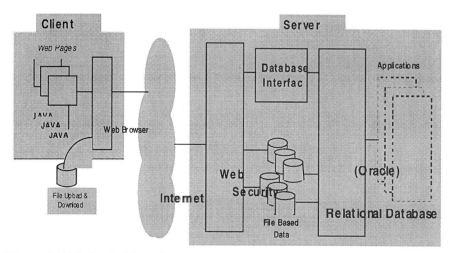

Figure 6 Web Market Interface

Related to this is the practical market reality of the energy spot market: the markets today are converging on the concept of an hourly spot market (which may evolve to finer time granularity in the future) The necessary linkages to the financial markets for purposes of hedging, futures and derivatives trading, and other financial risk management tools demands an organized spot market on an hourly basis. Whether done in a top-down mandated fashion as is the case in California or as an adjunct to existing exchanges such as CPEX or financial exchanges such as NYMEX, the power exchange of physical energy delivery is necessarily regional and linked to ISO definition and protocol. Thus the ISO scheduling periods and markets for ancialaries have to be coordinated with the exchange markets. Thus the California day ahead and hour ahead markets. The ISO concept of managing transmission scheduling, security, and congestion, have to be reconciled with the OASIS reservations concept in the future. If the California, New England, and New York markets work OASIS will have to adapt to them - the business needs of the power marketing industry will dictate that.

ISOs are formed, at first, by utility groups coming together. As such, the engineering problems and mind sets take priority. This results in an initial emphasis on transmission operations and security functions - which leads to the definition of, in effect, a regional transmission operations/coordinating center. In other words, an ISO EMS system which is largely business as usual for the task forces working on the problem. However, the realities of the transmission scheduling, metering, and settlements business problems overwhelm the analysis and control systems problem in complexity and technical challenge. The ISO ends up with a transaction processing system equivalent to a good sized bank ATM system - taking in 000's of schedules hourly; confirming every one against security checks (the network analysis for the ISO and account balances for a bank); and archiving all the information for future reconciliation. Also, great care for data integrity, protection against system failure, and account privacy must be maintained. Of the total cost of a comprehensive ISO in a retail choice environment, perhaps 75% is associated with the business systems. EMS engineers and EMS consultants generally fail to appreciate the difficulties of this problem and the costs involved with implementing solutions. Large IT firms with backgrounds in gas, trading, and transportation systems have a far better understanding and appreciation for the problem. Just because the only mathematics employed is basic addition, multiplication, subtraction, and maybe division, does not make the systems easy, simple, or inexpensive. And, because non-ISO personnel have to interact with these systems their performance and usability has to be far better than systems developed for the ISO Staff themselves to use. A failure to correctly plan for the business systems can only lead to major shortfalls in final results and a failed market as a consequence.

Beyond this, the ISOs are all going immediately to Internet technology (albeit on secure Intranets) as the basis for participant interaction. This is the equivalent of a bank putting *all* its ATMs on an Intranet at once. The result - the NY and CA ISOs are not only putting in place the largest commercial markets to spring up from ground zero but they are doing it entirely with cutting edge technology, and for an application where "system not responding" is a physically unacceptable behavior. Thus the attention to the technology architecture and a willingness to "overkill" the various hardware platforms is absolutely necessary to success. (again, contrast this with OASIS on a PC!)

PRACTICAL REQUIREMENTS FOR ISO SYSTEMS 183

The final major problem area to be discussed is "settlements." The settlements problem arose initially in the gas industry. Unlike any physical commodity market or financial market, the gas and electric industry share the problem that the actual "delivery" or transaction is only approximated by the scheduled transaction or trade. The actual energy consumed is measured after the fact and inevitably will differ from the scheduled transaction by some amount that cannot be known beforehand. This leads to the settlements problem of reconciling the meter readings with the schedules and then allocating the variations, and the prices for those variations, to the participants according to defined terms of trade. This can simply be described as the balancing problem - the terms of trade have to define who is responsible to physically provide the balancing energy - the seller, the ISO, the wireco, the buyer, or indirectly a provider of regulation - and then who is financially responsible for it and on what basis. Each scheduled energy delivery has to identify the responsible physical and financial providers and then the ISO has to reconcile the meter reads with the schedules and provide settlements statements to all the parties. This problem first arose in the gas industry at the wholesale level and solutions used in the gas industry have readily been adapted to the electric problem.

It is only this fall dawning on the power marketers/ESPs that they too will need settlements functions. They cannot simply accept power exchange and ISO settlements statements without their own verification processes any more than any business accepts invoices without verification against purchase order and delivery "receivers". A much larger problem not yet fully appreciated by the utility distribution companies (UDCs) is that they too will require their own settlements systems. They too will get statements from the ISO for settlements which incorporate their liabilities for default provider deliveries, losses, services, and balancing energy. The ISO is unaware of which customers have how much load served by the UDC - it is only aware of the scheduled deliveries by the ESPs. Thus absent a comprehensive settlements capability the UDC runs the risk of financial losses associated with the misallocation of ESP shorfalls and failures to its own balancing energy obligations. So every UDC requires its own settlement systems capable of analyzing meter reads and ISO settlements. Not yet contemplated or planned is some integrated settlement system on a regional basis such as exists in the financial markets for equity instrument trades which can reconcile the statements of providers (ESPs), market makers (ISOs) and buyers (UDCs).

In conclusion, the major practical issues around implementing an ISO have more to do with the business systems than the control and analysis systems. The systems have to be robust and perform at the state of the art for electronic commerce systems, with the added difficulty of performing in near real time to allow the operation of hourly markets. The analysis systems face the problems of user buy-in to the methods, data models, and test results.

III MARKETS OF THE FUTURE

6 AGENT BASED ECONOMICS
Gerald B. Sheblé

Department of Electrical & Computer Engineering
Iowa State University
Ames, IA 50011

INTRODUCTION

This chapter describes the application of artificial life techniques (ALIFE) to the study of auction markets for electric power optimization. Artificial life techniques include: artificial neural networks (ANN), genetic algorithms (GA) and genetic programming (GP). All ALIFE techniques are based on biological models of evolution and of neurological functions.

ALIFE techniques are being used to evaluate market operation efficiency and bidding strategies. This chapter reviews preliminary work in the area of electric power markets. The role of ALIFE techniques is compared with experimental economics to properly place the most fruitful application of computerized emulation. A definition of agent based computational economics (ABE) is offered to limit the framework within which ALIFE can achieve reasonable results. This introduction then outlines the expected electric marketplace including financial and physical aspects of production and delivery. An brief introduction to genetic algorithms is presented since it is fundamental to the research subsequently described.

The main three sections describe three approaches to ACE for electric power. The first section describes work to build a market simulator for market evaluation or for training. A blend of classical and ALIFE techniques were used within this simulator to provide active computer players to pit against the human player. The second section describes the use of GA to select bidding strategies to maximize profit. The bidding strategies are fixed in structure but the parameters are adapted as the market changes. The GA selects the best strategy based on the performance over many emulations of market behavior. The last section discusses how to generate strategy structures based on fuzzy logic rules. The intent is to grow expert rules which can determine the best bid price as the market evolves.

The intent of this research is not to answer all questions regarding market and electric power de-regulation. Instead, a set of approaches making incremental improvements are presented. The application of these techniques is valid not only for training, but also for market evaluation and for policy evaluation at the government level. The revolution of de-regulation is started, now the evolution can proceed.

Experimental or Artificial

Economists have attempted for several centuries to analyze the structure of markets, the efficiency of markets, and the evolution of markets. The traditional approach used game theory to find the steady state solution of the dynamic system of buyers and sellers. Such approaches have yielded significant research for the analysis and for the efficiency of simple markets. However, the size of markets and the complexity of markets, which could be analyzed, did not provide a sufficient basis to analyze the micro interactions amongst the market players. The latest approach to analyze markets is to establish an economic laboratory, which is then used to emulate a market with human subjects. This latest approach is termed Experimental Economics as established by Dr. Vernon Smith [57, 58, 69] and associates [16, 70].

Experimental economics emulates a market to analyze the structure and the efficiency of the market. An example experimental economics laboratory would be to construct a room in which a reasonably large number of subjects can play the market. One such examplew, as used by Vernon Smith, consists of forty (40) player personal computers (PPC) connected to an auction mechanism personal computer (AMPC). One PPC is used for each player to enter bids, to calculate the potential profit if a bid is selected, and to keep track of the players profits during the emulation process. First, the AMPC interrogates the PPCs to retrieve the bids entered by the players. Second, AMPC matches the bids according to the market rules. Third, AMPC sends messages to the PPCs to acknowledge the matching of a bid. Fourth, AMPC sends messages to all PPCs to give statistics on the latest round of bidding. Finally, AMPC keeps track of all bids that have been matched and are now contracts to determine the profitability of each player. An additional personal computer may be attached to the AMPC to keep track of the actually winnings of each player and other statistics to assist with the eventual market analysis. An example computer configuration is shown in Figure 1.

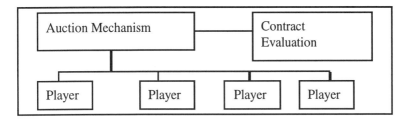

Figure 1. Computer Configuration

The PPC contains software for communication and for calculating the profit of each bid. The profit calculation thus requires a corporate model of the firm for each player who is the agent. The player is shown the cost of production if the associated firm is a supplier. The player is shown the revenue of sales if the associated firm is a buyer (retailer). The tally of the profits (or losses) gathered after each round is completed. An example entry display is shown in Figure 2 as used by this author. The top left shows the bids as sent to the AMPC. The lower left shows the acceptance or rejection of the last submitted bid. The lower right shows the messages received from the AMPC regarding the bids just submitted. The top right shows the statistics for all bids matched during the current series. The difference between a series and a round is the difference from a previous market closing and a currently open market as detailed below.

Markets are efficient because the valuation of the product is openly known. The valuation may be known from trade papers, daily newspapers, online communication services, etc. Each player has to know the current valuation of the product to formulate an individual valuation of the product for the player. An open ring market is the most common market structure for commodities. An open ring market is a raised arena enabling a group of two hundred to four hundred (200-400) traders to negotiate contract trades. Communication is primarily by hand signals, but voice communication is also used. The effective trader combines the visual and the audio price bids to determine a price valuation for the trader's position on the commodity. This indirect price signal is a major factor of the market efficiency to reduce the search costs for acceptable commodity transfers. A computer-based exchange as defined above does not have this visual and audio feedback for each player. This feedback is achieved by segmenting the bidding process into rounds.

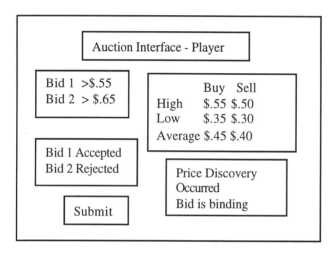

Figure 2. Player Screen

A typical computer based auction mechanism is shown in Figure 3. As shown in this mechanism, the player has several attempts at bidding before the matched bids are binding. This enables each player to gain information on what other successful players are valuing the commodity. The human auction as implemented by Vernon Smith uses a time clock to fix the bidding duration instead of fixing the number of bids, which can be submitted. The bids are binding only when the clock times out and the last bids entered were accepted. This interval of bidding is called the sequence period. The completion of a sequence period is called a round. Several rounds are conducted during a trading session. If a player is not satisfied with the contracts after each round, the player may request the market to reopen to search for a better position. Note that a fee would be charged of those wishing to reopen the market to limit "frivolous" searches of the market. Various other options may be implemented with double auctions [61, 75, 91]

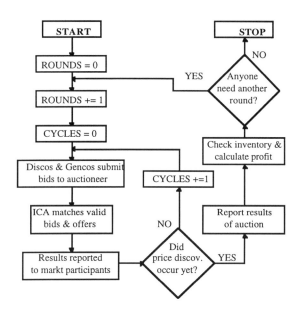

Figure 3. Computer Auction Mechanism
©IEEE 1996

A typical AMPC display is shown in Figure 4. This display shows the bids entered by the players, the bids accepted this round, the bids accepted on previous rounds, the profit of each player, the activity of each player. Additional statistics and graphs may be shown to examine the dynamic behavior of each player.

The above economic laboratory provides many insights and very significant analysis of market structures, market efficiency, and market evolution. However, the cost of running experiments is not insignificant. The time to run a sequence of experiments for a given market structure may be overwhelming. Based on previous experiments, ten market emulations of ten rounds consume between forty (40) and ninety (90) hours of work. This does not include the work to setup the software for the market rules, corporate models, etc. If ten sessions were required to gather sufficient statistics to analyze a given instance of research, then four hundred (400) to nine hundred (900) hours of work would be required for each player. Including the support staff of one instructor, one data base

192   POWER SYSTEMS RESTRUCTURING

programmer, and one software engineer, a total of forty three individuals would be need for each session. The total labor hours ranges from 400*43 to 900*43 hours of work. While most experimental economic labs use students as players to save costs. A company approach to such simulations would incur significant labor expenses to provide knowledgeable traders for experiments. Assume that engineers are used for each position at an annual cost of $250,000 per engineer, equivalent hourly rate of $150 per hour, then the cost of a single session would be $2.58 to $5.8 million. Even though such emulations are beneficial, it may be argued that the expense of such emulations may not be justified.

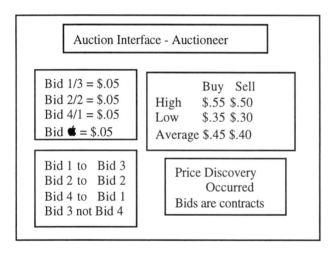

Figure 4. AMPC Display

Agent based computation economics (ACE) is a new method to approach the experimental economic laboratory with adaptive computer programs replacing the human players [2, 3, 28, 53, 94]. The adaptive computer programs are agents for the companies needed to trade a commodity. Naturally, the adaptive agents are not as conceptually powerful as human beings, in most cases, but when the resulting market sessions yield similar results, then the approximation can be justified due to the significant cost savings. The cost of such experiments is primarily the setup cost. The operational cost is one instructor, a data base programmer and a software programmer. The setup cost for adaptive agent emulation is greater than the setup for the experimental economic laboratory since the adaptive agents require significantly more programming. However, the advent of artificial life techniques is reducing this expenditure significantly. Artificial life techniques include artificial neural networks, genetic algorithms, genetic programming, and evolutionary programming.

Agent Based Computational Economics (ACE) Definition
[90]

Agent based computational economics (ACE) is roughly characterized by participants at several recent conferences as the bottom-up study of evolutionary economies modeled as decentralized systems of autonomous interacting agents. A central concern of ACE research is to understand the apparently spontaneous appearance of global regularities in economic processes, such as the unplanned coordination of trade in decentralized market economies that economists associate with Adam Smith's invisible hand. The challenge is to explain these global regularities from the bottom up, in the sense that the regularities arise from the local interactions of autonomous agents channeled through actual or potential economic institutions rather than through fictitious top-down coordinating mechanisms such as imposed market clearing. These new models combine biological models with microanalysis to look at economics in a new light.

This section focuses on terms for the reader such as "agent," "bottom-up model," "gene," and "genetic algorithm." The remaining sections present adaptive agent applications: the EPRI Auction Market Simulator (EPRI-AMS), the Kumar and Sheblé auction market simulator, the Sheblé and Richter genetic algorithm selection of rule sets (S&RI), and the Sheblé and Richter genetic algorithm selection of fuzzy rule sets (S&RII). A discussion of how closely ALIFE replaces the experimental economics laboratory is not presented. The intent is not to establish that the adaptive computer agents are replacing the human being. Instead the intent is to question if the results from adaptive agent emulation enable realistic problems to be partially answered when analyzing market efficiency, market structure and market rules.

Artificial life originated with the work of John Von Neumann with experiments on self-replicating automata. However, ALIFE was not an active research area until the establishment of ALIFE as a serious field of experimentation. The first ALIFE conference, organized by

Chris Langton, is often considered the start of ALIFE [90]. ALIFE is considered a "bottom-up" emulation of basic phenomena associated with living organisms. The genetic biologist concepts of self-replication, evolution, adaptation, self-organization, parasitism, competition, and cooperation are used within the computer. ALIFE is insufficient to replace the traditional biological and social sciences. However, it does complement the analytical, laboratory and field study of organisms by emulating the synthesis of life-like behavior within computers. ALIFE is most capable of enhancing understanding of biological and social systems. Fortunately for the field of optimization, ALIFE enables us to use nature as a guide to develop solution algorithms for difficult optimization problems. High-dimensional solution domains, extreme non-linearities, and multiple optima characterize problems best suited for ALIFE with a possibility of multiple global solutions.

ALIFE emulates systems, which are adaptive units whose response in the absence of a centralized coordination algorithm is the ultimate decentralized control system. The individual units respond to other units in parallel depending on the states and the actions of neighboring units. The overall system response is a result of the competition between the units subject to the environmental constraints and performance objectives. The complexity of the system is not obvious due to the simplicity of the individual units. Often, the interaction of units is redefined as the emulation progresses according to the environment. It is often the development of "niche" solutions, which are so interesting since their existence is not evident from the structure of the units or of the environment. It is most common for the environment and for the units to be mixed and recombined during the emulation process. The changes in the environment are referred to as the changes in the landscape. As a consequence, ALIFE is the emulation of continuously "evolving systems whose global behavior arises from the local interactions of distributed units [90]." The concept of "bottom-up" is the modeling of the individual units instead of the overall system. Since genetic algorithms or evolutionary programs are used, the units of the systems are often bit strings analogous to DNA or robots.

ALIFE has benefited extensively from the advent of very fast microcomputers, inexpensive memory (relatively) and object-oriented programming languages such as C++ (one could also include

PASCAL, FORTRAN extensions to OOPS) and JAVA. Engineers and economists who implement ALIFE emulations have been able to extend previous evolutionary research. The normal interaction is typified by cooperative and predatory characteristics with a clear relationship to the price and quantity relationships.

ALIFE research is to dissect and to identify the "apparently spontaneous appearance of regularity in economic processes, such as the unplanned coordination of trading activities in decentralized market economies that economists associate with Adam Smith's invisible hand [90]." It is important to note that rationality is "generally viewed as a testable hypothesis, or at least as a debatable methodological assumption, rather than as an unquestioned axiom of individual behavior [90].

New Environment for Power System Operation and Planning

Through the passage of the Energy Policy Act (EPAct) by the US Congress in 1992, the Federal Energy Regulatory Commission (FERC) is encouraging an open market system. The basic premise of the regulatory policies is to create a competitive environment where generation and transmission services are bought and sold under demand and supply market conditions. By increasing competition through deregulation of the transmission network, it is hoped to increase system efficiencies and improve benefits to electric consumers. Attitudes toward re-regulation vary between states. Many states are adopting a wait and see attitude, some are performing investigations, and others are almost ready to restructure their electric marketplace. California had been the first to announce plans to adopt an open market structure [8]. For decades, electric consumers in the US had only one source for their electricity—the local vertically integrated utility. Under the Energy Power Act, entities that did not own transmission-lines were granted the right to use the transmission system. This was termed *open access*. US electric utilities began to see limited competition in power production. Several other countries have already moved towards open competition: New Zealand, Australia, Norway, Sweden, Argentina, Chile, and The United Kingdom (UK) to name a few.

The open market system will consist of several companies as emulated within this chapter. Under the framework described by Sheblé [79, 80, 81], companies presently having both generation and distribution facilities would, at a minimum, be divided into separate profit and loss centers. These companies will be typified as generation companies (GENCOs), distribution companies (DISTCOs), transmission companies (TRANSCOs), energy service companies (ESCOs), ancillary service companies (ANCILCOs), independent system operators (ISOs), regional transmission groups (RTGs), national electric reliability council (NERC), national and state regulation commissions (FERC, et. al.), and possibly others. Power is generated by generation companies (GENCOs), transported via transmission companies (TRANSCOs) and distribution companies (DISTCOs). It is expected that the customer interaction, to be competitive will be handled by energy service companies (ESCOs) which include load aggregators for this chapter. In earlier research by this author [81], it was proposed that NERC would set the reliability and security standards, and that FERC would set the operation of the exchanges as presently performed by the Commodities and Futures Trading Commission (CFTC). The interconnection between these groups is shown in Figure 5 [50, 51, 52, 71]. Note that both FERC and NERC are shown with double plus signs to indicate their increased roles. Also note that an independent contract administrator (ICA) is shown. North States Power Company as an alternative to the ISO first used this acronym. Since the ISO does not operate the circuit breakers or other equipment directly, the term can be confusing. The ISO operates the system directing operation to maintain the system security and reliability. Since many TRANSCOs and DISTCOs will actually operate the equipment, the term ISO can be misleading. Additionally, the RTG plans the system to maintain the system security and reliability. This author [79, 83] has proposed that the ICA is a combination of ISO and RTG to provide a seamless transition from the future planned system to the present operational system. There is also the separate power exchange, referred to within this chapter as the energy mercantile association (EMA), that is an independent and disassociated agent for market participants and hence, is assumed to be the auctioneer in the electric marketplace.

Configuration of the transmission system and the fact that electricity flow is subject to the laws of physics, have some speculating that we will see the formation of regional commodity exchanges that would be oligopolistic [14] in nature (having a limited numbers of sellers). Others postulate that the number of sellers will be sufficient to have near perfect competition. Regardless of the actual level of the resulting competition, companies wishing to survive in the deregulated marketplace must change the way they do business and will need to develop bidding strategies for trading electricity via an exchange.

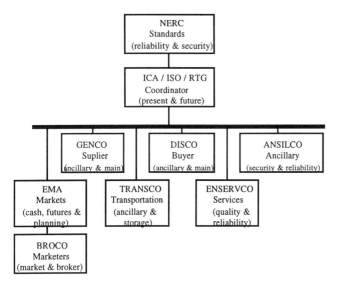

Figure 5. New business environment
©IEEE 1996

The buyers and sellers in this research interact through a central coordinator, an energy mercantile association (EMA), who matches the bids subject to all operational constraints. The ICA is responsible for ensuring that the energy transactions resulting from the matched bids do not overload or render the electrical transmission system insecure. GENCOs and ESCOs coordinate only via the prices transmitted to a central auctioneer. The ICA may submit information to the independent system operator (ISO) or to individual system operators for implementation. The key element is that the ICA is responsible for maintaining the security and reliability of the system. The ICA monitors and responds to the power system limits and transmission capacities. GENCOs and discos are required to cooperate with the ICA in maintaining system reliability. Supplying crucial generator parameters to the ICA during the bidding process is part of this cooperation.

The proposed simulation model provides a single central entity, the ICA responsible for merchandising as well as technical functions. This will facilitate the bidding process as the bids are directly evaluated in conjunction with the operating conditions and constraints. The issue of acceptance and rejection of bids is resolved disregarding the order of transactions. This is one of the key features of the model that will reduce the legalities involved in real-time implementation of a deregulated market operation. Furthermore, the practical implementation of the proposed model will reduce burden on the communication requirement. This is highly desirable due to intensive communication needs arising from the increased number of transactions in the new marketplace.

There are many technical and legal problems need to be resolved for the suggested industry transition to be successful [80, 92]. The proposed simulators can be used to address such problems, at least in part. The issues, such as pricing of transmission losses, priority of transactions, emergency power, etc. can be studied and analyzed by experimental simulation of electric markets. The ultimate goal of any simulation is to discover the problems associated with the new environment via computer/operator emulation instead of experimenting with the actual power grid.

Market players will be required to submit price-based bids to perform electric power transactions in this new environment. This is quite different from what the traditional system operators were accustomed in a vertically integrated industry. The proposed simulators can be used to train market players how to bid in the new environment. Market players include agents for any of the above companies but are primarily centered on the actions of the GENCOs and the ESCOs. Generation companies (GENCOs) and energy service companies (ESCOs) participating in an energy commodity exchange must learn to place effective bids in order to win energy contracts. Microeconomic theory states that in the long term for a perfect marketplace, the firm selling in a competitive market should price their product at its marginal cost of production.

A thorough discussion of price-based operation in an auction market structure is presented in reference [83]. Reference [50] presents a model for energy brokerage/auction system with reserve margin and transmission losses. This chapter uses the described framework in reference [50] and the proposed model in reference [51] to develop an auction market simulator. The developed simulator scheme is used to simulate auction market in the 24-bus, 10-generator IEEE Reliability Test System [42].

Electric Marketplaces

The research presented in this chapter assumes an electric marketplace similar to commodity exchanges like the Chicago Mercantile, Chicago Board of Trade, and New York Mercantile Exchange (NYMEX) where commodities (other than electricity) have been traded for many years. NYMEX has recently added electricity futures to their various energy contract offerings. The importance of these auctions is twofold. First, they provide an enhanced marketplace where the commodity is well defined, where the enforcement of the contract is very strong, where the delivery of the commodity is detailed, and where the valuation of the commodity is open for public scrutiny. The value of public scrutiny is that these marketplaces, where only ten to twenty percent of the commodity is traded, can serve as the proper valuation for bilateral contracts.

Economists have developed theoretical results of how markets are supposed to behave under varying numbers of sellers or buyers with varying degrees of competition. Often the economical results pertain to aggregation across an entire industry and require assumptions that may differ from reality. These results, while considered sound in a macroscopic sense, may be less than helpful to the individual company not fitting the industry profile who is trying to develop a strategy that will allow the company to remain competitive.

Auctions have existed within the electric power industry for several decades as exhibited by the Florida coordination Group [93]. Others have also proposed markets extending those successes. The research described in this chapter is an extension of those auction market proposed [25, 55, 59, 62, 93] and on the concepts of spot pricing of electricity [10, 19, 38, 77].

Post [67, 68] described and compared the various types of basic auctions including the multiple auction, which allows more that one bidder to be awarded a bid, and the double auction where several buyers and sellers submit bids and offers for one unit of a good. If and when a buyer accepts a seller's offer, or a seller accepts a buyer's bid, a binding contract is made. Post pointed out that "few theoretical results of double auctions exist since modeling strategic behavior on both sides of the market is difficult." Work by Fahd and Sheblé [29, 30] demonstrated an auction mechanism. Sheblé [79, 83] described the different types of commodity markets and their operation and outlined how each could be applied in the evolved electric energy marketplace.

The framework described in this chapter allows for a cash market, a futures market and a planning market. The cash market is for trading power for each 30-minute period in the next 30 days. The futures market allows electricity trading from 1 to 18 months into the future. Futures contracts are non-firm for a specific month. This futures market provides a means for electricity traders to manage their risk. The other market is a planning market that can be used to develop capital to build new plants and would allow trading more than 18 months into the future. Sheblé and McCalley [80] outline how cash, future, planning and swap markets can handle real-time control of the system (e.g., automatic generation control) and risk management.

The framework for the research described is adopted from this author's research [1, 48-52, 79-84]. This research is centered on double sided multi-lateral auctions (multiple buyers and sellers). Such auctions are an extension to the implemented auction in the United Kingdom and that being implemented in California which are one-sided, bi-lateral (sellers to aggregated buyer). A double-sided auction maximizes the number of bid awards for each time period of bid matching. This can be shown to be equivalent to the pool operating convention used in the vertically integrated business environment. This framework has since been implemented for New Zealand and is under development for Australia. This author [79] described the different types of commodity markets and their operation and outlined how each could be applied in the evolved electric energy marketplace. Under this framework, companies presently having both generation and distribution facilities would, at a minimum, be divided into separate profit and loss centers.

The major trading objectives in the auction market are hedging, speculation, and arbitrage. Hedging is a defense mechanism against loss and/or supply shortages. Speculation is assuming an investment risk with a chance for profit. Arbitrage is the crossing of sales (purchases) between the markets. This chapter assumes that there are four markets commonly operated: cash, futures, planning and swaps as shown in Figure 6.

A. Cash Market The two contracts normally traded in cash markets are spot and forward. The spot contracts reflect the current price of transactions. The forward contracts reflect future system conditions. Transactions are executed immediately for spot contracts. In the forward market, prices are determined at the time of contract but the transactions occur at some future time. The auction mechanism for cash market is developed in reference [91].

B. Futures Market A futures market is a place where buyers and sellers can meet readily. A futures market creates competition because it unifies diverse and scattered local markets and stabilizes prices. The contracts in futures market are risky because price movements over time can result in large gains or losses. There is a link between cash market and futures market, which allows price volatility.

The components of futures contract include trading unit, trading hours, trading months, price quotation, minimum price fluctuation, maximum daily price fluctuation, last trading day, exercise of options, option strike prices, delivery, delivery period, alternate delivery procedure, exchange of futures for, or in connection with physicals, quality specifications, customer margin requirements. Note that delivery, delivery period, alternative delivery procedure, and quality specifications would have to be specified in greater detail than when the industry was vertically integrated.

C. Swap Market In the swap market, contract position can be closed with an exchange of physical or financial substitutions. The trader can find another trader who will accept (make) delivery and end the trader's delivery obligation. The acceptor of the obligation is compensated through a price discount or a premium relative to the going price.

D. Planning Market The growth of transmission grid requires transmission companies to make contracts based on the expected usage to finance projects. The planning market would underwrite the usage of equipment subject to the long-term commitments to which distribution and generation companies are bound by the rules of network expansion to maintain a fair market place.

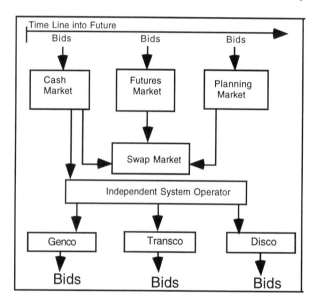

Figure 6. Electric energy markets
©IEEE 1996

In this work, we concentrate on the electric power auction in forward market and futures market. The financial drain inflicted on traders when hedging their operations in the futures market is slightly higher than the one inflected through direct placement in the forward market. Hence, an optimal mix of futures commitments and forward contracts is difficult to assess and depends on hedging, constraints imposed by different contracts, and cost of different contracts. 1. The futures market is a monthly market up to 18 months. The forward market is an hourly market for a one (1) month period. Table 1 explicitly depicts this time horizon of the different markets.

This research presented an approach to allocate and implement the future contract commitments via forward market simulation. The proposed contract allocation is based on linear programming formulations. These formulations can easily be extended to include stochastic variables, such as fluctuating market prices by using decision analysis methodology [41].

Electric Market Emulation Research

Work by Kumar and Sheblé [50, 52] brought the above ideas together and demonstrated a power system auction game designed to be a training tool. That game used the double auction mechanism in combination with classical optimization techniques. In other sections, trading agents use genetic algorithm (GA) coupled with various price-forecasting techniques to select appropriate bidding strategies for the current market conditions. The bidding strategies adapt, or evolve, as other traders change their trading behavior. The results have been written up for the electricity marketplace, but may be used directly for other types of trading.

Some work has been done in developing bidding strategies for other electric systems. Finlay [31] analyzed bidding strategies for the restructured Power Pool of England and Wales system, and showed mathematically that there exists an optimal bidding strategy for its bidders. Finlay's work differs in that his objective was not to maximize the profit of the individual generation companies, and the system itself is different from those proposed in the USA.

Developing bidding strategies with evolving trading agents for the deregulated electrical utility industry is a new field of research. Apart from the electrical utility industry, interest has grown in recent years for using evolving, or adaptive, agents to simulate trading behavior. Research with adaptive agents has proved to be a useful means of exploring trading markets outside of the electrical industry.

LeBaron [54] uses evolving agents to learn to play financial markets. Tesfatsion and others [5, 7, 87, 88, 90] describe research in which trading agents decide who to trade with based on an expected payoff. Ashlock, in reference [4], uses genetic programming combined with finite state automata to play a classic academic game involving bidding behavior and strategies. Other research investigate the evolution of trading structures as agents find easier interfacing methods to interact [6, 15, 21, 26, 27, 43, 44]. Each section below will define the difference that typifies an agent for the research summarized.

Table 1. Time horizon of forward and futures market

Day	Present Month	Future Month 1	Future Month 2	Etc.
First Day of Month	Cash Market	Forward Market	Forward Market	Forward Market
	Future Allocation	Future Purchase	Future Purchase	Future Purchase
Today	Cash Market	Forward Market	Forward Market	Forward Market
	Future Allocation	Future Purchase	Future Purchase	Future Purchase
Last Day of Month	Cash Market	Forward Market	Forward Market	Forward Market
	Future Allocation	Future Purchase	Future Purchase	Future Purchase
First Day of Next Month		Cash Market	Forward Market	Forward Market
		Future Allocation	Future Purchase	Future Purchase

In the double auction used for this research, the bids and offers are sorted into descending and ascending order respectively, similar to the Florida Coordination Group approach as described by Wood and Wollenberg [93]. If the buy bid is higher than the sell offer to be matched, then this is a potential valid match. The ICA must determine whether the transaction would endanger system security and whether transmission capacity exists. If the ICA approves, the valid offers and bids are matched and the difference in the bids ($/MW) is split to determine the final price, termed the *equilibrium price*. This is similar to the power pool split savings approach that many regions have been using for years. An example is given in Table 2.

In this example, there are three bids that are higher than the offers. If there are not a sufficient number of valid matches, then *price discovery* has not occurred. The auctioneer reports the results of the auction to the market participants. When all bids and offers are collected and insufficient valid bids and offers are found to exist, the auction has gone through one cycle. The auctioneer then reports that price discovery did not occur, and will ask for bids and offers again. The buyers and sellers adjust their bids and offers and another cycle of the auction is played. The cycles continue until price discovery occurs, or until the auctioneer decides to match whatever valid matches exist and continue to the next round or hour of bidding.

TABLE 2. Example of auction bid matching

Buy bids ($/MW)	Sell offers ($/MW)	Contract?	Equilibrium price ($/MW)
12.50	8.50	Yes	10.50
12.00	9.00	Yes	10.50
11.80	10.00	Yes	10.90
10.00	10.50	No	NA
9.50	11.00	No	NA

©IEEE 1996

After price discovery, the auctioneer asks if another round of bidding is requested. If the market participants have more power to sell or buy, they request another round. This process is continued until no more requests are received or until the auctioneer decides that enough rounds have taken place. See Figure 3 for a block diagram of the auction process.

Genetic Algorithms

A genetic algorithm (GA) is an algorithm that allows evolution of the contents of a data structure. GAs were developed by John Holland [34, 40] and are loosely based on the biological notion of evolution. The data structure being evolved contains a solution to the problem being studied. Populations of these solutions are initialized at the beginning of the algorithm. Each of the solutions is assigned a fitness based on its suitability for solving the particular problem being studied. If these solutions are initialized randomly, the chances of them being highly *fit* during the first *generation*, is not very high. At each generation, the GA will randomly chose members of the population to be "parents" favoring the highly fit members. The parents will then produce offspring via the *crossover* and *mutation* processes. The offspring replace the members of the population that have a low fitness. As the generations progress, there is a tendency for the contents of the data structures to adapt such that they become more suited to solving the problem.

Derived from the biological model of evolution, genetic algorithms (GAs) operate on the Darwinian principle of natural selection [34]. Populations of data structures appropriate for the optimization problem are "randomly" initialized. Each of these candidate solutions is termed an individual or a creature. Each creature is assigned fitness, which is simply a heuristic measure of its quality. Then during the evolutionary process, those creatures that have a higher fitness are favored and allowed to procreate.

During each generation of the evolutionary process, creatures are randomly selected for reproduction with some bias toward higher fitness. After parents are selected for reproduction, they produce children via the processes of *crossover* and *mutation*. The creatures formed during reproduction explore different areas of the solution space than did the parents. These new creatures replace lesser-fit creatures of the existing population. The basic algorithm can be written as shown in Table 3.

Table 3. Generic genetic algorithm process

Step	Action
1	Randomly initialize a population and set the generation counter to zero.
2	Until done or out of time, do the following, else do 9.
3	Calculate the fitness of each member of the population.
4	Select parents using some fitness bias.
5	Create offspring from the selected parents via crossover.
6	Mutate these new offspring.
7	Replace the lesser-fit members of the population with the newly created offspring.
8	Increment the generation counter and go to the beginning of step 2.
9	Generate output

Reproduction is the mechanism which the most highly fit agent in a population are selected to pass on information to the next population of agents. The fitness of each agent was determined by calculating the profit each agent made and then ordering the agents by decreasing profit. Then each agent was assigned fitness according to its rank. The agents that were kept for reproduction were sometimes determined by roulette wheel selection. This type of reproduction is called rank-based fitness and can be found in [12, 22, 23, 34, 36, 37, 63, 76] along with other methods.

The parents are required to be in pairs for reproduction, and the result is one or two children. Copying the contents of parent 1 into child 1 creates children and of parent 2 into child 2 until a randomly selected crossover location is reached. At this point, bits are copied from parent 1 into child 2, and from parent 2 into child 1.

Following the crossover process, the children are mutated. Mutation introduces new genetic material into the gene at some low rate. If the gene to be mutated in the child is represented by a binary string, mutation involves flipping the bit (0 goes to 1, 1 goes to 0) at each location in the string with some probability. If the gene is represented by an integer, mutation might involve adding an integer that will result in a different valid integer occupying that gene location (loci).

Mutation is generally thought of as a secondary operator. This operator ensures that no string position will ever be fixed at a certain value through the course of the search. Mutation operates by toggling any given binary weight string position using the probability of mutation.

Crossover is the primary genetic operator that promotes the exploration of new regions in the search space. Crossover is a structured, yet randomized mechanism of exchanging information between strings. Crossover begins by selecting two buyers/sellers previously placed in the mating pool during reproduction. A crossover point is then selected at random and information from one buyer/seller, up to the crossover point, is exchanged with the other buyer/seller. Thus creating two new buyers/sellers for the next generation.

Elitism ensures that the best buyers/sellers are never lost in moving from one generation to the next. The elitism subroutine takes the best buyers/sellers in the current generation and saves them into tile generation.

EPRI AUCTION MARKET SIMULATOR (EPRI-AMS) [84]

The auction market simulator (AMS) described in this section was developed under a contract with the Electric Power Research Institute (EPRI). The author is grateful for the support provided when market based research was not favored. This AMS evolves between 3 to 20 agents for the buyers and 3 to 20 agents for the sellers. The market is an extension of the auction rules used by the Florida Coordination Group [93] with the additional rules of transmission limitations as defined by available transmission capacities (ATCs). Specifically, transactions above the ATC between individual buyers and sellers are not allowed. The agents evolve the bid value until equilibrium is obtained as evidenced by a flat bid price curve. However, since the GA is a probabilistic technique, convergence is not guaranteed. Indeed, the mutation operator may find a better solution many generations (iterations) beyond observations of the flat curve. It is important to note though, that convergence often occurs very quickly (less than fifty generations).

Note that the Electric Power research Institute (EPRI) has authorized the author to provide this simulator free. The software can be downloaded from the author's personal web page at Iowa State University.

Artificial Adaptive Agents

An artificial adaptive agent is an encoded price that defines what a buyer or seller should offer according to marketplace conditions. Each artificial adaptive agent represents one buyer or seller. The artificial adaptive agent tries to learn the optimal allocation of its resources to maximize its profits in the marketplace. The following discussion assumes that each artificial adaptive agent will try to determine the optimal bid price per megawatt for a buyer or the optimal selling price per megawatt for a seller in the electric marketplace. A genetic algorithm was used with each group of buyers or sellers to determine the bid price per megawatt or the selling price per megawatt. The genetic algorithm is the mechanism by which each artificial adaptive agent will adjust to the marketplace and optimally allocate its resources.

Artificial Adaptive Agents Implementation

The artificial adaptive agent's implementation consists of initializing buyers and sellers; selection of the most fit to survive, a reproduction mechanism including crossover and mutation.

An explanation of each part of the artificial adaptive agent's algorithm implementation follows:

The buyer initialization is explained for one artificial adaptive buyer agent of the population. A buyer member of the population consists of a DNA string with bit length of 10. This array represents the buyer's bid. This array is randomly filled with 0's and 1's. The mapping of the bit string to a bid is determined by a linear transformation between his customer's paying price and a certain percentage of profit.

The seller initialization is explained for one artificial adaptive seller agent of the population. A seller member of the population consists of a DNA string with bit length of 10. This array represents the seller's asking price. This array is randomly filled with 0's and 1's. A linear transformation between his customers' cost and a certain percentage of profit determine the mapping of the bit string to a selling price.

The auction mechanism consists of matching the buyers and sellers of electricity as described above. The auction mechanism will set up the transaction that match the highest priced buy quotation with the lowest priced sell quotation. The transaction price is set at the midpoint between the buyer's bid and the seller's asking price.

Experimental Results

Artificial adaptive agent's results are given in Table 4 and Table 5. The buyers and sellers were all given similar values for parameters and the equilibrium point should have the same profits for buyers and sellers. It is interesting to note that did not happen in the computer simulation. The buyers have some advantage over the sellers that allowed them to make more profit. The code has been checked and no obvious mistakes have surfaced.

An artificial adaptive agent buyer should bid $6.81 per MW to maximize its return in the simulated open market environment. An artificial adaptive agent seller should bid $6.80 per MW to maximize its return in the simulated open market environment. It is interesting to note that the seller that tried to ask for $6.86 per MW ended up not being able to sell any electricity in the simulated open market environment and actually lost money because of fixed cost.

Table 4. Artificial adaptive agent buyer's bid and profit

Buyer	Bid Price/MW	MW Bought	Profit
1	$7.68	10	$53.65
2	$6.84	10	$53.10
3	$6.81	10	$50.91
4	$6.81	10	$50.77
5	$6.81	10	$46.53
6	$6.81	0	-$1.00

©IEEE 1996

Summary

This simulation has presented artificial adaptive agents for the buying and selling of electricity in an open market environment. The interesting result of this experiment was the adaptive artificial agent buyers figured out a method that exploited a problem in the computer simulation program and found a bidding strategy that allowed the buyers to make more profit than the sellers. Various agent encodings were tried to demonstrate aggressive and conservative behavior. The differences in profitability are striking, even though both should have been similar. If a genetic algorithm artificial adaptive agent can figure out and exploit an advantage in a trading network to the company's philosophy, just imagine what it could do in the highly complex power environment.

Table 5. Artificial adaptive agent seller's asking price and profit

Seller	Bid Price/MW	MW Sold	Profit
1	$6.80	10	$21.46
2	$6.80	10	$17.22
3	$6.80	10	$17.08
4	$6.80	10	$15.89
5	$6.80	10	$15.34
6	$6.86	0	-$1.00

©IEEE 1996

AUCTION MARKET SIMULATOR FOR PRICE BASED OPERATION [52]

This simulator used classical optimization techniques [9, 11, 13, 20, 45, 56, 65, 85], adjusted to not overbid the price for each cycle of the auction. The auction included all of the power flow equations including line limits. A reduced DC power flow [93] model accelerates the process. The use of sensitivity factors [93] could also simplify the optimization at the price of more optimization steps. This would be equivalent to using Available Transfer Capability (ATC) limits.

This proposed simulator allocates the implementation of futures contracts via cash or forward market substitution as through a swap market in addition to the spot market. The simulator is developed on the assumptions below; however, the approach described is flexible enough to allow these assumptions to be modified if desired.

Assumptions

The players are obligated to settle futures contracts in on-peak periods of weekdays via forward market. The allocation of futures contracts in forward market simulation is shown in Figure 7.

212 POWER SYSTEMS RESTRUCTURING

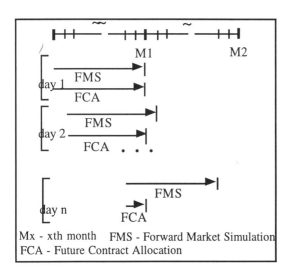

Figure 7. Market simulation and futures contract allocation
©IEEE 1996

The overall process of forward market simulation consists of periodic bid development by GENCOs and ESCOs followed by bid matching by auctioneer. This process is repeated several times until the price discovery has occurred in the auction market place. The *price discovery* is defined as the cycle at which the amount of binding contract exceeds a set minimum percentage value. The contracts established at the price discovery cycle is defined as the *closing contract*. The overall auction process cycle is shown in Figure 3. Thus, the major elements of auction market are (1) rules defining the auction mechanism and contract evaluation, (2) information available to the market players, (3) GENCO bids, and (4) ESCO bids.

Rules

The auctioneer will establish interchange schedules between the participating players on hourly basis by using the model developed in reference [51].

The bidding process is performed a fixed number of times within an hour. The auctioneer may decide when the last bids are binding to allow price discovery to occur without excessive gaming of the market. The players do not know which bid will be binding until the auctioneer declares such conditions. However, they do know that there is a maximum number of trials per bidding period. The auctioneer may start another round of bidding if any of the parties wish to bid.

A bid is a specified amount of electricity at a given price. Hence, the players should come up with the price of electricity for the amount they wish to transact (price per block for a given number of blocks).

The players are obligated to buy or sell the binding bids declared by the auctioneer. Hence, all local operational constraints (such as ramp rate constraint, emission constraint, fuel constraints, minimum down times, minimum up times, start-up procedure curves, etc.) must be considered by the players while generating bids for the next trading session or round.

Information Available

The auctioneer would post following information on an electronic bulletin accessible to all agents.

1. High bid
2. Low bid
3. Average bid
4. Number of bids accepted

A chronicle of the above mentioned information would describe the trend of the auction market. The trend analysis may be used to explain the expected behavior of players.

GENCO Bids

GENCOs would develop bids primarily based on the plant I/O curve. Typically, the incremental cost curve (ICC) of a power plant is monotonically increasing. A piece-wise linear ICC is shown in Figure 8. In order to develop a bid, GENCOs need to construct the curve illustrated in the figure. Most of the ICCs are flatter in the lower range of operating points as opposed to much more steeper in the higher range. Thus, more discretizing segments are needed as one moves to higher range of operating points for block bidding.

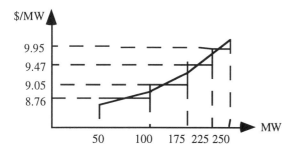

Figure 8. GENCO bids at different operating points
©IEEE 1996

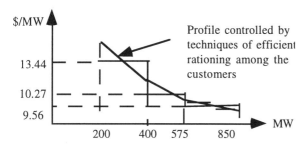

Figure 9. ESCO bids at different operating points
©IEEE 1996

It is very evident that the GENCOs need to generate their ICC accurately. All local constraints such as fuel constraints, emission constraints, and ramp rate constraints must be included in the optimization problem. Moreover, the bid development should be done by using not only the incremental cost curve, but also by deciding business strategies with due consideration to moves made by the other players in market place. Information posted on the electronic bulletin should be properly used to do a trend analysis and to conjecture performance of other agents.

ESCO Bids

ESCOs would develop bids primarily based on their decremental revenue curve (DRC) as shown in Figure 9. At present, the concept of DRC is not very prevalent. But in the future, ESCOs would have to use strategies based on their DRCs to operate efficiently. There are two major components that would affect the decremental revenue curve, (a) Delay Scheduling curve [50] and (b) interruptible and curtailable rate programs [18, 24, 32, 77, 86, 89].

The most important element that affects the decremental revenue curve is the ability to shift the load through different time periods to maximize the overall profit in procurement and delivery. Dynamic programming using any combination of the following can solve this optimization problem: (a) Self-generation, (b) Storage (cold water, etc.), and (c) Demand side management (DSM by moving load and losing some). Under interruptible and curtailable rate programs, a disco offers a bill discount to the participating customers in the exchange for the right to curtail their service under prescribed conditions within prespecified monthly frequency of interruptions.

The overall scheme for auction market simulator

The overall scheme for auction market simulator is depicted in Figure 10. Initialization is done on the basis of operating point and contract agreements. The players use the auction mechanism developed above to match the submitted bids.

The players use future contracts in generating bids to maximize profit by using formulation given in equations (1) - (6). The auctioneer ensures that the current day futures contract allocation is feasible, and then simulates the forward market for the current month. The auction cycle proceeds until the price discovery occurs. The convergence in forward market is defined as the auction cycle in which the number of closing contracts (as defined earlier) in different periods exceeds a set minimum percentage value. At each convergence, the operating points are updated in accordance with the established contracts and the simulation proceeds for the next day.

Simulation Results

The developed auction market simulator emulated the 10-generator, 24-bus IEEE RTS system. The test system data has been taken from the report prepared by the IEEE PES Reliability Test System Task Force [42]. For simulation purpose, the test system was divided into the 5 GENCOs and 5 ESCOs. The correspondence of the test system with RTS test system is described in Table 6.

The cost functions of GENCOs are assumed to be quadratic curves. Revenue curves of ESCOs are also modeled as quadratic curves. However, in this example, these curves are generated such that the data allows enough consumer surpluses for facilitating the trade game. This approach provides a basis of comparison between auction market outcome and the conventional pool dispatch.

An example of 3 period simulation of forward auction market is used to illustrate the power transaction in the proposed framework. Hydro units being the cheapest always enter the basis of the proposed LP model. Hence, they are ignored for this simulation example. Table 7 shows the future contract agreements of the GENCOs. Initial operating points of various units for the three periods are shown in Table 8.

The simulation process for the agents follows a two step procedure at every auction cycle. First, the futures contract amounts are allocated among the periods by using the deterministic formulation as described in earlier section. A multi-area economic dispatch program generates the price estimates. At the second step, the agents update their operating points based on the accepted futures contracts. Then, the bids are generated on the basis of operating points and desired profits. Using these bids, the overall process of auction market simulation for the 3-period is implemented. The simulation results are summarized in Tables 9, 10 and 11.

Note that the market-clearing price is higher than the lambda pool dispatch. This is due to high value of disco bids originally designed to have enough consumer surpluses in the system data. Hence, this should not be confused with the high price of transaction. The intent of this example is to show the overall process. Observing that all the market-clearing prices are correlated with system lambda recognizes the similarity of the proposed approach with the conventional pool dispatch.

AGENT BASED ECONOMICS 217

Table 6. GENCO and Disco Buses in IEEE RTS-96
©IEEE 1996

GENCO	Bus Number	Disco	Bus Number
1	1, 2	1	4, 5
2	13, 23	2	3
3	22	3	6, 8, 9, 10
4	7, 15, 16	4	20
5	18., 21	5	14, 19

Table 7. Future contract agreements
©IEEE 1996

GENCO	Futures Contract MW
1	70
2	105
4	45

Table 8. Initial operating points of units
©IEEE 1996

Generator type - GENCO.	Period 1 (MW)	Period 2 (MW)	Period 3 (MW)
U20 - 1	80.00	77.68	47.31
U76 - 1	165.00	165.00	165.00
U197 - 2	395.53	305.45	265.00
U155 - 2	308.19	249.42	175.57
U350 - 2	73.02	57.28	50.00
U12 - 4	60.00	60.00	60.00
U155 - 4	308.19	249.42	175.50
U100 - 4	125.00	125.00	125.00
U400 - 5	333.00	333.00	333.00

Figure 10. The overall scheme for auction market simulator
©IEEE 1996

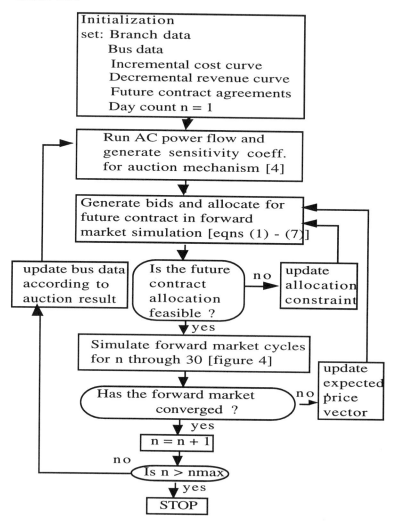

Table 9. Futures contract allocation
©IEEE 1996

GENCO	Period 1 MW	Period 2 MW	Period 3 MW
1	12	26	32
2	30	35	40
4	12	18	15

Table 10. Unit operating points after binding contracts in RTS
©IEEE 1996

Generator type - GENCO.	Period 1 (MW)	Period 2 (MW)	Period 3 (MW)
U20 - 1	80.00	80.00	80.00
U76 - 1	268.70	200.42	175.00
U197 - 2	547.52	471.02	336.16
U155 - 2	310.00	310.00	269.46
U350 - 2	99.57	86.21	62.65
U12 - 4	60.00	60.00	60.00
U155 - 4	310.00	310.00	269.46
U100 - 4	145.00	135.00	140.00
U400 - 5	353.00	353.00	343.00

Table 11. Transaction parameters
©IEEE 1996

Transaction parameters	Period 1	Period 2	Period 3
Market Clearing Price ($/MW)	10.58	10.41	10.13
Lambda for Pool dispatch ($/MW)	9.63	9.39	8.97
Total amount of sell bids accepted	292.06	360.81	314.60
Total amount of buy bids accepted	285.88	353.46	304.33
Transmission losses due to transactions (MW)	6.17	7.35	10.26

Summary

This AMS was developed for price based operation. The proposed approach allows the implementation of futures contract allocation via forward market simulation. This simulator can be used to experimentally study various aspects of the deregulated environment. The proposed simulator can also be used to explain the price-based bidding to the system operators.

GENETIC ALGORITHM EVOLUTION OF UTILITY BIDDING STRATEGIES [71]

This section describes an environment in which energy service companies (ESCOs) and generation companies (GENCOs) buy and sell power via double auctions implemented in a regional commodity exchange. The electric utilities' profits depend on the implementation of a successful bidding strategy. In this research, a genetic algorithm evolves bidding strategies as GENCOs and ESCOs trade power. A framework in which bidding strategies may be tested and modified is presented. This simulated electric commodity exchange can be used off-line to predict whether bid strategies will be profitable and successful. It can also be used to experimentally verify how bidding behavior affects the competitive electric marketplace.

Marketplace

This research assumes the existence of regional commodity exchanges in which buyers and sellers participate in a double auction. This framework is described above. For the results presented in this section, TRANSCOs are considered to be exogenous to the market, DISCOs and GENCOs are allowed to interact as described in the above environment. Although this author's framework makes use of the futures markets, only the cash market was modeled for the results presented here.

Genetic Algorithms for Agent Evolution

In the research described in this section, parameters used to develop GENCOs' bids are evolved using a GA. Each member of the GA population corresponds to a GENCO participating in an auction. There are three distinct evolving parts, or genes, for each of the GENCOs. First, the number of 1 MW contracts to offer at each round of bidding is evolving. This gene is filled with integer values. Valid integers are between 0 and a maximum value that corresponds to that GENCO's maximum capability divided by the number of rounds of bidding. Secondly, bid multipliers for each round of bidding are evolved. These bid multipliers are used in combination with the GENCOs' costs and their expected equilibrium price to develop a bid. Binary strings that are mapped represent this gene to a value between the GENCO's cost and forecasted equilibrium price during the bidding process. Thirdly, each GENCO has a gene that selects which prediction technique to use to forecast the equilibrium price. This is an integer-valued gene with valid integers being from 0 to 4, since there are 5 classical prediction techniques from which each GENCO may choose. Additional forecasting methods can be incorporated easily. See Figure 11 for a representation of the trading agent's data structures.

Roulette selection was chosen to select parents each generation of the genetic algorithm. Roulette selection is a parent selection method that chooses more highly fit creatures with a greater probability than the lesser-fit creatures. This fitness bias is more pronounced (especially when population sizes are small) than other parent selection methods like tournament selection or rank selection. However, the authors intend to do sensitivity testing on all of these methods in the future.

Based on sensitivity tests, three-point crossover was selected to create the children. Crossover is used on both the no. of contracts desired and the bid multiplier. The bid multipliers for each GENCO are concatenated together into one string prior to crossover, and three crossover locations are selected randomly from a uniform distribution over the gene's entire length.

The standard bit-flip mutation operator is used on the binary strings representing the bid multipliers. The number of contracts gene has the possibility of being mutated by two different mutation operators. The first mutation operator (mutation_A) adds an integer to the existing integer. If the result is not a valid integer, the value is wrapped around, i.e. if the result is greater than the maximum, then the maximum is subtracted from it. Further investigation might reveal better results if the gene is set to its maximum rather than wrapping it around. The second mutation operator (mutation_B) shuffles the values among the different loci. This way if a good number is found in one locus, it can spread to other locations more

222 POWER SYSTEMS RESTRUCTURING

quickly. Mutation on the prediction technique selection gene involves randomly selecting one of the valid predication techniques.

Figure 11. Data Structure of Evolving Agents
©IEEE 1996

Agent 0	Rounds of Bidding -------------------->				
MWs each round	12	4	20	.	14
Mult. each round	01011	01101	101101	.	00101
Prediction Technique	0				

Agent 1	Rounds of Bidding -------------------->				
MWs each round	15	7	1	.	6
Mult. each round	10011	01100	101100	.	10111
Prediction Technique	2				

o
o
o

Agent N	Rounds of Bidding ------------------------>				
MWs each round	15	3	19	..	20
Mult. each round	00011	11101	11001	..	00111
Prediction Technique	3				

The fitness of each creature is exactly equal to its profit after participating in an auction. A generation level for each GENCO can be determined by the number of contracts that the GENCO was able to obtain during the auction process. Profit becomes the total cost to generate at that level, minus the total revenue. Total revenue is equal to the sum of the contract price multiplied by the number of contracts over all rounds of bidding.

At each generation, one half of the population is replaced with the children. Although the parents were not taken strictly from the top half of the population, it is always the creatures on the bottom of the population that are replaced each generation.

Developing the Bid

This subsection explains how to develop a bid from the previously shown data structures. The number of 1 MW contracts is taken directly from the "MWs each round" gene. The bid multipliers can be used in the following two different methods. In the first of these, it can be mapped into a range where all 1s would be the expected equilibrium price (EEP), and all 0s would be the cost of the generator. With the second method it can be multiplied by the EEP such that the bid will be within some tolerance of that EEP. For the results shown in this section the first method was used.

There are two methods included for determining the EEP. The first method uses no prediction techniques, and simply assumes that the current round's price will be the same as the previous round's equilibrium price. This is a fair assumption given a stable market where the prices do not fluctuate very much. The other method uses prediction techniques to forecast the price. (A gene on in the GENCO's data structure determines which method to use.) The GENCOs are able to make use of any prediction method, examples might be:

1. moving average (MVA)
2. weighted moving average (WMVA)
3. exponentially weighted moving avg. (EWMVA)
4. linear regression (LR)

Any other technique to predict the clearing price can be added. Another technique, which the author has tried, is charting techniques and neural network techniques [39, 66, 78].

The MVA, WMVA, EWMVA and LR are standard and can be found in any statistics textbook [33]. The number of inputs and inputs to be considered can be adjusted until the best predictions are observed (minimum mean squared error). It was originally intended that each GENCO would have a unique set of parameters for its own prediction technique, but this was found to be too computationally expensive. Therefore, each GENCO that uses the same prediction technique gets the same forecasted equilibrium price.

See Figure 12 for a block diagram of how the genetic algorithm, the price prediction and the auction processes fit together to evolve the GENCOs that are trading electricity.

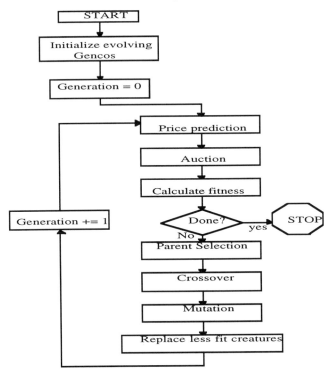

Figure 12. The GA agent evolution process.
©IEEE 1996

Results

The algorithm was initialized with a population size of 24, i.e., 24 GENCOs. Each generation 12 new creatures replaced the 12 worst fit of the population. Roulette selection was used to select the parents. A mutation rate of twenty percent was used at each locus of the new creatures for the standard mutation operation. Mutation_B was used 50% of the time, and the Mutation_A was used the other 50%. Three-point crossover was used during reproduction. Fitness was taken as the raw profit each GENCO received.

For simplicity one disco was bidding against all of the GENCOs. The disco bid a constant amount each round. Neither transmission constraints nor system stability violations were considered. The minimum generation of each GENCO is 200 MWs; the maximum is 480 MWs. Trading occurs for 24 rounds each generation. A maximum of 20 cycles is allowed for price discovery, at which time the round number is incremented. Each of the GENCOs has the same generation cost curve, represented by a quadratic input/0output curve.

For the case shown in the paper, the expected equilibrium price was taken to be the previous round's equilibrium price. The bid multiplier was used to bid between the GENCO's cost and the expected equilibrium price.

The case shown is a sellers market, where the electricity demand is twice the supply. The disco wants 960 MW each round, but the maximum total supply is 480 MW/round. The disco bids $20 each round. The GENCOs are allowed to evolve for 400 generations.

The graphical results in Figures 13 through 17 show that the auction is working as expected. The average, maximum and minimum GENCO fitness is plotted for 400 generations. The number of MWs actually purchased by the disco is plotted. The average, minimum and maximum bids of the GENCOs are plotted at each of the rounds for each generation, as well as the resulting equilibrium price. The number of MWs sold by the best GENCO each generation is plotted. The best GENCOs bid multipliers are plotted at the first generation and the last generation.

The fitness of the best GENCO/agent each generation is increasing. The best GENCO is taking advantage of the high demand by increasing the number of MWs it is offering for sale. However, many of the lesser-fit GENCOs are not finding that it is beneficial to sell as much as possible. Consequently the disco is purchasing less than half of the electricity it would like. The equilibrium price tends to stay in the $17-$18/MW range. It is smooth over the generations, but fairly mobile over the rounds during each generation. This is because the equilibrium price is the difference between sell offer and the buy bid, and the buy bid is constant. The offer prices of the lesser-fit GENCOs prevent the equilibrium price from reaching $20/MW. The average bid is about $15/MW for this case. The bid multipliers of the best GENCO evolved so that the bids were closer to the expected equilibrium price as opposed to the GENCO's cost.

Summary

The auction simulation is working as desired. The evolution of the GENCOs is also functioning as expected. The framework used and developed for this research should be a helpful tool for those who will be participating in the competitive electric marketplace of the future. It would also be helpful for those wishing to develop bidding strategies for other types of markets. Several other cases were run that have not been included in this section. Multiple runs verified that the bid multipliers were functioning. The bid multiplier gene evolved to increase fitness while the other genes were held constant. The prediction techniques must be properly tuned to get good forecasts. If the market is not volatile, using the previous equilibrium price works well. Market specific information including knowledge about the volatility can help to fine-tune the parameters for better performance. The GA is able to make use of a more complex data structure. Separate multipliers for each round of bidding evolve to result in a better solution than that derived from a single multiplier for all rounds.

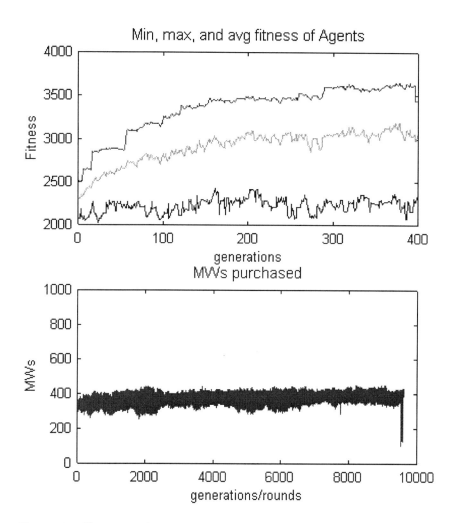

Figure 13. Fitness and purchasing results.
©IEEE 1996

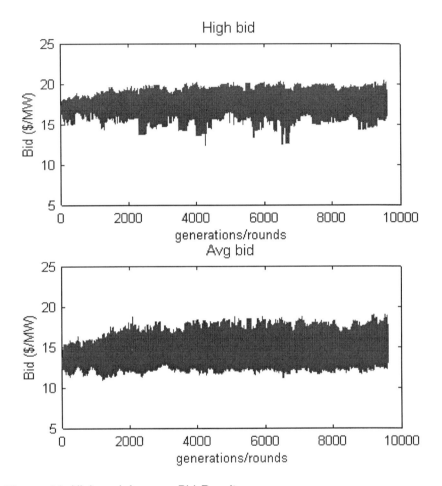

Figure. 14. High and Average Bid Results.
©IEEE 1996

Recent research includes a more detailed model that considers startup and shutdown costs, ramp constraints, minimum up and down times for the generators. Separate cost curves for each generating unit will be included. Future research will also include power marketers that both buy and sell power. Future research will verify that the results are valid when using "intelligent" buyers, in addition to the "intelligent" sellers, i.e. including multiple discos with their own bidding strategies that react to the GENCO's bids.

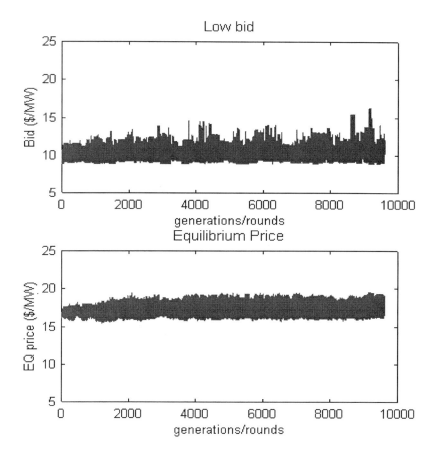

Figure 15. Low bid and equilibrium price results.
©IEEE 1996

230 POWER SYSTEMS RESTRUCTURING

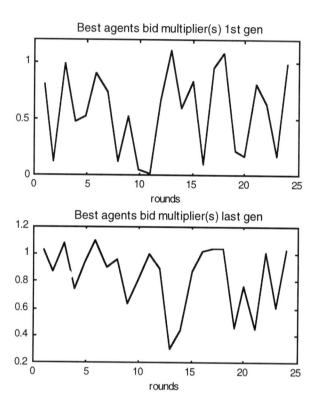

Figure 16. Bid multipliers of best agent.
©IEEE 1996

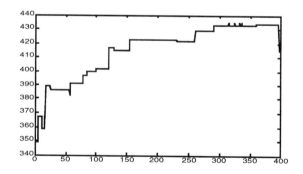

Figure 17. MWs sold by best agent.
©IEEE 1996

COMPETITIVE FUZZY BIDDING STRATEGIES FOR THE SUCCESSFUL GENERATOR [73]

In this section, the author builds on previous research that has done in this area. The bids of the agents participating in an auction are determined by fuzzy rules [35, 60, 64]. A collection of these fuzzy bidding rules can be fine-tuned off-line to perform well under specific circumstances. The agent can then select the bidding rules most appropriate for the current market conditions and use them to develop a profitable bid.

Marketplace

In the double auction used for this research, the bids and offers are sorted into descending and ascending order respectively, similar to the Florida Coordination Group approach as described above. If the buy bid is higher than the sell offer to be matched, then this is a potential valid match. The ICA must determine whether the transaction would endanger system security and whether transmission capacity exists. If the ICA approves, the valid offers and bids are matched and the difference in the bids ($/MW) is split to determine the final price, termed the *equilibrium price*. This is similar to the power pool split savings approach that many regions have been using.

If there is an insufficient number of valid matches, then *price discovery* has not occurred. The auctioneer reports the results of the auction to the market participants. When all bids and offers are collected and insufficient valid bids and offers are found to exist, the auction has gone through one cycle. The auctioneer then reports that price discovery did not occur, and will ask for bids and offers again. The buyers and sellers adjust their bids and offers and another cycle of the auction is played. The cycles continue until price discovery occurs, or until the auctioneer decides to match whatever valid matches exist and continue to the next round or hour of bidding.

After price discovery, the auctioneer asks if another round of bidding is requested. If the market participants have more power to sell or buy, they request another round. Allowing multiple rounds of bidding each time period allows the participants the opportunity to use the latest pricing information in forming their present bid. This

232 POWER SYSTEMS RESTRUCTURING

process is continued until no more requests are received or until the auctioneer decides that enough rounds have taken place.

Genetic Algorithm Evolution of Bidding Strategies

Richter [72] used a GA to evolve a structure containing bid multipliers, some were shown in the previous section. The bidding strategies that come from the evolved structures are fairly simple. The bid multipliers multiply the expected price of the electricity (obtained via some prediction scheme) and the result is used as the bid for that particular round of bidding. The results are promising, but the strategies are somewhat limited because they do not make use of inputs that are available during a particular round of bidding. Evolving the bidding strategies selection is like evolving the steps of a dance or a list of things to perform blindly to make a successful bid under a particular set of circumstances. Using this fixed-dance approach, means that the evolved rules are not very adaptive, i.e., they don't react to the environment. Each set of rules is evolved for a specific set of circumstances. If the circumstances vary from that, the set of rules may yield disappointing results. So the question becomes: How can we evolve bidding strategies that can take advantage of currently available information?

Fuzzy Bidding

Lotfi Zadeh made the field of "fuzzy logic" popular during the 1960s. Fuzzy logic provides a methodical means of dealing with uncertainty and ambiguity. It allows its users to code problem solutions with natural language syntax which with people are comfortable. In fact many of us regularly use fuzzy terms to describe things or events. For instance, if we were asked to describe a person, we might use terms like "pretty tall", "big nose" and "somewhat overweight". These terms can mean be defined differently by different people. There is a certain amount of ambiguity or uncertainty associated with any description involving natural language terms such as these. Most of the things we deal with daily in this universe are ambiguous and uncertain. "The only subsets of the universe that are not in principle fuzzy are the constructs of classical mathematics." [46]

Fuzzy logic allows us to represent the ambiguous or uncertain with membership functions. The membership functions map the natural language descriptions onto a numerical value. Membership to a particular description or class is then a matter of degree. For instance, if we define a person's height as described in Figure 18, then we can see that a person that is 6 feet in height is tall with a membership value of one. This membership value is also known as a truth-value. From the same figure, we can see that a person who is 5 feet 8 inches is tall to a lesser degree and short to a certain degree.

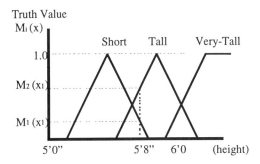

Figure 18. Fuzzy membership functions.
©IEEE 1996

Using similar reasoning, we might say that electrical demand is high in a region if it goes above 100 MW, and normal if it is between 50 MW and 75 MW. What if the demand is 90 MW? Using traditional logic we would classify it neither high nor normal. However using fuzzy logic, we might find that this demand is actually both high and normal, each to a certain degree (based on its membership function). Similarly we could have fuzzy membership functions for other inputs like fuel costs, risk aversion, level of competition, etc.

Once defined, these inputs can then be used in a set of fuzzy rules. For instance, a simple rule might be as follows:

IF demand is HIGH, then bid should be HIGH

where a "high" bid would be defined using another membership function. Multiple input conditions can be considered by combining rules with the "and' and "or" functions. For example a rule (not necessarily a good one) might be:

IF (demand is LOW) AND (risk aversion is HIGH) THEN (bid should be LOW)

Although it may not be necessary, we could have an output for all combinations of inputs. A three input fuzzy rule system where each input is broken into five classifications might be represented as in Figure 19. We could have more or fewer inputs, and we could use different classifications. Figure 20 shows the fuzzy system architecture. The inputs are fed into the rule base. The output (i.e., the bid values in the example) of each rule can be classified by a fuzzy membership function in the same manner as the inputs. The output of each rule may be assigned a certain weight depending on how important we determine that rule or corresponding inputs to be. We can then sum the weighted output of the rules and determine an overall fuzzy output. However, when the time comes to place the bid we can't just say, "bid high". We need a way to convert the fuzzy output to a single number. This is called the defuzzification process.

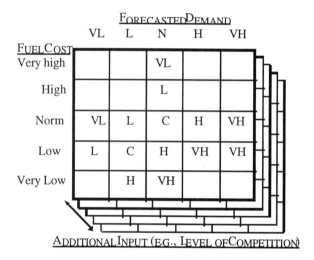

Squares contain the rule on how to bid relative to cost. Each squa[re] need not have an output. A particular input maybe classified in r[ore] than one square at a given instant. Some conditions might be ver[y] unlikely to occur. [V=very, L=low, H=high, C=cost, N=normal]

Figure 19. Three input fuzzy rule set.
©IEEE 1996

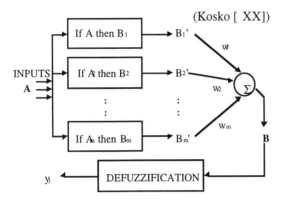

Figure 20. Fuzzy system architecture
©IEEE 1996

According to Kosko [46] defuzzification formally means to round off a fuzzy set from some point in a unit hypercube to the nearest bit-vector vertex. Practically, defuzzification has been done by using the mode of the distribution of outputs as the crisp output, or by the more popular method of calculating the centroid or center of mass of the outputs and using that as the crisp output.

Simulating the Auction and Evolving Fuzzy Rules

The research described here builds on a technique used by the authors and described in a previous section. The authors use a genetic programming [47, 74] to evolve data structures representing bidding strategies that can adapt to current market conditions. A group of GENCOs will compete to serve the electrical demands of the ESCOs. For the time being, we will aggregate the buyers into a single large ESCO. Each of the GENCOs will have it's own evolving data structure consisting of a fuzzy set of rules, and weights associated with each of those rules. The weights allow some rules more importance than others do. For the sake of simplicity, transmission constraints will be considered off-line and after-the-fact, but could be incorporated in the auction model. In previous work, the authors allowed each of the individual GENCOs to have their choice of price forecasting technique. This created much overhead, and for simplicity, where applicable, each GENCO will be receiving the globally forecasted data.

SUMMARY

The preceding sections have introduced the application of ALIFE onto electric power markets. This is barely a beginning on the possibilities of applications. Techniques to evaluate the dynamic response of markets, individually and interactively, are sorely needed. Tools to evaluate market rules have been started as shown above but need additional effort to answer all questions from a technical viewpoint. Tools to evaluate strategy are sorely needed to evaluate corporate performance in an unknown market place. Tools to generate strategy are needed, not to replace the human mind but to complement the work performed. Tools to evaluate competitor's strategies and to identify competitor's weak points are also needed. Such efforts are underway as a vital component of this author's research. However, it has yet to be clear that there is a light at the end of the tunnel. As once said, this is not the end of the evolution, nor the beginning of the middle implementation, but we are nearing the end of the beginning.

References

1. Anwar, and G. B. Sheblé, "Application Of Optimal Power Flow To Interchange Brokerage Transaction", Electric Power System Research Journal, vol. 30, no. 1, pp. 83-90, 1995.
2. Arthur, W.B., "On Learning and Adaptation in the Economy," The Santa Fe Institute, Report SFI 92-07-038.
3. Arthur (1993), "On Designing Economic Agents that Behave Like Human Agents," Journal of Evolutionary Economics 3, pp. 1--22.
4. Ashlock, "GP-automata for Dividing the Dollar," Mathematics Department,Iowa State University, Ames, IA 1995.
5. Ashlock, M. D. Smucker, E. A. Stanley, and L. Tesfatsion (1996),"Preferential Partner Selection in an Evolutionary Study of Prisoner's Dilemma," BioSystems 37, pp. 99--125.
6. Axelrod (1984), *The Evolution of Cooperation*, Basic Books, New York.
7. Axelrod (1987), "The Evolution of Strategies in the Iterated Prisoner's Dilemma," in L. Davis, Ed., Genetic Algorithms and Simulated Annealing, Morgan Kaufmann, Los Altos, CA.
8. Barkovich, and D. Hawk, "Charting a new course in California", IEEE Spectrum, Vol. 33, July 1996, p26.
9. Bard, "Short-Term Scheduling of Thermal-Electric Generators Using LaGrangian Relaxation," Operations Research, 36, 756-766, 1988.

10. Baughman and W. W. Lee, "A Monte Carlo Model for Calculating Spot Market Prices of Electricity," 91 WM 179-2 PWRS.
11. Bellman, *Dynamic Programming*, Princeton University Press, Princeton, New Jersey, 1957.
12. Belew, and L. B. Booker, (Editors), *Proceedings of the Fourth International Conference on Genetic Algorithms*, Morgan Kaufmann Publishers, Inc., San Mateo, CA, 1991.
13. Bertsekas and P. Tseng, "Relaxation Methods for Minimum Cost Ordinary and Generalized Network Flow Problems," Operations Research, 36, 93-114, 1988.
14. Binger and E. Hoffman. *Microeconomics with Calculus*, HarperCollins. 1988.
15. Brock and S. N. Durlauf (1995), "Discrete Choice with Social Interactions," Working Paper No. 95-10-084, Santa Fe Institute, Santa Fe, NM.
16. Bulfin, R. L. and Unger, V. E., "Computational Experience with an Algorithm for the Lock-Box Problem," Proceedings, Association for Computing Machinery, 16-19, 1973.
17. Bullard and J. Duffy (1994), "A Model of Learning and Emulation with Artificial Adaptive Agents," Working Paper, Department of Economics, University of Pittsburgh.
18. Chao, "Peak Load Pricing And Capacity Planning With Demand And Supply Uncertainty", Bell Journal of Economics, Vol. 14, No. 1, 1983, pp. 179-190.
19. Caramanis, N. Roukos, F. C. Schweppe, "WRATES: A Tool for Evaluating the Marginal Cost of Wheeling," paper presented at the 1988 IEEE PES Summer Power Meeting, New York, New York.
20. Cooper, and D. Steinberg, *Methods and Applications of Linear Programming*, W. B. Saunders Company, Philadelphia, Pennsylvania, 1974.
21. Cowan, and J. H. Miller, *Economic Life on a Lattice: Some Game Theoretic Results*, The Santa Fe Institute, Report SFI 90-010.
22. Davis, (Editor), *Genetic Algorithms and Simulated Annealing*, Pitman, London, 1987.
23. Davis, (Editor), *Handbook of Genetic Algorithms*, Van Nnostrand Reinhold, New York, NY, 1991.
24. David and Y. Z. Li, "Effect of Inter-temporal Factors on the Real Time Pricing of Electricity," 92 WM 117-2 PWRS.
25. Doty and P. L. McEntire, "An Analysis of Electric Power Brokerage Systems," IEEE Transactions on PAS, vol. PAS-101, no. 2, February 1982, pp.389-396.
26. De Vany (1996), "The Emergence and Evolution of Self-Organized Coalitions," pp. 25--50 in M. Gilli (ed.), Computational Economic Systems: Models, Methods, and Econometrics, Kluwer Scientific Publications, New York.

27. Durlauf (1996), "Neighborhood Feedbacks, Endogenous Stratification, and Income Inequality," in W. A. Barnett, G. Gandolfo, and C. Hillinger (eds.), *Disequilibrium Dynamics: Theory and Applications*, Cambridge University Press, Cambridge.
28. Epstein and R. Axtell (1996), *Growing Artificial Societies: Social Science from the Bottom Up*, MIT Press/Brookings, Cambridge, Mass.
29. Fahd, D. A. Richards, and G. B. Sheblé, "The Implementation Of An Energy Brokerage System Using Linear Programming", IEEE Trans. on Power Systems, Vol. 7, No. 1, pp. 90 - 96, 1991.
30. Fahd, and G. B. Sheblé, "Optimal Power Flow Emulation Of Interchange Brokerage System Using Linear Programming", IEEE Trans. on Power Systems, Vol. 7, No. 2, pp. 497-504, 1992.
31. Finlay, *Optimal Bidding Strategies in Competitive Electric Power Pools*, Masters thesis, University of Illinois, Urbana-Champaign, IL, 1995.
32. Gedra and P. P. Varaiya, "Markets and Pricing for Interruptible Electric Power," 92 WM 169-3 PWRS.
33. Gilchrist, *Statistical Forecasting*. New York, NY: John Wiley & Sons, 1976.
34. Goldberg, *Genetic Algorithms in Search, Optimization, and Machine Learning*, Addison-Wesley Publishing Company, Inc., Reading, Massachusetts, 1989.
35. Suran Goonatilake, John A. Campbell, and Nesar Ahmad. "Genetic-Fuzzy Systems for Financial Decision Making," Lecture Notes in Computer Science, vol. 1011, pp. 202-223. Springer-Verlag: New York. 1995.
36. Grefensteete, (Editor), *Proceedings of the First International Conference on Genetic Algorithms*, Lawrence Erlbaum Associates, Inc., Hillsdale, New Jersey, 1985.
37. Grefensteete, (Editor), *Proceedings of the Second International Conference on Genetic Algorithms*, Lawrence Erlbaum Associates, Inc., Hillsdale, NJ, 1987.
38. Jean-Michel Guldmann, "A Marginal-cost Pricing Model for Gas Distribution Utilities," pp. 851-863, Operations Research, Vol. 34, No. 6, November-December, 1986.
39. Hakin, *Neural Networks: A Comprehensive Foundation*, Englewood Cliffs, NJ: Macmillan Publishing Company, 1994.
40. Holland, J.H., *Adaptation in Natural and Artificial Systems*, MIT Press, Cambridge, MA, 1975.
41. R A Howard, "Decision Analysis: Practice And Promise" Management Science, 1988: 679- 675.
42. IEEE Power System Reliability Subcommittee, "Reliability Test System", presented at IEEE PES Winter Meeting, 96 WM 183-7 PWRS, 1996.
43. Ionnides, "Evolution of Trading Structures," Department of Economics, Tufts University, Medford, MA 02155, 1995.

44. Ionnides (1996), "Evolution of Trading Structures," Working Paper 96-04-020, Santa Fe, NM.
45. Jensen and J. W. Barnes, *Network Flow Programming*, John Wiley & Sons, New York, New York, 1980.
46. Bart Kosko. *Neural Networks and Fuzzy Systems: A Dynamical Systems Approach to Machine Intelligence.* Englewood Cliffs, NJ: Prentice Hall, 1992.
47. Koza, John, *Genetic Programming*, MIT Press, Cambridge, Massachusetts, 1992.
48. Kumar, K. H. Ng, and G. B. Sheblé, "AGC Simulator For Price Based Operation Part I: A Model", accepted and to be presented at IEEE PES Summer Meeting, 96 SM 588-4 PWRS, 1996.
49. Kumar, K. H. Ng, and G. B. Sheblé, "AGC Simulator For Price Based Operation Part II: Case Study Results", accepted and to be presented at IEEE PES summer meeting, 96 SM 373-1 PWRS, 1996.
50. Kumar, G. Sheblé, "Auction Game in Electric Power Market Place, "Proceedings of the 58th American Power Conference, 1996, pp. 356-364.
51. Kumar, G. Sheblé, "Framework for Energy Brokerage System with Reserve Margin and Transmission Losses," 1996 IEEE/PES Winter Meeting, 96WM 190-9 PWRS. New York: IEEE
52. Kumar, G. Sheblé, "Auction Market Simulator For Price Based Operation," presented at the 1997 IEEE PES Summer Power Meeting, Berlin, Germany, in press, 1997.
53. Lane, *Artificial Worlds and Economics*," The Santa Fe Institute, Report SFI92-09-048.
54. LeBaron, "Experiments in Evolutionary Finance," University of Wisconsin - Madison, Madison, Wisconsin, 1995.
55. Li and A. K. David, "Optimal Multi-Area Wheeling," 93 WM 174-3 PWRS.
56. Luenberger, *Introduction to Linear and Non-Linear Programming*, Addison-Wesley Publishing Company, Reading, Massachusetts, 1973.
57. Kevin A. McCabe, S. J. Rassenti, and Vernon L. Smith, "Auction Design for Composite Goods: the Natural Gas Industry," Journal of Economic Behavior and Organization, 14 (1900) 127-149.
58. Kevin A. McCabe, S. J. Rassenti, and Vernon L. Smith, "Experimental Research on Deregulated Markets for Natural Gas Pipeline and Electric Power Transmission Networks," Research in Law and Economics, Vol. 13, pp.161-189, JAI Press, Inc.
59. Manne and J. S. Rogers, "A Structure for Modeling Bulk Power Transfers among Systems in a NERC Region," sp 91-211.
60. Yusuf M. Mansur. *Fuzzy Sets and Economics*, Edward Elger Publishing Limited: Vermont. 1995.
61. Milgrom, *Auctions and Bidding: A Primer*. Journal of Economic Perspectives, Vol. 3, No. 3. Summer 1989, pp. 3-22, 1989.

62. Jo Min, "Unbundling the quality attributes of electric power: models of alternative market structures," uer-165, California Energy Studies Report, April 1986.
63. Michalewicz, *Genetic Algorithms + Data Structures = Evolution Programs*, Springer-Verlag, New York, NY, 1992.
64. Yusuf M. Mansur, *Fuzzy Sets and Economics*, Edward Elger Publishing Limited: Vermont. 1995.
65. Nemhauser, *Introduction to Dynamic Programming*, John Wiley & Sons, Inc., New York, New York, 1966.
66. Papalexopoulos, S. Hao, T. Peng, "An implementation of a neural network-based load forecasting model for the ems," Presented at the 1994 IEEE/PES Winter Meeting, 94 WM 209-7 PWRS. New York: IEEE, 1994.
67. Post, S. S. Coppinger, and G. B. Sheblé, "Application Of Auctions As A Pricing Mechanism For The Interchange Of Electric Power", IEEE Trans. on Power Systems, Vol. 10, No. 3, pp. 1580 - 1584, 1995.
68. Post, *Electric Power Interchange Transaction Analysis And Selection*, Master's thesis, Iowa State University, Ames, IA, 1994.
69. Rassenti, V. L. Smith, and R. L. Bulfin, "A Combinatorial Auction Mechanism for Airport Time Slot Allocation," The Bell Journal of Economics,13, 402-417, 1982.
70. Rassenti and R. L. Bulfin, "A Generalized 0-1 Programming Problem:Algorithm and Application," presented to the Canadian Operations Research Society/TIMS/ORSA Meeting, Toronto, Canada, April 1981.
71. Charles W. Richter, Jr. and Gerald B. Sheblé. "Genetic Algorithm Evolution of Utility Bidding Strategies for the Competitive Marketplace," Presented at the 1997 IEEE/PES Summer Meeting, in press, 1997.
72. Charles William Richter, Jr., *Developing Bidding Strategies For Electric Utilities In A Competitive Environment*, Master's thesis, Iowa State University, Ames, IA. 1996.
73. Richter and G. B. Sheblé, "Competitive Fuzzy Bidding Strategies For The Successful Generator," presented and published in the proceedings of the North American Power Conference,Boston, 1997.
74. Charles W. Richter, Jr., Tim T. Maifeld, and Gerald B. Sheblé. " Genetic Algorithm Development of a Healthcare Expert System," Proceedings of the 4th Annual Midwest Electro-Technology Conference, pp. 35-38.Ames, IA. 1995.
75. Rust, R. Palmer, and J. Miller, "A Double Auction Market for Computerized Traders," The Santa Fe Institute, Report SFI 89-001, 1989.
76. Schaffer, (Editor), *Proceedings of the Third International Conference on Genetic Algorithms*, Morgan Kaufmann Publishers, Inc., San Mateo, CA, 1989.

77. Schweppe, M. C. Caramanis, R. D. Tabors, and R. E. Bohn, *Spot Pricing of Electricity*, Kluwer Academic Publishers, Boston, 1987.
78. Gerald B. Sheblé, "Solution of the Unit Commitment Problem by the Method of Unit Periods," presented at the 1989 IEEE Power Engineering Society Summer Meeting and accepted for publication in the IEEE Transactions on Power Systems, 1989.
79. Sheblé, "Electric energy in a fully evolved marketplace," Presented at the 1994 North American Power Symposium, Kansas State University, KS, 1994.
80. Sheblé and J. McCalley, "Discrete auction systems for power system management," Presented at the 1994 National Science Foundation Workshop, Pullman, WA, 1994.
81. Sheblé, "Simulation Of Discrete Auction Systems For Power System Risk Management," Frontiers of Power, Oklahoma, 1994.
82. Sheblé, M. Ilic, B. Wollenberg, and F. Wu, *Lecture notes from: Engineering Strategies for Open Access Transmission Systems*, A Two-Day Short Course presented Dec. 5 and 6, 1996 in San Francisco, CA.
83. Sheblé, "Priced Based Operation in an Auction Market Structure", presented atthe 1996 IEEE/PES Winter Meeting, Baltimore, MD, 1996.
84. Gerald B. Sheblé, "EPRI Auction Market Simulator (EPRI-AMS)," published as EPRI Technical Report.
85. Simmons, *Nonlinear Programming for Operations Research*, Prentice-Hall, Inc., Englewood Cliffs, N. J., 1975.
86. Siddiqi and M. L. Baughman, "Reliability Differentiated Real-Time Pricing of Electricity," 92 WM 115-6 PWRS.
87. Smucker, E. A. Stanley, and D. Ashlock (1994), "Analyzing Social Network Structures in the Iterated Prisoner's Dilemma with Choice and Refusal," Department of Computer Sciences Technical Report CS-TR-94-1259, UW-Madison.
88. Stanley, D. Ashlock, and L. Tesfatsion (1994), "Iterated Prisoner's Dilemma with Choice and Refusal of Partners, " pp. 131--175 in C. Langton, ed., *Artificial Life III, Proceedings*, Volume 17, Santa Fe Institute Studies in the Sciences of Complexity, Addison-Wesley, Reading, MA.
89. Chin-Woo Tan, "Prices for Interruptible Electric Power Service," ISU presentation, July, 1991.
90. Tesfatsion, "A Trade Network Game with Endogenous Partner Selection," Economic Report Series, Department of Economics, Iowa State University, Ames, IA, 1995.
91. Thompson, and S. Thore, *Computational Economics: Economic Modeling with Optimization*, San Francisco, CA: Scientific Press, 1992.
92. Vojdani, C. Imparto, N. Saini, B. Wollenberg, and H. Happ,"Transmission Access Issues," Presented at the 1995 IEEE/PESWinter Meeting, 95 WM 121-4 PWRS. New York: IEEE, 1994.

93. Wood, and B. F. Wollenberg, *Power Generation, Operation, and Control*, New York, NY: John Wiley & Sons, 1984.
94. "A New Laboratory for Economists," Science and Technology Section of *Business Week*, pages 96-97, March 17, 1997.

7 ONE-PART MARKETS FOR ELECTRIC POWER: ENSURING THE BENEFITS OF COMPETITION

Frank C. Graves[*], E. Grant Read[**],
Philip Q. Hanser[*], and Robert L. Earle[***]

[*]The Brattle Group, Cambridge, MA
[**]Canterbury University, Christchurch, New Zealand
[***]The Brattle Group, Washington, D.C.

In order to ensure adequacy of generation supply, the utility industry has traditionally been required to carry two to three years of planning reserves, e.g., 20 percent over projected peak demand. Closely related, they have often used two-part (capacity/energy) pricing to buy and sell generation (real power) output. This paper argues that continued use of this approach, especially continuing to require planning reserves under power pool or NERC or other mandate, will undermine the benefits of power industry restructuring. In contrast, a market with no administered capacity requirement, but a one-part commodity price reflecting both marginal operating costs and capacity scarcity, will have many benefits. In particular, it will induce efficient capacity planning—which has been the real problem in the past (not inefficient dispatch) and which is where the real opportunities for future efficiency gains lie. It will also encourage demand-side participation in peaking "reserves", and forward contracting for risk protection and expansion financing, both of which also reduce generation market power. Independent system operator (ISO) planners and regulatory agencies should concentrate more attention on encouraging demand-side participation and forward contracting, and less on the sufficiency of physical reserves or on customer protection against possible high market prices.

Introduction.

Traditionally, the utility industry has administered the capacity market by requiring utilities to carry 2-3 years of planning reserves, *e.g.*, 20 percent over projected peak demand, as a condition of participation in power pools or regional reliability coordination councils. Closely related, they have often used two-part (capacity/energy) pricing to buy and sell generation (real power) output.

This paper argues that continued use of this approach, especially continuing to require planning reserves under power pool, North American Electric Reliability Council (NERC), or other mandate, will undermine the benefits of power industry restructuring. In contrast, a market with no administered capacity requirement but a one-part commodity price reflecting both marginal operating costs and capacity scarcity will have many benefits. In particular, it will induce efficient capacity planning—which has been the real problem in the past (not inefficient dispatch) and is where the real opportunities for future efficiency gains lie. It will also encourage demand-side participation in peaking "reserves", and forward contracting for risk protection and expansion financing, both of which also reduce generation market power.

This conclusion runs counter to much past utility economics literature that advocated two-part pricing, and to well-intentioned concerns about preserving generation reliability, but the assumptions of that literature and of the coordinated generation planning tradition are no longer appropriate to the future power markets (at least in the United States).

Notwithstanding its advantages, a commodity power market with one-part pricing has its own difficult implementation issues, including externality concerns and antitrust questions of how to tell when prices reflect more than just capacity scarcity and have become monopolistic rents. The latter determination will be especially difficult if, as many industry analysts expect, unregulated power markets become quite volatile and peak-intensive, recovering capacity costs in a few percent of hours per year at prices 5-50 times as large as typical variable costs. Concerns about *ex post* anticompetitive behavior in such periods may be alleviated if *ex ante* there are active, competitive forward markets that customers can access to insure themselves against price spikes. Hence, a key question is under what circumstances such forward markets will evolve. It is not clear that effective forward markets can exist if there is significant market power in the spot/physical market. Evidence from other commodity markets on this question is mixed, but it is clear that high levels of forward contracting reduce incumbents' incentives to force spot prices up. Game-theoretical modeling also suggests that forward markets can create a prisoners dilemma situation that drives oligopolistic producers towards efficient competition.

Besides diluting anticompetitive behavior, forward contracting is essential as a means for investors to cope with the risks of building and maintaining new peaking capacity that may have only rare, but significantly profitable use. Another aid for inadequate spot market competition and the risks of obtaining peak reserves is demand-side participation in pools. This could help cure the problem associated with must-run generation inside of many urban load pockets, allowing competition where bid-price restrictions would otherwise be required. Demand-side participation is more likely to be induced with one-part than two-part pricing, and no planning reserves obligation.

One thing which will certainly destroy the forward contract market is the expectation that regulators may force generators to supply consumers at "reasonable prices," even if they are not contracted. In that case, why should customers contract? If owning a backup plant means that a generator is forced to sell at the regulator's price, he/she will only build what the regulator requires...hence undermining the market. "Standard offer" service obligations, designed to protect customers from the risk that market prices will rise after stranded cost allowances have been set, are an example of such a policy that may interfere with competition.

There is good news for two-part pricing, despite the above argument that organizing the industry in traditional, two-part markets with administered capacity is fast becoming undesirable. Once efficient one-part spot markets have evolved, any buyers or sellers preferring the certainty of pricing they enjoyed under the two-part regulatory tradition can recreate it with call options written against the spot market. These are superior to administered capacity provisions for which quantity and cost recovery are determined by regulatory fiat, because each customer can design them to suit his or her risk preferences, *e.g.*, by buying options that are more or less out of the money to cover part or all of expected requirements.

This chapter first examines the economics literature on two-part pricing and argues that, at least in the United States, because of structural changes, much of it no longer applies to generation. Second, the undesirable consequences of capacity market management are examined and contrasted to the likely outcome from one-part pricing. Both empirical analysis and theory indicate that the benefits of restructuring are likely to come not from more efficient dispatch (allocational efficiency), but better deployment of capital and innovations in demand management (dynamic efficiency). Third, the volatility of prices that may result from one-part pricing is discussed. There, it is suggested that the high spikes of these prices provide the signals and incentives for the construction of new capacity or adjustment on the demand side, although forward markets should ameliorate price volatility's impacts. Fourth, some lessons are drawn from results in New Zealand. There, despite some initial problems, one-part pricing has not resulted in the disasters predicted for it beforehand. Fifth, it is briefly described how options can replicate the effects of a capacity purchase without necessarily distorting the energy market. Finally, the chapter concludes with some policy recommendations for power pool planners and regulatory agencies.

Economic Heritage of Two-part Market and Pricing.

Some economic literature has shown that, under certain circumstances, two-part pricing of electric power can be the second-best ideal (typified by Boiteux[1]). In brief, these analyses typically conclude that energy should be priced at short-run marginal cost, while capacity-related fixed costs should be recovered through a demand charge that achieves full recovery of the remaining costs (revenue requirement).

Today, however, those circumstances favoring two-part pricing are not likely to apply to some, perhaps most, prospective deregulated power markets. For instance, this literature tends to assume that there are significant economies of scale in generation, so that incremental consumption should not be discouraged by one-part prices. But generation technology has now evolved to where there are far fewer consequential economies of scale; 400 MW gas combined cycle gas turbines (CCs) are now cheaper to build (per kW) than traditional baseload technologies like coal and nuclear generation. This literature is also strongly tied to the revenue requirement/regulated monopolist tradition, seeking to prevent

ONE-PART MARKETS FOR ELECTRIC POWER 247

monopolistic prices while achieving cost-of-service average prices that minimally distort power use. But of course we are hoping to deregulate generation and let the market determine prices.

There may also be implicit assumptions that are equally important and increasingly inappropriate for a competitive generation market. Even when these are not present in economic derivations, they are common in engineering studies designed to formulate the industry's capacity management policies. For instance, they generally assume:

- Known/fixed capacity requirements, determined by loss of load probability or reserve margin targets—but empirical estimates of the marginal value of lost load (VOLL) designed to equate the social value of avoidable curtailments to the marginal cost of additional generation capacity are extremely imprecise, and clearly indicate that these values are highly situation-specific (*e.g.*, depending on lead time for notification, length of outage, *etc.*). Such uncertainty and diversity makes this an ideal problem for markets, rather than regulators, to solve.

- Static demand curves, sometimes with zero elasticity—yet real-time pricing (RTP) experiments in New York, Georgia, and elsewhere show strong price responses. Moreover, few customers have ever been given a strong price signal to which they could respond; competitive generation creates the opportunity to tap previously unexplored price elasticity, via spot prices that are likely to be volatile and peak-intensive.

- Uniform annual reliability/expected availability of existing generation—unlikely to apply to generators facing a volatile spot market. Traditional modeling of availability (e.g., in LOLP studies) treats outage as a Poisson process, randomly occurring throughout the year. Similarly, optimal pricing models often treat the supply curve as fixed with respect to outages, regardless of price. To the contrary, in a competitive generation market with one-part commodity prices, generators' incentives to be online during peak hours should be much higher, even if average availability does not improve. Thus administering two-part markets requires a degree of confidence in our information about supply and demand curves that we probably should not have.

Another concern about continued reliance on administered capacity markets is that they may become unworkable and even have anti-competitive impacts. Unworkability arises because in the past, planning reserve obligations have been imposed on utilities with long-lived obligations to serve and protected franchise monopolies. In this context, it was meaningful to speak of a long-term requirement, as well as reasonable and feasible to achieve cost recovery of the carrying costs of those planning reserves with average cost demand charges. This approach could be sustained under *wholesale* access if the reserve obligation were made a contractual obligation of the distcos. But it is untenable under *retail* access. Retail customers will want the opportunity to be served with self-selected reliability, not with pool-specified reliability. Of course, it is true that we cannot (easily) direct power towards certain customers and away from others. In this regard, power flowing in accordance with Kirchoff's Laws displays externalities that could result in free-ridership. But future smart appliances and meters that can be curtailed in response to price thresholds may eliminate much of this problem.[2] Moreover, this problem is not nearly as unique to electricity capacity as those concerned about externalities have suggested. For instance, any time someone builds a house it slightly improves (reduces) the price of housing for all other buyers, yet this is not deemed to discourage home building.

Similar issues arise because suppliers will want the opportunity to rely on spot power procurement *alone* to cover some of their customers' needs. That is, they will want the right to offer service without holding physical capacity, provided they are willing to financially indemnify their customers for nondelivery. (As explained further below, this becomes the only workable definition of reliability in a deregulated power market.) This is the way all other commodity markets enforce performance, *e.g.*, in their futures markets. Or, suppliers (*e.g.*, cogenerators) may only want to offer service in off-peak periods when the need for planning reserves is effectively zero. In principle, this could be accommodated: the planning reserve obligation on load-serving entities (LSEs) could be converted to a very complicated function, conditional on the precise time, location, and duration of service. In practice, this unbundling would itself become a source of controversy and litigation.

The antitrust implications of enforced capacity obligations arise if pools oblige LSEs to carry planning reserves as a condition of entry, and as a result fewer suppliers are able or willing to do so. This will reduce competition. Insistence by pool governing bodies on planning reserves could be challenged as a conspiracy in restraint of trade, if the incumbents are perceived as having created entry barriers that only they can readily surmount with their existing generation capacity reserves.

ONE-PART MARKETS FOR ELECTRIC POWER   249

Finally, the concept of "capacity" itself is not clearly defined, with the imprecision/gaps becoming increasingly important in the restructured power markets. In the traditional pricing models (and even some pool reserve requirements), all MWs are treated as equivalent. In fact, features that can vary enormously across same-sized units include:

- Minimum run constraints,
- Black start capability,
- Ramping speed and range,
- Availability (forced outage), and
- Emergency capability

Thus it is not just the number of MWs but performance capabilities that determine value. In principle, we could imagine performing an analysis of the marginal contribution to pool reliability of each LSE's reserves, but development of an effective ancillary services market is a better way to solve this problem, with only financial indemnification for non-delivery of real power being the relevant measure of "reliability" in the one-part commodity market.[3] This, of course, will not guarantee that power is always delivered; occasionally, indemnification for non-delivery will occur. However, the required amount of compensation will be market-driven, hence efficient, and thus will send the right sized signal for avoiding the problem in the future.

Undesirable Consequences of Capacity Market Management.

The danger in sticking with the conventional capacity/energy distinction is that it may encourage preservation of two markets that are not necessarily in equilibrium with each other, or that are in equilibrium but at undesirable levels. Pool rules that require traditional loss of load probability (LOLP)/reserve margin planning reserves for would-be marketers or LSEs will lead to:

- Administrative, not economic capacity targets.
- Artificial size of the supply pool with resulting distorted (too low) energy prices.

- Continued (over) reliance on supply-side approaches to balancing the market, rather than letting demand vary in response to a one-part commodity price that reflects marginal operating costs *and* capacity scarcity.
- A bias against adoption of supply-side technologies which do not provide the "normal" mix of generation and reserve capacity.

Figures 1 and 2 illustrate the difficulty. In Figure 1, the equilibrium prices and quantities in a on-price market are depicted. In Figure 2, the equilibrium conditions in the two-price market are shown.[4] If adequacy targets are administratively set, say by the state, then there may be a chance it will be set too high with the result that the single price energy market will need to be split into an administratively set price for capacity and an energy only market. In Figure 1, the increase in the price of electricity above GEN 6's short-run marginal cost cleared the market. In the two-price market, the additional capacity required by setting a reserve margin is labeled as GEN 7 in Figure 2. Note that GEN 7 never is inframarginal as a unit, thus never earns a margin to cover its fixed cost. As a result, GEN 7 will not remain in the market unless it receives a separate payment to cover its fixed costs—a capacity payment. Note also that all other generators' margins have been lowered by virtue of adding GEN 7. However, since they do have positive margins, they could potentially bid into the capacity market at a price lower than that for GEN 7. Because the reserve requirement mandates the existence of GEN 7, the resolution is to administratively set a capacity payment based on what GEN 7 requires in order to be financially viable and which all generators then receive. Of course, GEN 7 increases reliability but, if it was required in order to satisfy planning reserve standards, then it has not had to pass an economic test. This in turn will necessitate an administered cost recovery payment which will be inefficient, and ultimately more expensive.

If the quantity of MWs in the market continue to be administered, *e.g.*, by mandating traditional planning reserves, the role of economic forces in shaping the *capacity* decisions in the market will be severely reduced. All that may be achieved by restructuring is (slightly) improved dispatch from creating large regional pools.

ONE-PART MARKETS FOR ELECTRIC POWER 251

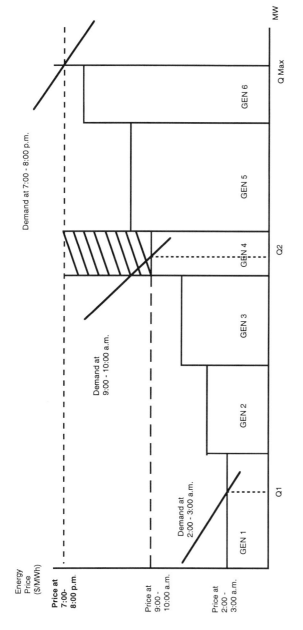

Figure 1: One-Part Price Market

Note that at Q1, GEN 4 is out of the money; at Q2, it just covers its variable cost; at Qmax, it earns a surplus above marginal costs which will contribute to its coverage of fixed costs.

252   POWER SYSTEMS RESTRUCTURING

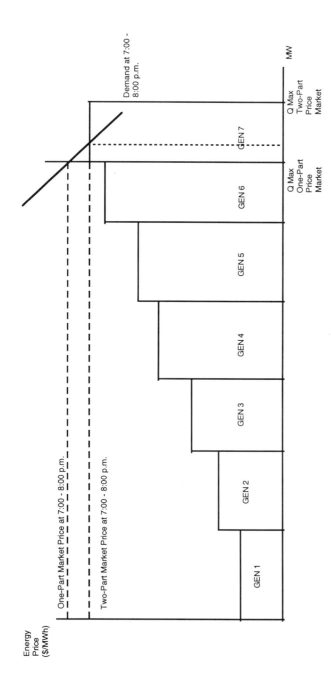

Figure 2: Two-Part Price Market

The U.S. already has fairly efficient dispatch, simply from pool practices and existing wholesale trading between utilities. The reason that restructuring and generation deregulation is an issue at all is because the regulatory process has *not* resulted in good *capacity* planning or more exactly, demand/supply balance. (Observe the share of non-utility generator (NUG) contracts and nuclear units responsible for the stranded costs of the industry!) Thus, it is essential that the focus be more on achieving competitive capacity planning, and demand/supply interactions, rather than competitive dispatch.

Recent analytic studies confirm the suspicion that existing dispatch is fairly efficient. Simulations of a few very large regions of the country (each well over 100,000 MWs, *e.g.*, the entire Western Systems Coordinating Council (WSCC)) show that only a *few percent* reduction in dispatch costs would ensue from region-wide dispatch as a single control area, *even assuming transmission capacity were costless and infinite*. Figure 3 summarizes the range of dispatch improvements in 1994, 1997, and 2000, for three regions of the country.[5] While this is a valuable savings, it is not anywhere near as large as the 10-15 percent savings that have sometimes been promised as the immediate payoff of restructuring.[6] These results strongly suggest that policymakers should not be looking to restructuring to provide an immediate "competitive dividend." The true benefits of restructuring will come in 5 to 10 years, as the fixed costs and capacity requirements of the industry and the character of the demand curves change in response to price signals and risk that have never been felt before by suppliers or customers under regulations. That is, the benefits will be both in dynamic efficiency and allocational efficiency.

Likely Characteristics of One-Part Electric Prices.

How might one-part prices for wholesale generation behave? As suggested above, it is quite likely that they will be very volatile, for several reasons: the capital intensity of the industry, volatility of the upstream fuel costs (especially natural gas in the U.S.) or availability (especially in hydro systems), high customer VOLL, and scarcity conditions that are random, often sudden, and non-linear. Evidence for what this might look like can be found in the U.K. power pool. Figure 4 shows the weekly average of half-hour spot prices paid to producers (the Producer Purchase Price) since vesting. This price includes the marginal bid price plus a capacity scarcity term

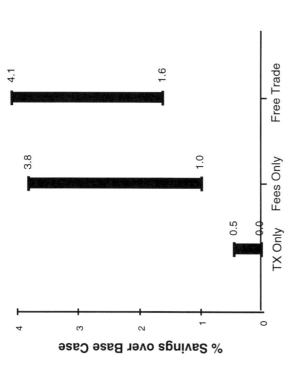

Figure 3: Cost Savings from Dispatch under Restructuring Not Likely to Be High

Base case is trade with existing transmission constraints and fees.

Range shown is for three regions (WSCC, NE, SE) dispatched for 1994, 1997, and 2000.

TX Only = No fees, but existing Transmission constraints.

Fees Only = No Transmission constraints, but Transmission fees (handoff charges).

Free Trade = No fees, No Transmission Constraints.

ONE-PART MARKETS FOR ELECTRIC POWER 255

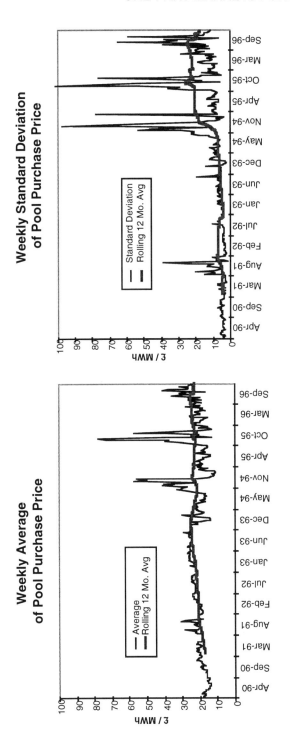

Since vesting, weekly average of half-hourly pool prices has not changed much, but volatility has more than doubled (with many weeks much higher) and is now around 100% of average price.

Figure 4: Power Price Volatility: U.K. Experience

256 POWER SYSTEMS RESTRUCTURING

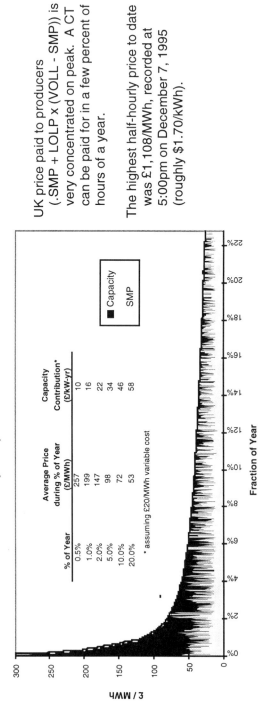

Figure 5: Power Price Peak Intensiveness: U.K. Experience

UK price paid to producers (.SMP + LOLP x (VOLL - SMP)) is very concentrated on peak. A CT can be paid for in a few percent of hours of a year.

The highest half-hourly price to date was £1,108/MWh, recorded at 5:00pm on December 7, 1995 (roughly $1.70/kWh).

Note: SMP = System Marginal Price = highest accepted energy bid price, without transmission contraints
LOLP = Loss of Load Probability

administered by the Pool based on a VOLL of roughly $4/kWh times the prevailing LOLP. We see that since privatization, the average price of power has not risen significantly but the volatility (as measured by standard deviation of half-hourly prices within each week) has risen dramatically. Moreover, the price has become very peak-intensive; Figure 5 shows that a combustion turbine (CT) could be paid for in 2-5 percent of hours/year, with prices in top half percent of hours averaging 257 £/MWh in 1996. Such volatile, peak-intensive prices are perfectly consistent with other capital-intensive commodity businesses (gas, metals, *etc.*) and with prevailing U.S. industry beliefs about VOLL. (Even though the U.S. pools may not follow the practice of adding a capacity term based on VOLL * LOLP, reserve targets are based on similar estimates. If we have been justified in requiring roughly 20 percent planning reserves, then we should expect demand-side bidders to set the market clearing price occasionally at such very high numbers as observed in the U.K. Wholesale power markets in the U.S. are already moving strongly in this direction. Day-ahead prices in July 1997 briefly exceeded $340/MWh in the East Central Area Reliability (ECAR) region. While, at first blush, these erratic, and occasionally very high, spot prices might appear undesirable, the opposite is actually the case:

- Acute price swings would motivate and reward time-of-use demand flexibility, though the focus would now be on peak demand reductions rather than reductions in the average use of energy that has been the focus of most administered demand-side management under regulation. A key driver in achieving this flexibility will be rapid penetration of real-time metering and behind-the-meter innovations in smart appliances and commercial equipment. This will result in flatter load shapes that reduce or defer the need for new capacity additions.

- Peak-intensive commodity prices also reward only those generators that are actually online in the few hours when prices spike. Under such a stimulus, forced outage rates around peak hours should drop radically, again delaying our need for generation capacity expansion.

- With locational spot prices that reflect capacity scarcity as well as energy variable costs, the transmission congestion rents (basis differentials) across bottlenecks will be much larger— large enough to elicit transmission expansion if/where that is the preferred solution instead of new generation. Transmission and generation should substitute for each other in this fashion, whenever expansion is an issue.

- Sharp peaks will elicit firm curtailable demand, prioritized by price levels, that effectively bring the entire curve of marginal willingness to be curtailed (*i.e.*, the entire VOLL curve) into competition with supply. This increases competition[7] and eliminates the need to estimate what the marginal VOLL is for generation LOLP planning.[8]

- The volatility of spot prices may be unappealing to many customers, who will enter into long term forward contracts that inoculate them from the extremes. These contracts provide financial stability for the seller as well as the buyer, enabling the seller to finance and operate new capacity profitably. Absent such forward contracts, a new entrant hoping to capture a piece of the spot market profitability faces very substantial risk, particularly at the high-price, low-volume peaking end of the market and, if making a significant capacity addition, may also drive the price of power down and fail to capture the desired profits. As explained below, this will help limit market power, if any, in the physical generation (spot) market.

Implementation Issues—Adequate Competition.

When prices are allowed to signal the need for capacity and the market allowed to fill that need, it is likely that spot prices occasionally will have some very high spikes. Contrary to the conventional wisdom, these spikes can be acceptable and desirable. Even if they are the result of the abuse of market power, there are a couple of principle ways of dealing with them. First, forward markets both hedge the risk of spiky markets and also lessen the incentive of the seller to manipulate prices. Second, the high prices serve both to communicate the need and provide the incentive for new capacity.

Spiky Prices Can be Acceptable.

Partly because extreme spot prices are so unfamiliar under our current system of administered, average cost prices, it is possible that they will occasionally have the appearance of being anticompetitive. Recall that they could be *much* higher than traditional electric prices; the British record to date is £1,180/MWh for a few half hours in December, 1995, or roughly

$1.70/kWh! Even higher prices have recently been observed in the highly competitive Australian market, where recent spot prices have averaged little more than 1¢/kWh (U.S. dollar) but VOLL is currently set to around US$3.30/kWh and likely to be raised to US$16.50. In part, the appropriate public policy reaction to these situations should be a function of the market conditions under which they occur. If there is little *a priori* reason to suspect market power, antitrust agencies should be more tolerant than if there is a structural reason for suspicion. This requires making a distinction between *ex ante* vs. *ex post* market power.

Scarcity premiums implicit in the spot price, *i.e.*, margins in excess of the variable operating cost of the marginal unit in the bid dispatch ladder, should not be deemed monopolistic unless 1) they are consistently above replacement cost, for assets likely to recover their total costs over a small, and highly uncertain, number of hours of use, and 2) there are no *ex ante* barriers to customers insuring themselves against *ex post* price spikes via forward contracts, *e.g.*, futures or options.

An example from another context may be helpful. If local prices of water spike after a hurricane, most economists would aver that we should be appreciative rather than resentful, even if the temporary premiums are very high. After all, there is a capital cost associated with holding water in inventory for this rare purpose, or a risk associated with foregoing some of your own water consumption in the interest of a fast profit. The high prices will encourage more people to self-insure (by storing water) or even to go into the emergency water business to drive down the price. There are no *ex ante* barriers to being insured against this, so *ex post* we should not object. Likewise for electric power contingencies.

Unfortunately, this replacement cost pricing test is very difficult to apply, because of the great uncertainty associated with how long or how often the opportunity to recover more than variable costs will prevail for the marginal generating units. For the majority of units, it will be profit maximizing to bid something very close to short run variable costs, assuming that the pool or power exchange (PX) clears all bids at the price of the last dispatched unit in each period. Units that are expected to dispatch for only a few hours may need to make a tough decision as to whether or not to embed startup costs in their bid, but any units very likely to be inframarginal for long periods gain little or nothing by bidding more than their variable costs. To do so risks impairing their position in the dispatch curve, hence reducing their hours of operation and profitability, since they are not setting the clearing price anyway. Even if a plant is marginal in one hour, it is likely to be inframarginal in the next if/when load is increasing. Thus only units at or near the very top of the ladder may be able to bid more than their variable cost; indeed they may have to do so to make up for the lack of inframarginal operating profits that would otherwise help cover fixed costs.[9]

For instance, a new CT costs around $350/kW to build and perhaps $8-10/kW-year to operate (ignoring fuel costs), entailing an annual carrying charge of $50-60/kW-year to cover fixed costs. If these amounts were to be recovered over only the few hours of the year in which the last percent of reserves were expected to be used, a bid of several dollars per kWh could be required. An owner might go a few years without experiencing the demand and system conditions in which such a bid would be taken, so bids in those years (or the number of hours in which the plant operates at those prices) might have to be well in excess of steady annual payment rates at cost-of-service levels for new capacity.[10] Identifying abusive versus reasonable pricing in this kind of environment will be very tough. Accordingly, it may be more useful to turn to the second test, of whether or not there are adequate forward markets to insure against ex post price premiums. If so, then spot price premiums should be tolerated.

Forward Markets.

The obvious distinction between electricity and water is that electricity is effectively not storable in bulk. However, power can be stored "synthetically" through forward contracts, *e.g.*, buying power in October and selling the same quantity in November is *financially* equivalent to storing and reselling. This suggests that forward markets might provide some of the same kind of *ex ante* protections as we find for other, storable commodities. So where might spot price abuses occur, and will forward markets be likely to emerge for those locations?

Bottlenecks behind contingent transmission constraints may create the opportunities for *ex post* price spikes we would like to insure against, and there are likely to be many such locations throughout the country that have occasional bottlenecks. Indeed, it would be a bad sign if there were very few such locations, as it would indicate that the grid was over-built; it is efficient that there be lots of locations where out-of-merit, multi-lambda dispatch occurs some of the time but not often enough to yet justify expansion. Many cities in the U.S. have this feature. For instance, San Francisco was recently identified as a "load pocket" in the California ISO planning studies. The San Francisco Bay Area has a peak load of between 5000 and 7400 MW and an import capability of roughly 4500 to 5300 MW. Generation within the San Francisco Bay Area must be used to serve the last 500 to 2100 MW of peak,[11] and ownership of that generation has been and will necessarily remain highly concentrated. (All were owned by Pacific Gas & Electric until its recent divestiture; now two of the plants are owned by Duke: Moss Landing (1474MW) in Monterey, and Oakland (165MW) in Alameda. Both are currently designated must-run by the ISO.)

The concern must be that horizontal concentration of generation in such load pockets could discourage third parties from writing forward contracts. Potential contract writers may fear that the local (nodal) spot market is not a "fair game." Specifically, the incumbent (or affiliate thereof) may be perceived as being able to write forward contracts that are relatively assured of ending up in the money, hence the incumbent can take less of a bid-ask spread on his transactions.

Evidence on the influence of horizontal market power in other commodities is mixed. Several markets with active financials have fairly concentrated supply markets, *e.g.*, coffee, and some metals, yet their futures contracts have escaped cornering. (Newbery reported that for eight traded commodities, producers from a single country, often coordinated internally, controlled over 50 percent of the available supplies, yet the corresponding futures contracts trade fluidly.[12]) Of course, these markets are not necessarily analogous to electric power, as cornering of commodities almost always involves inventory accumulations that are not feasible for electricity. There may also be a selection bias, whereby markets with successful futures contracts are precisely those that have not experienced pervasive spot price manipulations, notwithstanding high concentration of production ownership. Some futures markets have failed to take root because of market power in supply, *e.g.*, tin[13], and it has been speculated that Britain has not developed active power futures trading because of the horizontal market power problems in generation that have been documented there.[14]

The problem in a nutshell is that the futures price for a non-storable commodity should reflect the expected spot scarcity premiums. For a highly storable commodity, one can always buy the product when its spot prices are low (unmanipulated) and hold it or sell it forward. This cannot be done for power, so the forward price is a pure derivative of the spot market. This means that *ex ante* futures prices will be much lower per kWh than *ex post* spot prices in periods of spikes, but the futures prices will be paid on many kWh for which the spot price premium does not occur. If this is all that can be accomplished, then the forward markets do not reduce spot market power, they simply disperse the exposure across all other periods.

In fact, forward contracting creates more competitive protections than this. First, any forward contracting that the dominant players undertake reduces their incentives to manipulate spot prices at the expiration (delivery dates) of those contracts. It is only the net position, *e.g.*, spot sales in excess of forward sales, that can profit from price spikes. Moreover, the net position of a big supplier may occasionally be short (more presold than producible by that party), making spot price *decreases* preferred by that seller. Relatedly, forward sales reduce the amount of uncommitted capacity that the seller can manipulate, effectively reducing his or her market share in the price-setting generation. Concentration is down; competition is up. Moreover, there are game theoretic arguments for expecting that in a market with two or more dominant players, forward contracting will become a mode of competition between the oligopolists, driving them towards efficient spot pricing at delivery because of the very limited net positions they are likely to hold at the time. Allaz and Vila show that the situation becomes akin to a "prisoners' dilemma" for which the equilibrium, repeated (Cournot) game strategy is to not cooperate.[15]

Thus forward markets for power may play a key role in alleviating generation market concentration concerns, and those forward markets are more likely to develop the more the spot price of power reflects the full value of the service. That is, a one-part commodity market, with its attendant volatility and occasional substantial premiums, provides a more fertile ground for the emergence of forward markets than a two-part market in which the capacity component is regulated or obligatory under non-market standards. Of course, there remains a chicken-or-egg issue of whether enough market power will prevail to stifle participation in forward markets. The above results of Allaz and Vila suggest that this is likely to be a problem only where there is a near monopoly on local production, which can be alleviated by strict divestiture policies requiring a few new owners.

The Importance of Price Signals.

It is also likely that one-part spot prices exhibiting any kind of market power will quickly sow the seeds of their own destruction, if they are given the chance to do so: High spot prices in load pocket regions will encourage curtailable demand bidding into the pool. Indeed, such participation may be the best or even only comprehensive solution for inducing significantly more competition in these places, as generation expansion in urban areas is likely to be limited by environmental restrictions and the lack (or high cost) of good sites. Conversely, if bidding restrictions and cost recovery assurances (from the ISO) are imposed on "must-run" units behind transmission limits, then the price signal to induce demand-side participation, or to build generation, or to expand transmission across the bottleneck will be very weak. As a result, the dispatch will contain (many) out-of-merit plants not facing competition, and this inefficiency will be perpetuated indefinitely.

Note that transmission expansion is also impaired by such a policy. Transmission can be a substitute for generation capacity, once expansion is needed, if there is slack generation available elsewhere. The lack of a capacity scarcity premium in generation spot prices behind a transmission constraint will mean that there is no meaningful signal in congestion rents for solving this problem with new wires. The only congestion rents will be almost trivial, *e.g.,* for marginal losses and for occasional grid contingencies that are not systematic enough to have any associated expansion implications.[16]

This issue has arisen in California, where a recent study for the ISO concluded that under at least one of six possible operating conditions, more than 14540 MW of the fossil generation in the state (nearly a third of the state's generating capacity) should be deemed "must run for reliability purposes"—and this analysis did not even take into account the units that are occasionally needed for stability and security purposes.[17] Partly as a result, the vast majority of gas generating stations in California may be operated under a contract with the ISO that prohibits them from bidding above their variable costs during "constrained on" periods designated by the ISO, and which requires them to rebate 90 percent of their bid profits in other periods in exchange for operating under a contract that covers all of their fixed costs. Many of these units are gas-fired steam generators that are not particularly efficient and might even be uneconomic against new technology, absent their particular locational roles. Another undesirable consequence of this is that since under California's restructuring law power generated by nuclear plants, qualifying facilities, and spilled hydro must be accepted first in the PX, the remaining amount of generation open to competitive bidding is fairly limited.

Imposing price restrictions of this kind is of course a very sensible policy to pursue if one has no recourse but to rely on unconcentrated ownership and control of margin-setting physical supply before deregulating generation. But there are clear costs to this approach, despite the protections it offers—most notably the risks of extending the economic lives of units that might not necessarily be viable otherwise, and discouraging innovations and expansions that would otherwise ameliorate this problem. If, as in California, a very large number of "must-run" units are designated, then many of the potential fruits of restructuring may be suppressed.

Empirical Evidence for One-part Power Markets—the New Zealand Experience.[18]

In the U.S., we can only assess these restructuring policy issues by analogy to other commodity markets. In a few other countries, there has been deregulated production and trading of power for some time under one-part pricing, notwithstanding some concerns about potential market power. New Zealand provides one of the most interesting case studies. A similar regime is operating successfully in the Australian market.

The New Zealand system consists of two alternating current (AC) sub-systems, for the North and South Islands, connected by a 1200 MW submarine high voltage direct current (HVDC) link. The South Island system which is entirely hydro, with moderately sized reservoirs allowing storage of Spring/Summer flows to meet winter peaks, but relatively small inter-annual carry-over, typically meets South Island requirements and allows export to the North, where there is a mixed hydro/thermal system. On average, 75 percent of national requirements are met from hydro, 7 percent from geothermal, and the remainder from a variety of thermal plant, mainly burning gas.

Reservoir capacity is also relatively small, making the country quite vulnerable to shortages in a dry year, when inflows may be 20 percent less than normal. The transmission network is relatively sparse, covering an area the size of the U.K., but serving only 3.5 million people, with an annual load of around 3000GWh. As a result, transmission accounts for a relatively large proportion of the cost of delivered energy, especially because the major hydro resources are in the south of the South Island, while the population is concentrated in the north of the North Island. In such a small system, spinning reserve is a significant issue, particularly because it is necessary to guard against the failure of the HVDC link. In fact, there have been occasions on which more generation capacity has been devoted to spinning reserve than to energy production in the receiving island.[19]

Prior to 1987, the energy sector was totally dominated by the Government, and reform has proceeded in three broad stages. First, the Government's generation/transmission assets were corporatized as the Electricity Corporation of New Zealand (ECNZ), which operates under normal commercial law, and is expected to operate as a profitable commercial enterprise. Second, ECNZ's transmission assets were formed into a separate company, Trans Power, while the local distributor/retailers, which had formerly been under local government ownership, were corporatized and in many cases privatized. Finally, ECNZ's generation assets have been formed into two competing firms, both still in public ownership, with fringe assets being sold off. There is no explicit regulation of the sector. Apart from some surveillance by the Ministry of Commerce, the parties have been left to form and maintain satisfactory institutional and commercial arrangements, with all parties being free to appeal to the Courts and/or the Commerce Commission. There are no legal barriers to entry in generation and retailing, or in transmission/ distribution, although the latter sector is not expected to be particularly competitive.

Market arrangements have also evolved over the last decade. Prior to the reforms, ECNZ sold power to a large number of generally small, local distribution companies under a two-part tariff, 50 percent of which was accounted for by the capacity charge reflecting, in part, the traditional importance of transmission capacity constraints. After corporatization, transmission and energy pricing were separated. For energy pricing, ECNZ introduced a pseudo-spot market.[20] Under these arrangements, half-hourly short-run marginal cost (SRMC) based "spot" prices were determined a week in advance, using ECNZ's production costing models[21]. Annual "energy" contracts were then written in the form of "two-way" financial contracts, under which the parties each agreed to compensate each other on the agreed quantity, should the "spot" price turn out to be higher, or lower, than the agreed strike price. These contracts were negotiated with each party, and sculpted to meet their requirements, although ECNZ insisted on its customers contracting for between 90-110 percent of forecast demand. The goal of this market was twofold. First, it was designed to improve efficiency by ensuring that, even though the average price of power was largely determined a year in advance by the contract price, the market participants always faced an efficient price signal reflecting the SRMC of meeting their actual requirements. Second, it was designed to introduce power companies to what was then a radically new way of organizing the electricity market, as a preliminary to the development of a truly competitive spot market in which all parties would be free to enter into their own spot and contract arrangements, and take commercial responsibility for the consequences as advocated, for example, by Ruff.[22]

An important feature of those interim market arrangements was the imposition of a cap on spot prices, set to the SRMC of the most expensive thermal generator, which was around three times the average spot price. Conceptually, it was recognized that the spot market should really be uncapped, and that capping the market was equivalent to issuing all parties with "one-way" call options, or "capacity tickets," with a strike price set at the market price cap. In principle, it was argued that the price of such option contracts would set the "capacity charge," and that, in the long run, this should equal the cost of building and operating peaking plant, in this case open cycle distillate-fired turbines, able to generate at an SRMC equal to the price cap. Thus, under this interim "capped market" regime, contracts included an additional component, sometimes described as an "insurance premium" to cover this cost, with a similar "uplift" component being added on to pool prices. These arrangements were not ideal, but they did operate for several years, with apparent success in terms of facilitating retail competition[23] and inducing more flexible demand response, as evidenced by a flattening of the daily load curve. This is reflected at the retail level, where domestic customers may have a variety of tariff options, involving different day/night rates and/or interruptibility provisions.[24] Larger users, such as universities or factories, can opt to face spot market prices directly and may install sophisticated signaling and control systems to reduce usage at times of peak prices.

These market arrangements passed their severest test in 1992 when hydro inflows reached record low levels. Peaking plant which had seen little or no service in its twenty year life was called on to operate extensively, and spot prices were maintained at the market cap level continuously for several weeks. This induced national load reductions of the order of 15 percent, with most local power companies making significant profits under their contracts, effectively 'selling back' power which ECNZ had contracted to deliver. Far from 'profiteering' in this shortage situation, ECNZ paid out nearly as much to buy back that power as it did for extra fuel, and had very strong incentives to keep output up and prices down. (This experience may be compared with that in a similar crisis twenty years earlier when, in the absence of a spot market, regulation induced load reductions of only 7 percent.) It is now expected that there will be a much stronger demand response in any future crisis, because power companies and ultimate consumers have put systems and procedures in place to react more quickly and effectively. This suggests that the spot market can provide very effective signals for demand-side response in extreme circumstances. The implications for reserve margins are obvious.

These interim market arrangements remained in place for rather longer than had been planned, but a competitive spot market was introduced in 1996, following the breakup of ECNZ. The spot/dispatch market is now operated by the Electricity Market Company (EMCO). Generator offers, and load bids, are used to perform a day-ahead (indicative) market clearing to produce a pre-schedule and forecast prices. The market clearing/dispatch model represents all nodes in the national network, and determines nodal prices reflecting inter-regional dispatch and congestion for some 120 nodes where active trading occurs. Offers and bids may be freely changed up to four hours before real time, though, and the market is re-cleared to provide new dispatch and pricing forecasts on a regular basis. Real time dispatch must meet actual loads, but is determined, as far as possible, by re-solution of the market clearing model using the latest generator offers. The former price cap has been removed, and final prices are determined, *ex post*, by re-solution of the market clearing model to meet the actual metered load using the generator offers, and grid state, which pertained for that half-hourly trading period.[25] Spot prices have been fairly stable at levels similar to those prevailing prior to the establishment of the market. There is little day/night variation, as is expected given the ability of the two hydro reservoirs to absorb short-term demand/supply fluctuations.

The principal concern here is with the role of contracting in such a market, and particularly with its role as a substitute for regulatory intervention, and in fostering appropriate provision of various forms of reserve. During the reform process, there were real concerns about the economic wisdom, or political viability, of breaking up the relatively small ECNZ system, and fear that the resultant firms would be unwilling, or unable, to manage their reservoirs appropriately in the event of another hydro crisis. There was also considerable fear that an uncontrolled transition to market structures could result in significant short term damage to public welfare, and to an unfortunate regulatory backlash. Thus it was accepted that, even given a goal of totally free markets, transitional guarantees of some form were required. In fact the Wholesale Electricity Market Study[26] concluded that, without a reasonably high degree of long term contracting, the market would be subject to gaming by oligopolistic generators in the small New Zealand system, even if breakup were pursued to the level of individual hydro systems and major thermal plant. Conversely, it was shown that, by reducing market power, financial contracts would provide strong incentives to maintain adequate storage, and assist in providing reservoir coordination similar to that obtainable in an integrated system.[27] Such contracts were also seen as key to ensuring a smooth transition to a more competitive structure by providing consumers with protection against the risk of exploitation, while protecting the value of public assets. Thus it was proposed that agreement should be reached on a set of vesting contracts, withering over the initial years of the market, before any asset breakup or divestiture occurred.

It may be argued that this approach is "artificial," but it should be understood that, whatever approach would have been taken, the situation existing immediately after such a sectoral re-organization will necessarily be artificial, with market institutions, firms, supply/demand balance, plant mix, and contracting relationships all creations of the re-organization process, or of the prior regulatory regime, rather than of the market. In particular, it would have been equally artificial to create a market in which the incumbent suppliers inherit sunk cost plant, but no long term contracts. Thus it was proposed to establish the market with contracts in place, perhaps approximating the portfolio which might have arisen from a competitive market history. The aim was to leave the market free to determine signals for operation, consumption and investment, via spot prices, while determining an allocation of the value of the existing system between the stakeholders. The key principle was that any transitional guarantees should be implemented via contracts of the form that might have been expected to evolve naturally in a mature market, and that could evolve into appropriate market instruments. It was feared that once any form of regulation was established it would tend to perpetuate itself and expand, rather than allow the market to mature to the point where such intervention became unnecessary. Thus, rather than impose price controls on a "bottleneck" generator, for example, a policy was considered which would allow local prices to be uncapped, but require a negotiated option contract that negated local market power. This philosophy continued that of the interim market arrangements, which established a market apparatus which could be "freed up" later, rather than choosing a form which might have been mathematically equivalent, but would have tended to develop toward a regulated, rather than a market regime.

Even this degree of intervention proved unacceptable, and a compromise was reached whereby ECNZ was only partially broken up, but the market power of the larger firm resulting from that split was limited by a requirement to *offer*, annually, a substantial proportion of its capacity in the form of mid-term (one- to five-year) contracts at prices close to those prevailing prior to breakup. To date, the market has shown a reasonable, but by no means overwhelming, appetite for such contracts, with ECNZ selling perhaps 70 percent of its potential generation capacity in this way. This relatively low participation may be taken as evidence that the market does not really believe that ECNZ is in a position to sustain spot prices above current levels, because of an implied threat of regulatory intervention and/or anticompetitive pressure. The threat of intervention cannot be denied, but the requirement to offer contracts effectively disciplines spot prices, while competitive pressures are actually quite strong in the critical investment market. Although some critics maintained that independent investors would be reluctant to enter so long as the market is dominated by an incumbent such as ECNZ, it can be argued that, provided access to transmission and ancillary services was available to all on a neutral basis, the presence of a dominant incumbent is more likely to cause excessive entry.[28] This results from investors expecting the incumbent to accommodate entry by sacrificing volume to maintain prices and because consumers will be strongly motivated to foster diversity of supply by offering contracts to new entrants. These predictions now appear to have been correct, with significant entry occurring beyond, and putting much more downward pressure on prices than previous analyses might have suggested. Similar effects would seem to be evident in the U.K.

It should be understood that most contracting in the New Zealand market is still in the form of two-way options, as in the previous regime. This is probably not surprising, given the immaturity of the market, but may also reflect the fact that most parties are not attempting to match their load profile at all exactly, instead only contracting a year ahead for perhaps 75-80 percent of their expected load, a level which they are almost certain to consume at any price. Generators, purchasers and traders operating in the market are free to form a wide variety of financial contracts between themselves, though, according to their own circumstances. Dairy factories, for example, have highly seasonal loads but can only predict the onset of the peak season within a few weeks; they can purchase an option covering that window of uncertainty. Smaller purchasers who simply want to buy on a traditional 'tariff' basis can do so from traders who can diversify some volume risk. "Call options" (*i.e.*, price caps) have been offered for those who may only want to protect part of their load against very high price spikes, while "capped options" suit those who wish to simply reduce load at such prices. Such contracts are not likely to be actively traded, though, and WEMS laid great stress on ensuring that standard tradable contracts were available, in the form of a hedge against prices at each island's reference node, so as to ensure sufficient liquidity in the contract market.[29]

Still, there has only been limited trading in the first year of market operation. Initially, EMCO introduced a day-ahead contract market, but this soon collapsed due to lack of interest. This should not be surprising in a market where the major uncertainties relate to hydro inflow variations in the monthly to annual time horizon. *Ad hoc* trading occurs between participants when, for example, a trader on-sells "re-packaged" contracts to clients and although EMCO's plans to facilitate trading of standard ECNZ hedges have yet to materialize, monthly contracts are being traded on the futures exchange. Trading activity is sporadic, occurring mainly when there is a significant change in long term weather predictions, but active trading should not be expected in a sector where contracts are mainly bought by participants to cover expected retail sales, and retail market shares change relatively slowly. It is more significant that each of the two major firms involved have reasonably balanced plant portfolios. However, there is little of the speculative activity which creates much of the liquidity in other markets. This is partly attributable to unfamiliarity with a new market in a technically specialized sector, but it is probably unreasonable to expect independent speculators to get involved with a sector dominated by only two firms, each with significant stockpiles. The small New Zealand market clearly does not provide an ideal environment for the development of active trading in electricity contracts. Indeed further reform of both the generation and retail sectors is now under very active consideration to create a more dynamic competitive and contracting environment. Thus, the New Zealand experience suggests some caution with regard to predictions of very active electricity futures markets, although experience may be very different in larger, more competitive, thermal dominated markets, where the pattern of price volatility will be quite different. Nonetheless, these deficiencies should not distract from the very real achievements of the market in bringing forward new investment to discipline prices and providing opportunities for all parties to enter into contracts that match their particular requirements with regard to price and volume risk.

Finally, it is useful to consider the impact of these market arrangements on reserve provision of various kinds. Read *et al* [1997][30] explain how the provision of "contingency reserve" is actually incorporated into the spot market itself, with joint energy/reserve offers being used to form a joint energy/reserve schedule in a single market-clearing optimization, which also produces prices for energy at every node, and for two classes of reserve in each island. In principle, it has been suggested that contingency events should be handled by allowing "instantaneous' spot prices to reach very high levels during the event, and relying on market participants to hold reserve in order to capitalize on such occurrences. In practice, this ideal has been deemed unworkable, and the New Zealand spot market uses half-hourly pricing periods. But generators, and other reserve providers who are paid to be "on reserve duty," are effectively offering half hourly "hedging" contracts against a notional instantaneous spot price, to deliver the agreed increment of power at the half-hourly energy price, rather than to exploit the market power which they would otherwise have, for a brief period, to drive prices to extremely high level. The resultant reserve prices can be significant in New Zealand, averaging around 5-10 percent of the energy price, hence their inclusion in the spot market. Inclusion of reserve in the market seems to be having the desired effect in terms of innovation and refinement of reserve management mechanisms. In particular, prices are trending down as more interruptible load enters the reserve market. Legitimate concerns have been raised with regard to the market power which particular participants may possess in the reserve market, particularly in situations where inter-island transfer rates are limited by reserve requirements. In reality, by offering interruptibility at competitive prices, consumers are opting to handle the prospect of inter-island link failure themselves, rather than accept generator offers of secure transfer, backed by traditional spinning reserve. Thus this market is actually much more competitive than the underlying energy market.

Of more concern, in the long run, may be the provision of dry year backup of the type which might only be required once in twenty years. Critics of the market reforms advocated a U.S. style regulatory regime, whereby generators would be required to carry some kind of capacity margin. It was difficult to see how that could be achieved in a market context. Apart from the overheads of trying to administer such a regime, it seemed likely to militate against the very kind of innovation in unconventional technologies and demand side response which the market was designed to foster. It was recognized, though, that independent investors would have insufficient incentives to invest in plant which was so rarely called upon, particularly if there was any prospect of government intervention to limit "price gouging" in any future crisis. The appropriate form of contract to underpin such investment was seen to be a "capacity ticket," in the form of a call option with a strike price of, say, three times the average energy price. It was also suggested that retailers who have no local franchise and, unlike "distributors," few assets could be tempted to take their chances on the spot market rather than carry what amounts to fairly expensive insurance cover for quite rare events. Thus a transitional measure was proposed whereby retailers should be required to cover a specified proportion of their load with such capacity tickets, or equivalent financial guarantees, or at least to make some public declaration of their cover.[31] Ultimately, this approach was rejected in favor of simply leaving the spot market price uncapped so that those who do not contract for dry year capacity will be fully exposed to the costs of non-supply in such circumstances. It remains to be seen how much dry year/peak support plant will be built under the new regime, but that may be a moot point so long as entry continues at the current rate. In fact, the current level of energy contracting, which includes cover of extreme price peaks, is probably not too far below the load levels which might occur in response to extreme prices, and there is certainly sufficient physical capacity now in the system to cope with any likely event. Thus there is no evidence, as yet, that any kind of intervention is required to ensure adequate provision of backup capacity.

A Restructured Role for Two-part Pricing.

The New Zealand experience is quite encouraging in that it appears to demonstrate that competition can flourish despite apparently worrisome initial concentration in generation ownership. Further, it provides support that forward contracting and demand-side bidding can jointly alleviate market power abuse that might otherwise occur. Substantial entry has occurred, despite a lack of traditional engineering "need" for new capacity. The New Zealand experience also reveals that the forms of forward contracting can be quite various, spanning many durations and degrees of risk protection. This diverse appetite is by itself an important insight and a good reason for pursuing one-part pricing, since regulated capacity markets under two-part pricing typically offer only one "flavor" of service.

One of the most prevalent new forms of forward contracting in the past few years in the U.S. has been option purchase agreements (OPAs), which provide the opportunity but not the obligation to buy and sell firm energy in the future at pre-specified "exercise" or "strike" prices.

- OPA contracts are obviously much more valuable if written for peaking periods than for off-peak periods, at a given exercise price
- The are also more valuable in a given period the lower the exercise price

Thus the value in OPA contracts is contingent on the precise terms and conditions of delivery, even if the same number of MWs are involved. Such contracts can be valued formally as call options against the projected spot market.[32] Figure 6 illustrates this idea.

Since OPAs are for future delivery, it is almost impossible to appraise the reliability characteristics (especially, the forced outage rate or LOLP) of the machine or system that might eventually supply the power, as those machines or capabilities may not even be known at present, and they may vary enormously in complex ways across offers for the same number of MWs, rendering them noncomparable. Those who have been involved in OPA analyses usually conclude that the term "capacity" is not well defined and that the most useful, standardizable alternative is financially indemnified, firm energy (perhaps with a penalty above replacement cost for non-delivery). With that standard in place, one can put OPA offers (or any other kind of forward contracts) on a common foundation and evaluate their insurance value against price spikes from future capacity shortages.

While there are many intriguing features of OPA contracts, their relevance for the present discussion is that power options written against a one-part commodity market recreate two-part pricing, but in a *customized* fashion:

- Each buyer (or seller) can move as much or as little cost per MW into the fixed, demand charge (option acquisition premium) or the energy charge (exercise price per MWh) by simply altering the terms of the option (*e.g.*, a higher exercise price implies a lower premium and a higher energy cost).

- Each buyer or seller can insure as little or much of his/her demand/load obligation as desired, with no need to put all pool-area customers on the same footing, by buying (or selling) options to cap their individual exposure to only a self-selected portion of total requirements.

- Such options could (in principle) be traded in a secondary market, especially if they were written against a public exchange. Financial indemnification as the standard for delivery firmness allows this, while tying performance to specific physical units makes this practically impossible.

Thus, two-part pricing can be recreated in a much more customer-sensitive manner when it arises as derivative contracts written against a one-part commodity market.

ONE-PART MARKETS FOR ELECTRIC POWER 275

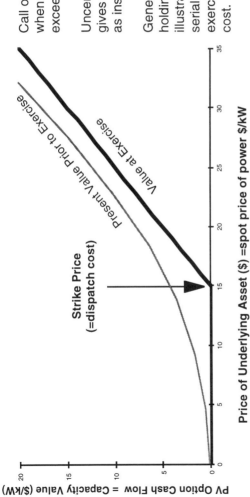

Call option on power is exercised if/when the underlying spot value exceeds the strike price.

Uncertainty over that future value gives options a greater present value as insurance against high prices.

Generating capacity is equivalent to holding a portfolio of call options (as illustrated above) on energy, with serial expiration dates, with the exercise price equal to the dispatch cost.

Figure 6: Capacity as a Call Option

Conclusions.

There are a few clear public policy implications of the above. In several venues around the country, ranging from NERC to power pool ISOs, to Congress and state regulatory commissions, policy makers concerned about supply reliability under restructuring are contemplating imposing legal obligations on suppliers to carry planning reserves. This should be discouraged. Instead, the strategy should be to:

1. Encourage one-part pricing and avoid the temptation to regulate capacity planning reserves or to guarantee "reasonable" prices under supplier-of-last-resort regulation. There is no good model for policy among the U.S. retail restructuring experiments to date: California is not requiring planning reserves, but it may place so many units under must-run cost-based contracts that competition is felt only in off-peak periods, when no signal for capacity expansion or conservation will be felt. Conversely, the eastern tight power pools do not (yet) have must-run, cost-based protocols in mind for their load-pocket capacity, but they seem intent on administering a planning reserve obligation on LSEs.[33]

2. Encourage demand-side participation as a precondition to opening up power markets that might otherwise require must-run, cost-based contracts for plants behind transmission bottlenecks. Such contracts prevent abuse but discourage entry and would inoculate those (often inefficient) plants from competition.

3. Encourage forward markets with as simple protocols as possible and rapid resolution of regulatory policy uncertainty. Try to create at least duopolies in the ownership of bottlenecked capacity, using divestiture, and then observe whether the producers contract themselves forward and largely out of their incentives to manipulate spot prices.

4. Tolerate *ex post* spot price premiums unless there are strong, *ex ante* concerns about sustainable market power. Many such price shocks will be transients, others will be reasonable by replacement cost standards, and the few others will rapidly sow the seeds of their own destruction.

Above all, do not try to create a regime that mixes competition and regulation in generation. This could easily prove less efficient than a more enlightened form of conventional regulation, *e.g.*, comprehensive real-time pricing, which would at least send market-like commodity prices to all customers in all locations. A hybrid system is more likely to bring the worst of both worlds rather than the best.

---

[1] See Jacques H. Dreze [1964], "Some Postwar Contributions of French Economists to Theory and Public Policy," *54 American Economic Review* (June):1-64, for a good summary in English of Boiteux's work.

[2] It is conceivable that certain customers may lack the means to purchase devices that would insulate them from expensive on-peak consumption of commodity power. Their needs for reliability could be covered with special subsidies that do not alter the mode of restructuring recommended herein.

[3] At the time of this writing yet another article (Robert J. Michaels [1997], "MW Gamble: The Missing Market for Capacity," *The Electricity Journal* (December): 56-64) has appeared espousing the "uniqueness" of electricity and the risks of inadequate future capacity under a deregulated market. We feel that it is ironic that Michaels claims that the old, regulated system generated an "efficient mix of energy and capacity, regulated to be priced as if it were in a competitive environment." To the contrary, regulators are pursuing restructuring precisely because past capacity decisions based on uniform reliability criteria have not produced an economical supply mix. The financial indemnification approach to motivating reliability is exactly alien to the rules and incentives used by commodity futures markets to assure liquidity and compensation for contract default.

[4] The curves labeled demand in the figures should be interpreted as consisting of demand plus losses and operating revenues.

[5] Simulations were performed with the IREMM model of production costing and trading. It uses derated capacity and monthly load duration curves, with local areas dispatched first to serve their own loads and trading occurring with marginal resources across all interconnected regions. Interconnections are based on NERC summer total transfer capability ratings, and wheeling charges are based on FERC filings or equivalent cost-based rates derived from FERC Form 1 transmission data. This work is discussed in detail in *Electric Utility Restructuring Impacts on Fuels—Analyzing Off-Peak Conditions* (Palo Alto, CA: Electric Power Research Institute, 1998).

[6] To the extent that the 10-15 percent savings occur, it will be because of stranded cost disallowances, which of course are not efficiency gains, just zero-sum wealth transfers.

[7] See Steven R. Backerman, Steven J. Rassenti, and Vernon L. Smith [1997],"Efficiency and Income Shares in High Demand Energy Networks: Who Receives the Congestion Rents When a Line is Constrained?," *Economic Science Laboratory* (January 21), The University of Arizona, Tucson; and Steven R. Backerman, Michael J. Denton, Steven J. Rassenti, and Vernon L. Smith [1997], "Market Power in a Deregulated Electrical Industry: An Experimental Study" (February), The University of Arizona, Tucson; and Jügen Weiss [1997], "Market Power Issues in the Restructuring of the Electricity Industry: An Experimental Investigation," Harvard Business School.

[8] For the integration of demand bidding with supply dispatch see Fred C. Schweppe, Michael C. Caramanis, Richard D. Tabors, and Roger E. Bohn [1988], *Spot Pricing of Electricity* (Boston: Kluwer Academic Publishers). For discussion of curtailable demand prioritization see Hung-po Chao and Stephen Peck [1997], "An Institutional Design for an Electricity Contract Market with Central Dispatch," *Energy Journal* 18(1): 85-110.

[9] As a bidding strategy, it would be feasible and profit-increasing to bid the variable cost of the unit most likely to be just above you in the dispatch ladder. For most

hours in most market areas in the U.S., this next unit is likely to be very close in price, as the supply curve is almost continuous. As peak conditions are approached, the steepness of the curve and the sparsity of competitors may make this a more meaningful profit-seeking strategy of the competing parties.

[10]This range assumes 33-50 percent debt financing and 20 year capital recovery, with the cost of equity at 14-15 percent. The "1 day in 10 years" LOLP standard that many pools have applied for planning reserves is often associated (depending on load shapes and other factors) with 4-20 hours per year of expected unserved load. It is over this many hours that the fixed costs of the marginal peaker would have to be recovered if it were to remain viable. Of course, a new peaker is very likely to have an efficiency much greater than the older units in a market area (i.e., much better heat rate), so it is unlikely that the full costs of a CT will set the marginal value of capacity in a region. That is, any new CT will offset some of its fixed costs with inframarginal profits. Nonetheless, the truly marginal unit will have to recover some of its fixed costs in its per kWh bid, and the premium over variable costs could be quite large.

[11]Steven Schleimer [1997], "Market Power Analysis in Support of PG&E's Application for Market-Based Rates Before the FERC," presentation to the Energy Modeling Forum, Stanford University, January 23.

[12]D. Newbery [1984], "Manipulation of futures markets by a dominant producer," in *The Industrial Organization of Futures Markets*, R. Anderson, ed. (Lexington, MA: Lexington Books).

[13]Hendrik Houthakker, private communication; Stephen C. Pirrong [1995], "The Efficient Scope of Private Transactions-Cost-Reducing Institutions: The Successes and Failures of Commodity Exchanges," 24 *Journal of Legal Studies* (January): 229-255; Stephen C. Pirrong [1993], "Manipulation of the Commodity Futures Market Delivery Process," 66 *Journal of Business* (No. 3): 335-369.

[14]Hendrik Houthakker, private communication.

[15]Blaise Allaz and Jean-L.C. Vila [1993], "Cournot Competition, Forward Markets and Efficiency," 59 *Journal of Economic Theory*: 1-16. . See also Richard Green [1996] "The Electricity Contract Market" Cambridge University, U.K., preprint May, 1996. It should be stressed that the above authors are addressing oligopoly situations; their results would not apply in a market with a single dominant player.

[16]The above discussion of one-part pricing applies to generation because scale economies are not a major problem. This is not the case for transmission, which means that you may still need two-part pricing with a large part of the transmission infrastructure cost being recovered via some form of "fixed charge" (possibly a long run take-or-pay TCC). At the retail, or even wholesale, level these charges may well be converted into a peak consumption charge, though, and this can cause unfortunate distortions. A pure one-part tariff system is not necessarily ideal at all levels in the system. See E.G. Read [1998b], "Transmission Pricing in New Zealand," forthcoming in *Utilities Policy*, for a detailed discussion of this issue. See also, Richard Green [1996], The Electricity Contract Market, Cambridge University, U.K. May 1996 (preprint).

[17]R.R. Austria, T.I. Leksan, W.R. Puntel, and J.R. Willi [1997], "Final Report Phase 1: Operating Reliability Requirements Study" (July 18), PTI Report No. R53-97 (Schenectady: NY: Power Technologies, Inc.).

[18]The nature and history of the New Zealand electricity system is discussed by J.G. Culy, E.G. Read and B. Wright [1996], "Structure and Regulation of the New Zealand Electricity Sector," in R. Gilbert and E. Kahn (eds.), *International Comparison of Electricity Regulation* (Cambridge University Press), p 312-365, while current arrangements and recent experience are covered by E.G. Read [1998a], "Electricity Sector Reform in New Zealand: Lessons from the Last Decade," forthcoming *Pacific*

Asia Journal of Energy, from which this summary is derived. E.G. Read [1998b], "Transmission Pricing in New Zealand," forthcoming in *Utilities Policy*. E.G. Read and D.P.M. Sell [1987], *A Framework for Electricity Pricing*, Arthur Young report released by the Electricity Corporation of New Zealand, Wellington.

[19] See Note 18 above.

[20] Read [1998b] argues that capacity elements are still important in the transmission pricing arrangements, reflecting the prevalence of scale economies in that sector, although those capacity elements should properly be expressed in the form of long term transmission congestion contracts.

[21] In fact the market structure adopted here was a natural extension to the way in which ECNZ's costing model had been constructed. Since hydro was typically "on the margin," the SRMC prices were themselves determined by a reservoir management model, which effectively calculated the "option value" of a unit of water as the expected value of that water, if stored to avoid the cost of thermal fuel, or shortage, at some future date.

[22] See L.E. Ruff [1992], "Competitive Electricity Markets: Economic Logic and Practical Implementation" in *Coping with the Energy Future: Markets and Regulations*, Proceedings of the International Association of Energy Economists, Tours, France, for a description of the kind of market structure that New Zealand may pursue eventually.

[23] See M.E. Bergara and P.T. Spiller [1997], "The Introduction of Direct Access in New Zealand's Electricity Markets" in A. Lapointe, P.-O Pineau and G. Zaccour (eds.), *Proceedings of the International Workshop on Deregulation of Electric Utilities*, Montreal, pp. 253-274.

[24] Most domestic water heating systems are electric, with a substantial storage tank and remote control systems allowing the local power company to determine when heating will occur. "Night storage" heating is also common, and can be controlled in a similar fashion.

[25] This supply/demand asymmetry in dispatch and final pricing is less than ideal, and seems likely to prove inadequate, should there be a repeat of the 1992 crisis. Normally though, this is a reasonable approximation, given that demand can only react to ex ante price projections, and this reaction is reflected in the loads to be met by real time dispatch.

[26] WEMS [1992] *Towards a Competitive Wholesale Electricity Market: Conclusions and Recommended Approach*, Wholesale Electricity Market Study, Final Report, Wellington, New Zealand.

[27] See T.J. Scott and E.G. Read [1996], "Modeling Hydro Reservoir Operation in a Deregulated Electricity Sector," *International Transactions in Operations Research*, 3(3-4): 209, 221.

[28] See WEMS [1992].

[29] Hedging between those reference nodes and local nodes may be obtained from Trans Power, but this is not compulsory, and many participants, who are only exposed to relatively minor price differentials due to losses, may operate without any such cover.

[30] E.G. Read, G.R. Drayton-Bright and B.J. Ring [1997], "An Integrated Energy/Reserve Market for New Zealand" in A. Lapointe, P.-O Pineau and G. Zaccour (eds.), *Proceedings of the International Workshop on Deregulation of Electric Utilities*, Montreal, pp. 275-291.

[31] The grounds that purchasers in the wholesale market may not really represent the interests of small consumers who have insufficient choice, or information, to judge the creditworthiness of competing suppliers in the event of a crisis. Such

arrangements were seen to be equivalent to the prudential requirements placed on financial institutions involved in what are essentially similar types of transaction.

[32]Frank C. Graves and James A. Read, Jr. [1997], "Capacity Prices in a Competitive Environment," Chapter 7 in *The Virtual Utility*, Shimon Awerbuch and Alistair Preston (eds.) (Boston: Kluwer Academic Publishers): 175-192.   Frank C. Graves [1997], "Capacity Prices in a Competitive Power Market," presentation, IEEE PES Summer Meeting (July), Berlin, Germany.

[33]*Reliability Assurance Agreement among Load Serving Entities in the PJM Control Area*, June 2, 1997.   William W. Hogan [1997], *Report on the Proposal to Restructure the New York Electricity Market* (January 31).

IV PLANNING IN THE NEW INDUSTRY

8 SYSTEM PLANNING UNDER COMPETITION

Raymond Coxe* and Marija Ilić**

*NEES Global Transmission, Inc.
Westborough, MA 01582 U.S.A.

rcoxe@ibm.net

**Department of Electrical Engineering and Computer Science
Massachusetts Institute of Technology
Cambridge, MA 02139 U.S.A.

ilic@mit.edu

INTRODUCTION

Having discussed the operation of electric systems under competition, we now consider the planning of electric systems in such environments. While the introduction of competition can significantly change the operation of an interconnected electric system in the near-term, the long-term effects on system planning may be even greater. In this chapter, we consider how the introduction of competition into an electric market affects the process of planning new generation and transmission investments.

We start by describing the role of the vertically integrated utility in a monopoly market, and the public service nature of monopoly utilities. We describe the traditional planning criteria and processes used in monopoly electric markets, and present a generalized formulation of how the monopoly utility will select from a mix of generation and transmission alternatives. We then discuss the role of transmission as the transportation function of the vertically integrated utility, and note the natural integration of transmission and generation planning in a monopoly utility.

With this background, we are finally ready to consider system planning under competition. After a discussion of some suggested conditions for workably competitive electric markets, we describe the need to separate the overall system reliability criteria (used by monopoly utilities) into generation-related reliability criteria and transmission-related reliability criteria. We then consider how the generation criteria can be complemented or supplemented with market prices and performance incentives when competition is introduced. We then reformulate the problem of evaluating generation investments for competitive electric markets.

We note that in competitive electric markets, the role of the transmission system is now to insure workable competition in addition to providing reliable energy transportation. We propose a generalized framework for transmission planning in competitive markets, while noting the shortcomings and assumptions of our model. This framework is illustrated with two specific, real world examples. Finally, we discuss some of the remaining challenges in adapting traditional licensing and siting criteria to the new realities of a competitive electric market.[1]

In many countries (including the United States), the electric supply industry is undergoing a significant restructuring that has only just begun. Fully competitive electric markets are a fairly recent phenomena, and the existing competitive markets have not operated through periods of resource shortages. As of late 1997, enough generating capacity has been present in the current competitive markets to maintain overall system reliability. Capacity margins have been generally maintained, even as energy prices in these markets have fallen. However, the presumption that long-term reliability will be fully sustained solely by competitive prices (with no other incentives) still attracts questions and challenges. This critical scepticism, and the need for new planning paradigms, was summarized in the executive summary of *Reliability Assessment by NERC*:

> It is apparent that the integrated planning process, which ensured coordination of generation and transmission plans in the past, is being dismantled as the industry restructures for an open market. In its place, market-based mechanisms are evolving to provide the necessary signals to spur investment in needed generation and transmission. However, these processes are neither fully developed nor mature. In the transition, it is not clear how the processes will evolve, but coordination of resource plans must be maintained. The risk is that an inadequate supply of either resources or transmission could result in an inability to deliver electricity to the customer. A new process is required to assure this coordination. It is possible that the new entities in the industry structure, like Independent System Operators (ISOs), Regional Transmission Associations (RTAs), or Regional Councils, may establish themselves as the coordinators of transmission and generation planning.

Clearly, the promise of market-based generation development has not yet been fully proven over the long term. In this chapter, we propose that competitive market prices and monetized performance incentives will, by themselves, attract enough investment in new generating facilities to sustain the reliability desired by electric consumers. Our fundamental premise is that unrestricted competition in generation will increase social welfare relative to regulation of generation, and that the transmission planning process can adapt to meet the needs of the competitive market. This conclusion has not yet been "stress-tested" through periods of capacity shortages and/or high prices in a fully deregulated market; more experience with success (and failures) in competitive energy markets is required to confirm our faith in purely commercial incentives.

A related caveat must be applied to our proposed framework for transmission planning in competitive markets. In this chapter, we note the need for a strong separation of the generation and transmission function (up to and including legal separation of generation and transmission assets). We also proposed centralized planning for most "network" transmission projects, and market-driven development of specialized transmission projects. We also believe that limited competition can (and should) be introduced into selected parts of the transmission sector for the construction, ownership, and maintenance of the transmission assets desired by regional electric consumers.

While this type of framework has been at least partially adopted in countries such as Argentina and Australia, other electric markets (such as the United States) confront difficult questions in the separation of generation and transmission and the introduction of competition in transmission asset ownership. The issues of planning, developing and financially supporting an integrated transmission system under competitive conditions are still being resolved, and probably constitute the most difficult challenges for a competitive electric market. For instance, the challenges of siting new transmission facilities (which are often controversial when the project is needed for reliability) will probably be greater for transmission projects that are solely motivated by commercial considerations. Our proposal in this area should be critically reviewed and refined by others, including our readers.

Finally, the process of planning and evaluating investments in mature competitive markets is challenging enough. Different companies use different processes; some investments succeed, and some investments fail. The uncertainties regarding generation investments in a newly competitive electric market are very large, and the theory of evaluating such investments is immature (and often overtaken by broader strategic considerations). Our suggestions regarding how a competitive generating company might value a particular asset may not always be followed in every instance; competing businesses rarely have the same valuation process or results.

CHAPTER SUMMARY

For a vertically integrated utility in a monopoly market, the *need* for new generation and transmission capacity is often determined by an engineering reliability criterion (often defined as a "loss-of-load probability"). The *type* of new generating capacity, and the associated transmission facilities required, is determined by a central planning process directed by the monopoly utility. This process is a natural byproduct of the monopoly utility's franchise right (and obligation) to be the exclusive supplier of electric energy within a region. A generalized formulation of this process is presented.

The introduction of competition in generation fundamentally changes this planning dynamic. In competitive electric markets, the timing and type of new generation investments is driven more by expectations of market prices, generating plant operating costs and resulting profit margins, and less by explicit reliability criteria. Profit-maximization is the primary business objective of competitive energy producers.

The primary consequence of this change is that electric energy production becomes a more commercial (and riskier) business enterprise. In competitive markets, generating plants are opened, operated and closed on the basis of demand and market prices, much as production facilities in other industries. In turn, the competitive market prices will be determined primarily by the decentralized decisions of competing producers instead of through regulation. Generation reliability is now explicitly valued, priced and determined through market forces and performance incentives, rather than through prescriptive criteria. Centralized generation planning by monopoly utilities is replaced by decentralized investment planning by individual generation companies in response to commercial incentives.

The introduction of competition also has significant impacts on transmission planning. The fundamental physical role of the electric transmission system in any electric market is to connect producers and consumers of electric energy. This physical role applies in both monopoly and competitive electric markets. However, the commercial and financial arrangements for the transmission system are greatly different for these two types of markets, especially in developing new transmission facilities.

In monopoly electric markets, transmission is planned as part of the integrated electric system of the monopoly utility. In these markets, new transmission projects are planned, financed and constructed as part of the utility's overall strategic resource plan. The utility will seek the overall investment plan that minimizes the total delivered cost of energy.

In competitive electric markets, this explicit tie between new generation and new transmission is severed. The transmission function now becomes a separate energy transportation business, with its own product (unbundled transmission service), pricing (unbundled transmission prices) and plans (transmission planning that now reacts to decentralized generation plans). The new mission of the transmission function is to serve as the agora[2] for all buyers and sellers of electricity. Multiple buyers and sellers now meet financially as well as physically at the transmission system to consummate their decentralized and competing transactions. In competitive electric markets, transmission planning is driven by a combination of traditional transmission-related reliability criteria, and new market-driven opportunities for energy trading. In this chapter, we propose a new model for transmission system planning in competitive markets that retains the benefits of centralized grid planning, while recognizing the expanded commercial opportunities for energy trading and introducing competition into selected parts of the transmission sector.

SYSTEM PLANNING IN MONOPOLY MARKETS

System planning in monopoly electric markets is typically done by a vertically integrated electric utility that is responsible for generating, transmitting and distributing electric energy to captive consumers[3]. Much of the historical theory of utility system planning in these markets is based on the characterization of the electric sector as a *public service activity* with natural monopoly[4] characteristics.

In monopoly electric markets, the local utility often has a *franchise right* for certain defined geographic areas. Under this franchise right, the local utility is usually the sole wholesale or retail seller of electric energy within the franchise area. The specific form of the franchise right may vary from region to region (or even town to town within a region), but the intended purpose is to endow the utility's activities with both the authorities and restrictions associated with providing a monopoly public service. Under the franchise, the utility usually has an obligation to provide any and all consumers with electric energy, and to provide it with a certain level of reliability.

Of course, customers of a monopoly utility require protection from monopoly prices. With no pricing constraints, a monopoly utility would charge a monopoly profit-maximizing price that is higher than the socially optimal (or competitive outcome) price, and thus undersupply the product. Prices charged by monopoly utilities are thus regulated to protect consumers from such monopoly pricing.

In exchange for assuming the obligation to serve all consumers and agreeing to charge only the regulated rates, monopoly utilities receive broad powers to execute their public service activities. These powers typically include the power to acquire land (even from unwilling sellers) in order to construct and operate generation, transmission and distribution facilities. As part of those powers, utilities in the United States can often appeal local zoning and land use decisions to a higher administrative body responsible for insuring that the utility can carry out its public service activities.

However, the most important rights granted to a monopoly utility are probably the commercial protections. These rights include protection from competition, and the right to charge rates that provide an opportunity to recover all "prudent" costs (including a "reasonable"[5] return on equity). The "rate-of-return" rate regulation used in the United States passes most of the benefits of good investments (whether due to good judgement or simply good luck) on to the consumers; it also protects the utility from the full consequences of bad investments (unless the investment are found to be "imprudent"). This ability to recover all prudently incurred costs, with a regulated return on equity, effectively protects the monopoly utility from many of the business risks of a normal competitive enterprise, while also constraining the utility's commercial opportunities. Under this framework, the business basis for a monopoly utility is success in the regulatory arena. Success in the commercial energy market is usually guaranteed by the utility's monopoly status.

This business arrangement is typical for other natural monopolies such as water and sewer services. The franchise right to be the sole provider of a public service normally has a corresponding obligation to reliably provide that service at regulated rates to all consumers within the franchise area. For a monopoly electric utility, the obligation to provide reliable service translates into an obligation to build or purchase new generation and transmission capacity as necessary to meet the expected electric demand with a certain level of reliability.

288   POWER SYSTEMS RESTRUCTURING

In the United States, the reliability criteria for planning vertically integrated utility systems in monopoly markets have been established by the various regional electric reliability councils and other utility groupings. Regional reliability criteria are generally based on the planning standards and criteria developed by the North American Electric Reliability Council and the regional reliability councils. In other nations, system reliability criteria have generally also been set by the monopoly utilities in cooperation with regulatory authorities.

Reliability criteria in monopoly electric market

In a monopoly market, the vertically integrated utility will have different criteria for different activities, but all of the criteria are directly related to the overall system reliability objective. For the monopoly utility, all reliability criteria are set exogenously (either by the monopoly utility, trade associations of such utilities, or governmental fiat), usually after limited discussion among a narrow group of participants[6]. Often, there is little opportunity (or reason) for explicit cost justification of the chosen criteria or an explicit valuation of reliability to the consuming public.

These reliability criteria often take the form of prescriptive "top-down" rules that explicitly direct the planning and operation of the integrated electric system. Illustrative examples of such prescriptive criteria would include:

- An operating requirement that enough operating reserve capacity be available and identified to replace the single largest unit within 10 minutes of a loss of that unit;

- A short-term maintenance planning requirement that the utility not take the two largest generating units out for maintenance at the same time;

- A substation design criterion that all substations with load in excess of a certain amount must be supplied by two separate transmission lines;

- A capacity planning requirement that the utility have a certain installed capacity reserve margin (e.g., 22%) over the expected annual peak load; and

- A longer-term resource planning requirement that the utility have enough resources under development to maintain an 80% probability of meeting the installed capacity reserve margin for the next five years, given uncertainty in future loads and resource availabilities.

Two striking features of this type of criteria are their prescriptiveness and the frequent absence of any cost-benefit test in implementing the criteria. Prescriptive criteria often leave little room for innovation or for evaluating their cost and benefits. An installed capacity rule that is uniform for all utilities (regardless of the type or reliability of their individual generators) will not directly promote the most reliable technology or encourage highly reliable operation of existing plants. Similarly, prescriptive criteria often fail to address whether the cost of meeting the criteria is justified in all (or even most) instances.

Prescriptive reliability criteria such as these are appropriate for monopoly markets. With no competitive price signals or commercial incentives to guide investment decisions, "top down" planning criteria are required. We will see later that in competitive electric markets, the generation-related reliability criteria can be complemented and supplemented with competitive market price signals and monetary performance incentives.

Generation planning in monopoly electric markets

In the United States, virtually all utilities adhere to the system planning criteria developed by the regional reliability councils. For resource planning, a typical standard would be as follows:

> Each *Area's resources* will be planned in such a manner that, after due allowance for scheduled outages and deratings, forced outages and deratings, assistance over interconnections with neighboring *Areas* and region, and capacity and/or load relief from available operating procedures, the probability of disconnecting customers due to *resource* deficiencies, on the average, will be no more than once in ten years.

This generation-related reliability criterion can be restated as a requirement that the probability of not serving firm load (the "loss- of-load probability" or "LOLP") be no greater than 10% per year (equivalent to 1 event in 10 years)[7].

Many utilities (or regional cooperative groupings of utilities) evaluate their integrated systems using mathematical reliability models that simulate the integrated bulk power system. These models forecast the LOLP for future years, given input assumptions about current and future loads, existing and planned generating units (including planned retirements), interconnections with other regions, planned maintenance schedules, forced outages, and load relief available from operation procedures. In years in which the annual LOLP exceeds 10model would identify a need for new generation resources.

We note that this planning process and this investment criterion is a fundamental reversal of the normal decisions to invest in new production capacity in a competitive industry. In a monopoly electric market, the market size for production capacity (generating units) is exogenously set, based on the LOLP criterion. The price of the product (electric energy) is considered secondary, and is the *result* of the utility's plans. In competitive electric markets, prices are set by competition, and those prices (relative to the capital and operating costs of new production plant) determine investments in new production capacity.

Once the need for new generating capacity has been identified, vertically integrated utilities typically select the *type* of new generating capacity based on their internal assumptions regarding the cost of various generating options and other alternatives. In the United States, these assumptions are often formally codified in an *integrated resource planning process* that attempted to identify the optimal set of new resources that would meet the franchise obligation to provide for reliable energy services at the lowest cost.

In summary, generation planning in monopoly markets is driven by generation-related reliability criteria (i.e., reserve margins) and centralized assump-

290   POWER SYSTEMS RESTRUCTURING

tions about generation costs. The monopoly utility determines when it requires new generating capacity, identifies the "least cost" mix of generating resources to meet that need, and acquires those resources (either through direct ownership or contracts with other generators)[8]. The costs of those resources, and any associated transmission facilities, are then recovered through the utility's bundled rates.

The system planning process for a monopoly utility

We now consider the economic evaluation of a range of hypothetical generating projects by a vertically integrated utility operating in a monopoly market. The specific formulae used by any utility are less important than the general framework and objective functions of the process. Therefore, we will focus on the mechanics of the evaluation processes and its critical assumptions.

Before we review the specific evaluation mechanism, we note that this illustrative example has been highly simplified and then presented using apparently rigorous formulae. The actual planning process for a monopoly utility would be more complex, and could consider costs and benefits that have been neglected here (including externalities such as air emissions). Furthermore, individual generating projects are always evaluated in the context of the other investment alternatives that may be available, such as demand side management (DSM) options[9].

We first set the stage by assuming the following characteristics of the monopoly utility's market:

Expected demand for electric energy.

- Peak demand forecasted for year y (in MW): $PL(y)$
- Total energy consumption expected in year y (in MWh): $E(y)$
- Demand in hour t of year y, relative to peak: $d(t)$ $(0 < d(t) < 1)$[10]

We thus have:

$$E(y) = PL(y) \sum_{t=1}^{8760} d(t) \qquad (8.1)$$

Existing generating units. We consider an existing stock of N generating units, each with the following characteristics:

- Maximum generating capacity k_j (in MW),
- Hourly production cost function given by $hc_j(q_j)$ (in dollars per hour), where q_j is the hourly output of unit j, in MWh,
- Scheduled maintenance requirements of mor_j (typically expressed as a fraction),
- Forced outage rate of for_j, (also usually expressed as a fraction)

In this simplified example, we will neglect interconnections with adjacent utilities[11]. Hence, in every hour t of every year y, the expected demand would be served entirely from the N existing generating units within the market, plus any new units selected as part of the planning process.

We also assume that the unit characteristics are stable over time. Hence, k_j, hc_j, mor_j, and for_j do not depend on y. Time-varying unit characteristics can be represented, but such detail is cumbersome in even the best models, and would unnecessarily clutter our generalized formulation. Other constraints beyond meeting expected hourly energy demand (such as a minimum operating reserve margin or dynamic constraints on the hour-to-hour operation of each generating unit) may be present. For this analysis, we neglect such additional constraints.

Finally, the reader should note that we have neglected the fixed cost component of the utility's existing assets. Many utility planning models assume that continued operation of the existing plants to a pre-specified retirement date is always economical, and hence, often do not consider early plant retirements. In particular, the capital costs associated with existing units are treated (for planning purposes) as an unavoidable sunk cost.

Project assumptions. We assume that the utility has M alternatives for new generating capacity[12]. Each alternative m has the following characteristics:

- maximum generating capacity of k_{N+m}[13], in MW;

- hourly production cost function in year y given by $hc_{N+m}(q_{N+m})$ (in dollars per hour), where q_{N+m} is the hourly output of unit $N+m$, in MWh;

- annual maintenance requirements of mor_{N+m};

- forced outage rate of for_{N+m}; and

- annual fixed cost equal to fc_{N+m} (in dollars per year).

As with existing units, we assume that these characteristics of the M alternatives are static (e.g., no change in the production function over time).

For each alternative, the annual fixed cost fc_{N+m} is calculated as the total cost annual cost that is *not* sensitive to hourly output. This cost includes both annual fixed operating costs (such as plant staffing, property taxes and certain maintenance costs) and the annual capital costs associated with construction of the plant. Thus, the total annual fixed cost can be expressed as:

$$fc_{N+m} = foc_{N+m} + cc_{N+m} \tag{8.2}$$

where

For the monopoly utility, the annual capital cost cc_{N+m} is calculated from the project's total construction cost, the utility's cost of debt and equity, the utility's capital structure (relative amounts of debt and equity), the book and

foc_{N+m} = annual fixed operating cost (staffing, property taxes, etc.) for alternative m, expressed in dollars per year;
cc_{N+m} = annual capital cost (interest and cavity return, depreciation, income taxes) for alternative m, also expressed in dollars per year.

tax life of the generating project, the income tax rate and the specific ratemaking practices of the utility[14]. In other words, for a monopoly utility, the annual capital cost is calculated to yield a specified return on the investments of the utility's shareholders and lenders. Later, we will see how this calculation is reversed for a generation company operating in a competitive electric market.

The projected construction cost of a new generation project will include the cost of any new transmission facilities required for the project. Generation projects far from demand centers may require significant transmission and thus incur a capital cost penalty relative to projects closer to the utility's loads. Correspondingly, generating projects that can defer transmission investment that otherwise would be required can receive a construction cost "credit" in the planning process. In this manner, transmission planning is intrinsically linked to generation planning, and the integrated planning process will identify an integrated plan for new generation and transmission facilities.

We now address the question of which, if any, of the alternatives should be built in a future year. We consider how the monopoly utility will determine the need for new capacity, and how the utility would select from the available alternatives.

The monopoly utility's expansion plan. Most generation planning processes focus on developing an "optimal" generation expansion plan. A plan p is defined by a unique set of new generating units that are added over the study horizon. For plan p, we denote the number of new generating units of type m operating in year y by $\delta_m(p,y)$. For example, if under plan 3, one unit of type 5 is added in each of years 2, 4, and 6 (for a total of three units of type 5), the annual values of $\delta_5(3,y)$ would be as follows:

Year (value of y)	Value of $\delta_5(3,y)$
1	0
2	1
3	1
4	2

The generating planning process will focus on finding the optimal plan p^* that has the lowest total cost to the utility's ratepayers while meeting the utility's overall system reliability criterion. The values of $\delta_m(p,y)$ associated

Year (value of y)	Value of $\delta_5(3, y)$
5	2
6	3
7 and subsequent years	3

with that optimal plan (denoted as $\delta_m^*(p^*, y)$) will specify the optimal set of new resources that should be acquired by the utility.

We assume that the utility has the following criterion for generation-related reliability:

> The utility will construct or purchase sufficient generating capacity such that the probability of not meeting load due to generation-related constraints is less than $LOLP_{crit}$[15] in any future year.

The starting point for the monopoly utility's planning process will thus be a reliability calculation regarding the need for future generating capacity. To rigorously calculate whether or not the generation criterion is met for a given plan p, the utility would compute the probability of serving the forecasted hourly load, given the projected peak demand $PL(y)$, the expected demand in each hour (given as the product of $PL(y)$ and $d(t)$), the stock of existing and planned generating units in each year for that plan (hence, the values of $\delta_m(p, y)$ for that plan), and the unit maintenance requirements (mor_j) and expected forced outage rates (for_j). The $LOLP$ in year y for expansion plan p is thus given as a function of those parameters:

$$LOLP(p, y) = LOLP[PL(y), D, K(p, y), MOR(p, y), FOR(p, y)] \quad (8.3)$$

where

$PL(y)$ = Forecasted peak load for year y (in MW)
D = Vector of hourly loads $d(t)$, relative to peak load in year y
  = $[d(1), ...d(t), ...d(8760)]$
$K(p, y)$ = Vector of generation capacities k_j for existing and planned generating units
  = $[k_1, ..., k_j, ..., k_N, \delta_1^*(p, y)k_{N+1}, ...\delta_m^*(p, y)k_{N+m}, ..., \delta_M^*(p, y)k_{N+M}]$
$MOR(p, y)$ = Vector of maintenance requirements mor_j for existing and planned generating units
  = $[mor_1, ..., mor_j, ...mor_N, mor_{N+1}, ..., mor_{N+m}...mor_{N+M}]$
$FOR(p, y)$ = Vector of forced outage rates for_j for existing and planned generating units
  = $[for_1, ..., for_j, ...for_N, for_{N+1}, ..., for_{N+m}...for_{N+M}]$

A computer program would compute the $LOLP(y)$ for each year given these inputs, and compare the calculated $LOLP$ to the criterion value (denoted by $LOLP_{crit}$). The program would identify the years in which additional generating capacity is required to keep the calculated $LOLP$ value above $LOLP_{crit}$.

The more sophisticated versions of these programs are capable of identifying maintenance schedules that minimize the $LOLP$, given the forecasted load and generating resources. Finally, the program may also include some modeling of interconnections with adjacent utilities and the reliability benefits of such interconnections[16].

Many monopoly utilities have used simplified implementations of the $LOLP$ criterion in their planning process. A very common simplification is to assume that the total installed generating capacity should exceed the expected peak load by a certain percentage (usually termed the "installed reserve margin"[17]). The assumption is that if the installed reserve margin criterion is met, the $LOLP$ criterion will also be met. This assumption is valid if the aggregated characteristics of the generating unit mix do not change over time.

For this simplified interpretation of the generation-related reliability criterion, the $LOLP$ criterion can be converted into a constraint regarding the total installed capacity (from new or existing units) that is required in each year:

$$\forall p, y: \sum_{m=1}^{M} \delta_m(p,y) k_{N+m} + \sum_{j=1}^{N} k_j \geq PL(y)[1+RM(y)] \quad (8.4)$$

where $RM(y)$ is the installed reserve margin requirement applicable to year y[18], and the other variables are as defined earlier. With these assumptions, the need for new capacity is solely a function of the forecasted peak load, and any changes to the installed reserve margin requirements.

As with other prescriptive planning criteria, the outcome of the capacity need assessment is not sensitive to the cost of the alternatives. If new capacity is required to meet the reserve margin criterion, the monopoly utility would then simply identify the alternative with the lowest net cost as discussed below. Not investing, and allowing generation reliability to decrease, is typically *not* an alternative for the monopoly utility. We will see how this absolute reliability criterion is complemented and/or supplemented in the competitive market by an investment criterion.

The next step in the evaluation process is to identify the optimal mix of new generating units (or other resources modeled as generating units). Most planning models used by monopoly utilities use the net present value of all ratepayer costs associated with a "satisfactory"[19] generation plan as the objective function to be minimized. In calculating this net present value of ratepayer costs, the fixed costs of existing generating plants are often ignored, since those costs are assumed to be common to all generating expansion plans[20].

Therefore, for any generating expansion plan p, the "plan index" $PI(p)$ will be calculated as the sum of (1) the net present value of all operating costs that depend on plant-specific energy production (mostly fuel costs) and (2) the net present value of all fixed costs (including capital costs) of the *new* generating units added under plan p. This net present value is calculated over an evaluation horizon of Y years[21], using an annual discount rate r. Mathematically:

$$PI(p) = \sum_{y=1}^{Y} \frac{OC(p,y) + FC(p,y)}{(1+r)^y} \qquad (8.5)$$

where

$PI(p)$ = Plan index of plan p
$OC(p,y)$ = Operating costs in year y of all generating units present in plan p
$FC(p,y)$ = Fixed costs in year y of all new generating units present in plan p

The best plan (denoted p^*) will have the lowest plan index (denoted $PI(p^*)$). The fixed costs for the new generating units in plan p are given by:

$$FC(p,y) = \sum_{m=1}^{M} \delta_m(p,y) \times fc_{N+m} \qquad (8.6)$$

The operating costs for all generating units (existing and new) in plan p are given by:

$$OC(p,y) = \sum_{t=1}^{8760} \sum_{j=1}^{N+M} hc_j(y, q_t(t,y,p)) \qquad (8.7)$$

where we have now recognized the dependence of q_j for a given generation unit j on all of the other generating units present in plan p. In other words, construction of a new generating unit will affect the output of the existing generating units and of any other new units.

The key parameters for the plan index are thus seen to be (1) the annual fixed costs of the new generating alternatives under consideration in the current plan and (2) the annual production cost of the individual generating units included in that plan, calculated for all hours in the year. To solve for the plan index (and thus identify the least-cost resource plan), we must identify the actual values of q_j for each unit in the plan.

Many planning models include a *production cost algorithm*. These algorithms identify, for a given set of generating units, the combination of generating unit output levels that minimize the total system production cost, while respecting unit and system constraints. Many early production cost algorithms used composite load duration curves to represent blocks of similar hours and thus eliminate the need to solve for hour-by-hour values of q_j. This simplification can significantly reduce the computational requirements for the program. Steady decreases in computational costs have resulted in hourly production costing programs becoming more common. For this discussion, we assume that an hourly production cost model is used that can optimize $q_j(t,y,p)$ over all units and in every hour t.

Virtually all production cost algorithms attempt to minimize the total cost of operating the integrated system over a given time horizon, subject to various constraints (such as maintaining a minimum operating reserve at all times).

Here, we assume that the operating constraints can be neglected, so that the cost-minimization problem can be restated as follows:

For all generating units j in plan p (existing and new), find the optimized values of $q_j(t,y,p)$ (denoted $q_j^*(t,y,p)$) that minimize $C(p,y)$ for each year y, subject to system and unit specific constraints.

In other words, for each plan p and year y, find the values of $q_j^*(t,y,p)$ that satisfy:

$$\forall j: \frac{\partial OC(p,y)}{\partial q_j^*} = 0 \tag{8.8}$$

subject to the following constraints:

1. Meeting load in every hour

$$\forall t, y: \sum_{j=1}^{N+M} q_j^*(t,y,p) = d(t) \times PL(y) \tag{8.9}$$

2. Capacity constraint for existing generating units (when those units are not on maintenance):

$$\forall t, y, p, j \leq N: q_j^*(t,y,p) \leq k_j \tag{8.10}$$

3. Capacity constraint on new generating units (again, for non- maintenance hours):

$$\forall t, y, p, m \leq M: q_{N+M}^*(t,y,p) \leq k_j \delta_m(p,y) \times k_{N+M} \tag{8.11}$$

4. Maintenance requirements of existing and planned generating units:

$$\forall y, p, j \leq N+M: q_j^*(t,y,p) = 0 \text{ if } t \in \text{ Maintenance hours for } j \tag{8.12}$$

The optimized values of $q_j^*(t,y,p)$ must also respect any internal transmission constraints within the utility's transmission system[22].

Note that since the hourly availability of an individual generating unit is uncertain due to the random nature of forced unit outages, the calculated values of $q_j^*(t,y,p)$ are expected values, and must reflect the forced outage rates of all generating units. Hence, the production cost algorithms usually contain an explicit recognition of forced outages at generating units. Much of the work on production cost programs over the past several years has focussed on improving the probabilistic representations of forced outages. Here, we do not explicitly model the forced outage rate forj, other than to note that the expected hourly production values $q_j^*(t,y,p)$ must reflect the expected actual availability of the individual units.

Remembering that the total annual fixed cost of the new units is equal to the sum of the annual fixed operating cost and annual capital cost, we now restate the plan index using the optimized hourly production levels $q_j^*(t,y,p)$, and the number of new generating units added of type m $[\delta_m(p,y)]$:

$$PI(p) = \sum_{y=1}^{Y} \frac{1}{(1+r)^y} \sum_{m=1}^{M} \delta_m(p.y) \times (foc_{N+m} + cc_{N+m})$$
$$+ \sum_{y=1}^{Y} \frac{1}{(1+r)^y} \sum_{allt} \sum_{j=1}^{N+M} hc_j(q_j^*(t,y,p)) \quad (8.13)$$

The monopoly utility will then identify the optimal plan p^* that produces the lowest value of $PI(p)$, and thus, the lowest total cost for meeting the electric energy needs of the utility's captive customers, subject to the constraints of Equations (8.9) through (8.12).

Many sophisticated software programs have been written to determine the optimal plan p^*, given the input values of the existing system data and the costs of the new generating units. We can recapitulate our generalized problem formulation by summarizing the input and output parameters of the planning process for a monopoly utility:

Input Parameters

For all hours t	:	$d(t)$
For all years y	:	$PL(y), RM(y)$
For existing units $(j = 1, 2, ..., N)$	:	$k_j, hc_j(q), mor_j, for_j$
For new units $(m = 1, 2, ...M)$	:	$k_{N+m}, hc_{N+m}(q), mor_{N+m},$ $for_{N+m}, foc_{N+m}, cc_{N+m}$

Typically, the planning processes also require input data on the utility's financial structure, in order to calculate the capital cost cc_{N+m} and discount rate r.

Output

For each new generating alternative m	:	$\delta_m(p^*, y)$ Quantity of new units of type m added through year y in the optimal plan p^*
For all existing generating units j	:	q_j^* Optimized hourly output of existing generating unit j in the optimal plan p^*
For all new generating alternatives m	:	q_{N+m}^* Optimized hour output of new generating unit m in the optimal plan p^*
For all years y	:	
Total operating costs of all units	:	$OC(p^*, y)$
Total fixed costs of new generating units	:	$FC(p^*, y)$

Over the years, many large and sophisticated computer programs have been developed to identify the optimal plan p^* from the input parameters described above. The most comprehensive of these planning models may includes many submodels, such as:

- an input module to collect, preprocess and check the large amount of input data required for the planning process;

- a reliability module that calculates the value of $LOLP(p, y)$ for each year y in plan p and compares this value to the specified reliability criterion $LOLP_{crit}$ (this model may only calculate and compare the installed capacity to the reserve margin criterion);

- a financial model that calculates the annual revenue requirements for a new generating project, based on the project's capital cost and construction spending curve, the utility's cost of debt and equity, the utility's target capital structure, the book and tax life of the project, the income taxes paid by the company and the specific ratemaking treatment used by the utility[23];

- a load forecasting model that uses historical demand data, forecasted population and economic trends, and the effects of planned energy efficiency programs to forecast values for $PL(y)$ and (occasionally) trends in $d(t)$;

- a production cost simulation routine that identifies $q_j^*(t, y, p)$;

- a generation expansion module that simulates and evaluates alternative plans, identifies the plan p^* with the lowest total net present value of revenue requirements, and reports the generation projects that are included in that plan (the values of $\delta_m(p^*, y)$ for each alternative m); and

- an output module that summarizes the output of the various modules in more convenient formats.

These models can become extremely large and complex, with many additional features (such as a probabilistic simulation of future loads, or a simulation of the entire utility's financial statements, price expectations and forecasted earnings). Over the years, a large amount of research has gone into developing improved production cost models and optimized generation expansion algorithms.

The output of this process is the utility's long-term strategic resource plan. For our example, a proposed generation project of type m would be pursued if it was included in the optimal system-wide generation plan p^* (i.e., if $\delta_m(p^*, y)$ were nonzero for any year y).

This planning process has many names, including "integrated resource planning" and "least cost resource planning". For many vertically integrated utilities in the United States, this planning exercise is a periodic event, with considerable resources devoted to the planning process. The resulting plans are often filed with the utility's regulators as an indication of the utility's long-term plans for generation and transmission development. In many states, the approval of utility's resource plan is the first step towards approval of the specific generation projects contained in that plan.

Transmission planning in monopoly electric markets

We now consider how the transmission planning flows from this generation planning process. In a monopoly electric market, the transmission function is carried out by the vertically integrated utility. As an integral part of such a utility, the transmission function is to physically integrate the utility's generation plants and its captive customers. A hypothetical mission statement for the Transmission Department of a vertically integrated utility might read as follows:

> Working with other departments in the company, the Transmission Department will plan, develop and operate a transmission system that transports electric energy from our generators to our customers in accordance with the company's least-cost integrated resource plan, while meeting applicable reliability criteria.

Note that this mission statement embodies several concepts specific to monopoly electric markets:

- The statement includes words such as "our generators" and "our customers", and does not refer to a separately priced energy transportation service. Transmission is only one of several functions in the production and delivery of the final product to the utility's captive consumers. The transmission system is planned and developed as part of an integrated electric system, according to company-wide strategic resource plan. In the United States, most utility resource plans have traditionally emphasized identifying the optimal generation resources. This focus was motivated by the high cost and siting restrictions associated with central station coal and nuclear plants. Such plants were historically the "technology of choice" for many United States utilities and their costs dominated the utility planning process. For vertically integrated utilities in the United States, generation plans often drive transmission plans.

- As a results of this integrated planning, new transmission investments are financed as part of the utility's overall investment program. Explicit prices for transmission service are neither developed nor needed, since the bundled rate paid by captive customers financially support all of the utility's costs, including the cost of transmission.

Thus, we see that the transmission system is planned as part of the total integrated electric system. The generation planning process discussed above represents the transmission improvements required for a specific generation project as an integral part of that project. Thus, the optimal transmission plan is the transmission component of the optimal system plan p^*. We will see how this close planning link between generation and transmission is supplanted in competitive markets with the separate generation and transmission prices.

Under this planning paradigm and "rate-of-return" regulation, the monopoly utility recovers, through bundled rates[24], their cost of producing or buying energy for the captive consumers and the cost of transmitting and distributing that energy to those consumers. These monopoly rates included an allowance for the return on the utility's equity investments in generating plants and transmission facilities. Thus, the financial returns for the generation sector were set mostly by regulation. For the monopoly utility, exogenous reliability criteria, centralized cost assumptions and regulated rates drive the generation planning process. Prices for electric energy are the *result* of the planning exercise, not the starting point.

In short, the franchise right to be the sole seller of electric energy within the franchise area is inextricably bound to the obligation to provide reliable service and thus to the utility's generation and transmission reliability criteria. These criteria in turn drive the utility's investment evaluation process. The normal profit maximization objectives of a competitive enterprise are thus replaced by the prudency and reliability objectives of a regulated public service monopoly. New generation and transmission plant is built when reliability criteria demand the investment, not when competitive prices justify it.

We can now see that system planning in a monopoly electric market is:

- driven by exogenous planning reliability criteria derived from the utility's public service obligation;

- focussed on identifying the set of resources that meet the reliability criteria at the lowest net present value cost to the utility ratepayers;

- vests all resource decisions with the utility and other participants in the resource planning process; and

- tightly couples the utility's generation and transmission investments.

In such a planning atmosphere, "following the rules" can become more important than obtaining the best outcome, particularly if following the rules entitles a participant to an equal sharing of the actual system reliability[25]. Moreover, if the financial penalties for failing to meet the rules are relatively low, the only deterrent to leaning on your neighbors is an honor system[26]. A system of prescriptive planning criteria, shared reliability and light financial penalties for failure to actually perform may be appropriate for regulated monopolies but is probably susceptible to the commercial pressures of a competitive marketplace. Participants in competitive markets will be motivated by commercial rewards and penalties, not peer pressure.

SYSTEM PLANNING IN COMPETITIVE ELECTRIC MARKETS

With the introduction of competition in generation, the overall system reliability criteria used by a public service monopoly must be reviewed. Specifically, we believe that overall system reliability criteria must be separated into generation-related reliability criteria and transmission-related reliability criteria. We also consider how each of these types of criteria are affected by the move to competition in generation.

We first define "generation-related reliability" as:

> Generation-related reliability is that portion of the regional system reliability directly associated with the availability of generating capacity, assuming unlimited transmission capacity.

In other words, the generation-related reliability of a region can only be improved by increasing the generating capacity that is available to serve that market[27]. Criteria regarding generation- related reliability would determine the quantity of regional generation that is required to serve load with a specified probability, assuming that sufficient transmission capacity within the region exists to allow all generators to operate freely.

Examples of generation-related reliability criteria include:

- An operating requirement that enough operating reserve capacity be available and identified to replace the single largest unit within 10 minutes of a loss of that unit;

- A short-term maintenance planning requirement that the utility not take the two largest generating units out for maintenance at the same time; and

- A capacity planning requirement that the utility have a certain installed capacity reserve margin (e.g., 22%) over the expected annual peak load.

Conversely, we defined "transmission-related reliability" as:

> Transmission related reliability is that portion of the regional system reliability directly associated with the availability of transmission capacity, assuming unlimited generation capacity.

Thus, criteria for transmission-related reliability would determine the quantity of local transmission that is required to serve load with a specified probability, assuming that sufficient generating capacity always exists to meet that load. Examples of transmission- related reliability criteria include:

- A substation design criterion that all substations with load in excess of a certain amount must be supplied by two separate transmission lines; and

- An operating criteria that the loss of a single transmission line does not force non-interruptible consumers to be disconnected.

This dichotomy of reliability criteria is necessary to understand how introduction of competition affects these criteria. We will now discuss how generation-related reliability criteria are complemented or supplemented by market incentives in competitive electric markets. Transmission-related reliability criteria are less directly affected.

Generation planning in competitive electric markets

With the introduction of competition, generation planning shifts from vertically integrated monopoly utilities to decentralized competitive generators. In competitive electric markets, generation companies invest in new generation plant when it is profitable to do so (just as in any other competitive market).

Since competitive generators do not have public service obligations or rights, they will be better motivated by profit concerns.

The usual measure of profit is return on investment. A generating company evaluating a new plant will forecast the expected revenues (from the sale of electricity services in the competitive market), net of expected operating costs (fuel, staffing, property and income taxes, interest on debt associated with the project, etc.) This stream of future net revenues will be compared to the net investment in the plant. If the resulting internal rate of return for the project exceeds the company's *hurdle rate* for such investments, the company will likely pursue the project. If the forecasted internal rate of return for the project does not exceed the hurdle rate, the project will likely be deferred. In competitive electric markets, the timing of generating investments now depends on projected returns, not anticipated reserve margins.

Before we consider how a competitive generation company would evaluate various projects, we briefly discuss some of the conditions for effective competition in generation within a regional electric market. While this discussion is qualitative rather than quantitative, it does illustrate some of the conditions that we believe are required for such a market

Suggested conditions for a workably competitive electric market. We propose that the following conditions are necessary conditions for workable and effective competition in the generation sector of a regional electric market:

- A sufficiently workable "open-access" regime is in place for regional transmission.

- A large number of retail electric consumers are free to choose their electric supplier (either directly or through equivalent financial instruments)[28].

- Within the regional market, entry and exit barriers for producers are small. In other words, sufficient sites for new generating capacity exist to allow new entrants to develop generation projects.

- The market is large enough to support several energy producers can achieve the minimum size required for a competitive entity.

- Generating capacity is not concentrated in too few generation companies.[29]

- Sufficient transmission capacity exists within the regional market so that most load can be served by a reasonably diverse mix of generators in most hours of the year and most generators can operate most of the time.[30]

This brief discussion of the conditions for workable competition in electric markets is primarily intended to stimulate our readers' own thinking about the nature of competition in electricity. Electricity is unlike most other commodities, since it is instantaneously transported and distributed over a shared, non-switchable network (the transmission and distribution grid) and cannot be stored in transit. Furthermore, the capital intensive nature of generation and transmission facilities, and their more complex siting issues means that new entrants may not be able to respond as rapidly to higher prices as new entrants in other industries. Hence, the usual criteria for effectively competitive commodity markets may need to be reconsidered for electric markets. As competition is introduced into more electric markets, and as competitive generation companies develop and implement their individual strategies, the conditions for workable competition will become clearer.

To meet our definition for a workably competitive electric market, a regional transmission network does not have to provide completely unconstrained access to all producers and consumers for all hours. Some transmission constraints are acceptable (and appropriate), as long as no single supplier or consumer can exercise market power for extended periods. Typically, the transmission network of a large vertically integrated utility will have multiple generators serving consumers throughout the region, and multiple paths to connect them. Hence, a workably competitive electric market is possible in these geographic areas, if several competing generation companies own and operate the generating plants of that utility[31].

We now consider how spot market energy prices (particularly during period of energy scarcity) can complement or supplement generation-related reliability criteria.

Generation reliability and spot market pricing in competitive electric markets. In a competitive market, the best measure of a company's success is its actual performance. The best motivators for actual performance are monetary: financial rewards for good performance and financial penalties for poor performance. Furthermore, actual performance in a commodity market is best measured by actual performance relatively to contractual obligations, with financial penalties extracted for any failures to deliver the product at the contractually agreed time. This focus on actual performance, rather than compliance with prescriptive rules, is facilitated by the existence of a spot market for the underlying product, where physical failures to perform can be remedied through spot market purchases or sales.

Other texts have discussed the theory and organization of spot markets, and the absolute need for effective spot markets to allow competition in generation. The existence of a spot market in electricity allows generators and consumers to share a common transportation network (the interconnected grid) and allows imbalances between contracted and actual production (or consumption) to be resolved physically and settled financially[32].

With the existence of a spot market, generation-related reliability criteria can be complemented or supplemented with market pricing mechanisms. By pricing spot market transactions to explicitly reflect the value of generation reliability (particularly during periods of generation scarcity), the decisions about how best to insure that reliability can be left to the competitive market participants who may transact at those spot market prices. The competing energy producers and consumers will assess how to best minimize their net costs, including the costs of any spot market transactions and any penalties for actual failure to perform. The commandment "you shall maintain a 20%

reserve margin" is replaced by the warning "the financial penalty for failing to produce enough energy during generation shortages is US $3000 per MWh". The cost and risk of the penalty can now be balanced by the competitive generator against the cost and risk of adding capacity to avoid the penalty.

Mechanisms for explicitly pricing generation reliability can take many forms, and have varying time horizons for establishing performance requirements, measuring actual performance against those requirements and assessing penalties. At one extreme, relatively long time periods can be used to measure an installed capacity requirement and assess financial penalties for failing to meet that requirement (e.g., pay $5 each month for every kilowatt shortfall in the installed capacity requirement). Other methods for monetizing reliability would typically use shorter time horizons for measuring performance and assessing penalties. For instance, on- peak spot market energy prices may contain an adder to the normal spot market price. This reliability valuation mechanism is used in Argentina. Spot market energy prices may also be exogenously increased during periods of *expected* generation shortages, as is done in England. Finally, spot market energy prices can be left entirely to competitive forces unless demand is actually curtailed, whereupon spot market prices increase dramatically. For example, in Victoria, Australia, spot market energy prices can rise to over US $3000 per MWh when load is being curtailed, but there is otherwise no reliability adder.

In a competitive electric market, competing generators will make decentralized commercial decisions about how to meet their contractual obligations to produce or acquire energy. These decisions will be made in light of the generator's individual expectations regarding future spot market prices and performance penalties. In essence, "top-down" generation planning is replaced by "bottom- up" reactions to expected market prices (with some flavoring from the spot market reliability pricing mechanisms). The decision to build or not build new generating capacity is now made purely for commercial reasons. We can now discuss generation planning in competitive markets.

Evaluating a generation project: the competitive generator's perspective. Evaluating a generating project in the competitive case is much more straightforward than in the monopoly case, but perhaps harder (since the stakes to the investors are typically higher). Instead of starting with exogenous generation-related reliability criteria, and identifying an optimal generation expansion plan after considering many resource plans that meet the criteria, the competitive generating company will evaluate the project by simply looking at its expected returns. If the project's capital and operating costs are sufficiently low compared to the expected price of the project's output, the project's returns will meet or exceed the company's target investment criteria, and the project would likely be pursued.

For our formulation of the investment evaluation problem, we assume that the competitive generation company will act as a price taker and will not assume any change in the price of electricity due to its individual investment. Thus, each generation project can be evaluated individually. Although the competitive generator will consider costs and revenues for extended periods (e.g., 20 years), the decision evaluation process often only considers near-term investment opportunities. Unlike the monopoly utility, the competitive generation company may not assume that its immediate actions directly affect the future price of electricity or the future value of investments. Here, we assume this behavior.

Finally, we also assume that:

- an ISO operates the grid and coordinates the final dispatch of all generators,

- a spot market resolves all final imbalances between planned and actual energy production, and extracts penalties for any failure to perform,

- all generators are owned by a diverse set of competing generation companies who each act as a price taker and sell their output at unregulated prices,

- the primary market for energy and options on energy (capacity) determines most of a generator's revenues, and

- the competing generators are responsible for their own commercial decisions, including the scheduling and hourly dispatch of their generation.

The competitive transactions determine the prices at which the physical energy production from a generator (energy) and options on that production (capacity) can be sold. We denote the competitive unit price of energy in any period t as $p(t)$, with a further assumption of a transparent spot market price that is identical for all producers and consumers. We also assume that the value of generation reliability is represented through an increase in the spot market energy price during periods of energy shortage[33]. Hence, during periods of generation scarcity, $p(t)$ for energy would be expected to increase, perhaps dramatically[34].

The analysis of the project investment next involves a comparison of the expected revenues from the project against the project's operating cost. The net revenue generated by the project is finally compared to the project's capital cost, to yield an internal rate of return for the project investment. If this internal rate of return exceeds the company's "hurdle rate" for similar investments, the project would likely be pursued. When considering several generating alternatives, each project with an acceptable internal rate of return would be developed, unless the company's funds were limited. For that case (the classical "capital rationing problem"), the company would develop the projects with the highest internal rates of return.

The problem of evaluating a generating alternative now does not include any explicit consideration generation-related reliability criteria, such as installed reserve margins. The competitive generating company will not invest simply to maintain the desired reserve margin (unless motivated by financial penalties for failing to do so), nor will it directly optimize its investment for the entire market. Rather, the competitive generator will optimize its own investment portfolio, seeking the highest returns possible on its investments.

To project these returns, the competitive generator must explicitly forecast the operation of the proposed project, and the price that would be received for the associated energy sales. Hence, much of the explicit analytical framework presented earlier regarding forecasts of total system operation is replaced with an effort to forecast the *price* of electricity.

Calculation of expected revenues in a competitive market. The expected revenue for the generating project in any period t is thus calculated as the product of the energy price in that period and the expected generation in that period. Mathematically, we have (in the simplest form):

$$Rev_m(t,y) = p(t,y) \times q_m(t,y) \qquad (8.14)$$

where $Rev_m(t,y)$ is the expected revenue in period t of year y, $p(t,y)$ and $q(t,y)$ are unit prices and production quantities, respectively, and we have retained the subscript m to identify the generating project under consideration.

The generation company must now forecast both $p(t,y)$ and $q_m(t,y)$ as a function of the market structure, the company's pricing strategy, the expected total demand, and the expected offer prices of other generators. The determination of $q_m(t,y)$ is analogous to the production cost problem in the monopoly market, with the change that the physical production of the unit m can now respond to the regional spot market price.

Forecasting any competitive market price is challenging, and projecting the price of electricity is particularly difficult. The physical characteristics of electric energy (particularly the inability to easily store it) lead to high price volatility, and increase the forecasting uncertainty. These challenges are compounded by the relatively recent experience with competitive electric markets, the variations in valuing generation-related reliability, and the range of forecasting techniques and philosophies. In the monopoly market, the conceptual equivalent of $p(t,y)$ was the system marginal cost that could be derived from the production cost simulation. No explicit recognition of the dynamic nature of competition was required, since no competitors existed. In the competitive market, the price forecast should reflect the expected dynamic behavior of all market participants.

Another complication in forecasting electricity prices is the difficulty of predicting the future demand of electricity. Forecasting macroeconomic trends and overall demand levels is difficult at best. Monopoly utilities have put considerable effort into developing their long-term load forecasts, and yet errors in those forecasts have often led to excess generating capacity and corresponding stranded costs. In competitive electric markets, forecasting total demand is probably one of the most difficult components of the price forecasting process.

The competitive generation company must also consider the ways in which energy can be sold. In a competitive market, generation companies will sell energy through a variety of contracts. Some sales will be through longer-term contracts (perhaps up to five years), to specific large retail consumers or retailers (who then sell to specific consumers). Other sales will be to a fluctuating mix of smaller retail consumers who have elected to buy from the generator's marketing division. Still other sales will be short term wholesale transactions with other generation and retailing companies, including spot market sales on a very short notice.

Forecasting the expected revenue from a generation project (or a mix of generation projects) requires that the generating company forecast prices and quantities for the types of contracts that are contemplated. The specific mixture of contracts and their relative prices will depend on the expectations of the individual market participants, and their risk profiles. If most of the plant's output is contractually sold before the investment is made, then those term contracts will dominate the generator's revenue. Projects with fewer long-term contracts will have a greater emphasis on the spot market, and expected spot market prices.

In our formulation, we denote the type of term contracts (and their associated quantities and prices) through the use of the index k. We also explicitly recognize net sales to the spot market as the difference between the generator's actual output and contracted sales[35]. Note that the relative prices and quantities of each type of contract will depend on the expectations of the individual market participants and their risk profiles. While the spot market price will be the same for all buyers and sellers, the term contract prices will not.

Furthermore, the price forecast will usually consider different energy prices in different time periods. Since most electric markets have limited capabilities for storing electric energy, energy prices can vary dramatically over time. For instance, in Victoria, Australia, hourly spot market energy prices can range from less than US $10 per MWh during minimum load periods to over US $3000 per MWh during periods of generation shortages. Bilateral contracts between individual buyers and sellers (whether physical or financial) often contain at least an on-peak/off-peak pricing component for energy. Thus, a rigorous evaluation of a generation project will require a forecast of energy prices in multiple time periods.

For the same reasons, the expected output of a generating unit will vary in time. During off-peak periods, a generation company may elect to not operate some of its resources, and to fulfill its contracts with energy purchased from other generators or the spot market. Higher cost resources (such as oil-fired peaking units) might only operate during peak hours on weekdays, and perhaps not at all during certain periods of the year.

To reflect this time sensitivity of both energy price and unit output, we have denoted the time periods of interest through the index t, with a total of T periods within each year of the evaluation. The most rigorous price forecast might include a projection of hourly or subhourly spot market prices. Here, our summations over time simply include all periods of interest. Evaluation of a generation project will require a forecast of both price and output levels for each of these time periods, and each of the contract types.

Finally, a fully rigorous evaluation of a generation project would consider variations in quantities and prices that might arise from transmission constraints within the regional electric market. If transmission from the generator is or could be constrained, the generator's prices and quantities might be affected. For instance, the spot market price received by a generator in a constrained exporting region might be lower than spot market prices in other regions, and this price decrease might be also reflected in the bilateral market. The forecasted prices and quantities should include an appropriate recognition of the potential transmission constraints within the market. In essence, many of the internal system constraints modeled in the production cost problem are now internalized for the competitive generator in the form of variations in p and q.

Hence, we recast Equation (8.14) to reflect the various types of contracts, and the residual spot market transactions:

$$Rev_m(t,y) = \sum_{\text{all } k} p_c(k,t,y) \times q_c(k,t,y) + p_s(t,y) \times \left[q_m(t,y) - \sum_{\text{all } k} q_c(k,t,y) \right] \quad (8.15)$$

where:

$p_c(k,t,y)$ = Contract price for contract k during period t in year y
$q_c(k,t,y)$ = Contract quantity for contract k during period t in year y
$p_s(t,y)$ = Spot market price during period t in year y
$q_m(t,y)$ = Output of generator m during period t in year y

For our analysis, we assume that an energy price forecast for all periods t of future years y is available for the evaluation of the project. This forecast would include a forecast of contractual prices $[p_c(k,t,y)]$ and projected contractual quantities $[q_c(k,t,y)]$[36]. Thus, we assume that all prices, and contractual quantities at those prices, are known for all values of k, t and y.

Finally, for simplicity, we assume that no separate market for capacity or other services exists. This constraint could be relaxed by explicitly pricing multiple products and summing project revenues over those products. However, since most of the revenue from a generator will be for sales of either actual output (energy) or options on that output (capacity), we can reasonably represent the project revenue stream by considering only physical energy sales (recognizing that some portion of the physical energy price should include an option value or capacity component). Hence, in our formulation, the only product sold by the competing generators is physical energy actually delivered to the grid.

Forecasting sales quantities. In our simplified analysis, we assume that the company will operate the generator at a level $q_m(t,y)$ that maximizes its profit, given the exogenous spot market price $p_s(t,y)$ and the hourly production cost function indicated earlier. In a fully competitive market, the competitive generator will find, for each hour, the output level at which the generator's marginal cost equals the spot market price,

$$\forall t, y: \quad \frac{\partial hc_m(q)}{\partial q_m(t,y)} = p_s(t,y) \quad (8.16)$$

subject to the capacity constraint,

$$\forall t, y: q_m(t,y) \leq k_m \tag{8.17}$$

the maintenance constraint,

$$\forall y: q_m(t,y) = 0 \text{ if } t \in \text{ Maintenance hours for unit } m \tag{8.18}$$

and any other operating constraints (such as minimum run times or minimum down times).

As before, we denote the *expected*[37] optimum output level for hour t in year y as $q_m^*(t,y)$. For hours in which the expected spot market price is too low, the value of $q_m^*(t,y)$ will be zero.

We can now recalculate the expected optimized revenue as $Rev_m^*(t,y)$ using the optimized production levels $q_m^*(t,y)$:

$$Rev_m^*(t,y) = \sum_{\text{all } k} p_c(k,t,y) \times q_c(k,t,y) + p_s(t,y) \times \left[q_m^*(t,y) - \sum_{\text{all } k} q_c(k,tky) \right] \tag{8.19}$$

This revenue is then compared to the expected operating costs for the optimized output level, the fixed operating costs (staffing, property taxes, etc.), and the income taxes due for the project, to yield the net free cash that could be returned to the project's investors:

$$NFC(y) = \sum_{\text{all } t} [Rev_m^*(t,y) - hc_m(q_m^*(t,y))] - [focm + trans_m(y) + it_m(y)] \tag{8.20}$$

where:

$NFC_m(y)$ = "net free cash" in year y generated by the project,
$hc_m(q)$ = hourly production cost for unit m, at hourly output level q,
$focm$ = annual fixed operating costs for unit m
$trans_m(y)$ = expected transmission cost for year y and
$it_m(y)$ = income tax associated with the project for year y[38].

Note that in the competitive market, the cost of transmission service is usually considered as an annual operating expense paid by the competitive generator, rather than a modifier to the capital cost of generation project (as was done in the monopoly utility case).

Before dividends can be distributed to the project's equity investors, the net free cash would be further reduced by any debt service (interest and principal payments) for the project. In this example, we assume that the project is financed entirely with equity and that all of the net free cash calculated above can be distributed back to the equity investors in the project.

Finally, this net free cash is compared to the capital investment required to develop, license and construct the project. The discount rate at which the net present value of the future free cash stream equals the project's capital investment (denoted as $Capinvest_m$), is termed the project's *internal rate of return* (irr_m). Numerically, irr_m is the discount rate that satisfies the following equation[39]:

$$Capinvest_m = \sum_{y=1}^{Y} \frac{NFC(y)}{(1+irr_m)^y} \qquad (8.21)$$

where Y is the business horizon over which the project is evaluated (e.g., 20 years). The values of $NFC(y)$ and $Capinvest_m$ are given; hence, irr_m can be determined. The best generating projects, with the highest free cash flow for a given capital investment, will have the highest values for irr_m.

The final evaluation step might be to compare the individual project's value of irr_m to the company's minimum acceptable rate of return for comparable projects, the company's other investment opportunities, and qualitative factors that have broader considerations for the company.

In considering the evaluation of a generating project in a competitive market, we note that many analytic tools are available to the competitive generating company. Tools such as option theory valuation, portfolio analysis and decision analysis have been used to evaluate investment decisions in competitive markets. However, these analytic tools have typically been applied to markets for goods that are storable and widely transportable. Electric energy is often neither. The art of evaluating generation investments in a competitive market is new, and relatively little quantitative data are available on competitive electric prices and their volatilities. Hence, the valuation of generating assets in a competitive market is an emerging art, not a rigorous science.

Finally, as regional electric markets initially introduce competition, competitive generating companies will often base their market entry investment decisions on very broad strategic considerations and qualitative expectations about future business prospects, rather than a narrow analytic valuation using rigorous formulae. For instance, a competitive generation company may have a strategic reason for gaining an immediate market presence, and may thus value near-term investment opportunities more highly. During the transition period as electric markets introduce competition, generation investment decisions may not be exclusively based on analytic formulations.

Of course, this discussion greatly oversimplifies the complex nature of decision making in a competitive business environment. The art of managing a competitive business has never been distilled into a rigorous science, and the valuation of generating projects in a competitive environment is no different. Competitive generating companies may not always evaluate a given generating project in the rigorous manner described above, nor will two companies evaluate the same project identically. Considerations such as regional market share, early entry opportunities, synergies with other projects, and corporate prestige will affect the evaluation of an individual generation project. Companies may pursue projects whose projected *irr* values are below the target *irr*, and they may neglect projects with attractive *irr* values. In competitive electric markets, the process for evaluating a generating project will be a decentralized process performed by individual companies with their own views and biases.

Discussion of differences. We can now summarize the two most important differences between the processes of evaluating a generation investment in a monopoly market and in a competitive market:

- *Spot market prices and performance penalties complement generation-related reliability criteria.*

    For the monopoly case, the key exogenous inputs are the peak demand forecast and the reliability criterion. For the competitive case, the key exogenous input is energy price. This difference is critical, as the future demand for electric energy and the desired generation reliability may not be treated explicitly in competitive markets, except through their impact on expected prices.

    This conclusion represented a fundamental shift in the basic ground rules of the electric supply industry: *In competitive electric markets, generation-related reliability criteria are complemented or supplemented by energy prices, particularly energy prices during generation shortfalls.* Competitive generation companies are best motivated by commercial incentives, not by exogenous reliability criteria that may not reflect their own commercial perceptions. In a competitive market, competing generation companies must be allowed to "place their bets" as they see fit, or to not place them at all. The externalities of a shared network are best internalized through appropriate prices, not mandatory criteria.

- *The commercial risks of an individual generating project lie with investors, not ratepayers*

    For the monopoly market, an individual generating project is evaluated as part of the utility's least-cost resource plan. This resource plan will be designed to minimize the cost of meeting the demand for electric energy over the planning period. The cost calculation includes the utility's regulated profit level (return on equity). Therefore, the utility will be both prevented from making above-normal profits associated with a good investment, and protected from below-normal profits associated with a prudent, but poor, investment. Historically, monopoly utilities in the United States have been allowed to recover most prudent investments, including a regulated equity return on the equity portion of those investments. In essence, profits have been tightly regulated, and prices to the final consumers have been at least partially determined from those regulated profit levels.

    In competitive electric markets, energy prices and profits are set through competition and financial risks (and rewards) are shifted to the competitive generation companies. The generation reliability risk still resides with all consumers (and may be greater or less than in monopoly markets), but the investment risk of owning generation assets is now clearly with investors rather than ratepayers.

In summary, there is no textbook formula for evaluating a specific generating project in a competitive market, nor will one likely emerge. The more structured and rigid process applied in monopoly markets will be replaced by decentralized and vastly differing decision processes of individual companies. Some decisions (and decision processes) will prove more successful than others. Good (or lucky) decisions will produce the best outcomes for investors, while the worst (or unluckiest) decisions will suffer in the competitive market. In a competitive market, the investment goals will be success, rather than prudence. The question of which companies can best manage the risks of fully competitive market will be continuously answered by the marketplace.

Transmission Planning in Competitive Electric Markets

Having reviewed generation planning under competition, we now come to the physical and commercial interface between generation and consumption: transmission. Unlike most products, electric energy requires a unique and dedicated set of physical facilities for transporting the final product from the producer to the final distribution system. These facilities constitute the transmission network.

This section explores the role of transmission in competitive electric markets, implications of these differences, and the challenges of transmission development under competition. A framework for developing new transmission facilities in competitive electric markets is proposed. This framework classifies new transmission projects as either network transmission projects (used by many electric producers and consumers and centrally planned) or dedicated transmission projects (benefiting specific users and developed in response to market signals). Specific, real-world examples of a network transmission project and a dedicated transmission project are given, to illustrate the development process and commercial arrangements for that type of project.

The role of transmission in competitive electric markets. With the introduction of competition in generation, the role of the transmission function is dramatically changed. In competitive electric markets, electric consumers are free to enter into unregulated contracts with electric producers. As discussed earlier, these term contracts, and the accompanying spot market for resolving imbalances, are now the financial basis of investments in generating plant.

For both term contracts and spot market transactions, energy buyers and sellers "meet" at the transmission system to consummate their competitive transactions. This meeting can be either:

- both physical and financial, in which the seller delivers physical electricity to the buyer at a defined location or

- purely financial, in which the buyer and seller settle a financial transaction (such as a "contract for differences" pegged to spot market price).

All competitive electric markets require that the transmission system function as the meeting place for these transactions. In these markets, the function of the transmission system is to impartially[40] enable competing transactions in electric energy.

These considerations lead to a new hypothetical mission statement for the transmission system in a competitive electric market:

> The Transmission Function will facilitate a competitive electric market by impartially providing energy transportation services to all energy buyers and sellers, while fairly recovering the cost of providing those services.

The transmission business unit now must (1) provide enough transmission capacity to allow competitive market transactions for electric energy, (2) allow impartial access to and use of that transmission capacity, and (3) charge unbundled transmission prices that recover the cost of constructing and operating the transmission system while sending appropriate price signals to generators and consumers regarding their use of the grid.

This revised mission is now suitable for an independent transmission enterprise. In fact, independence of the transmission system operator is a hallmark of competitive electric markets. *Every* competitive electric market features an "open access" regime for transmission. These markets have implemented a variety of measures to insure impartial access to the electric transmission system, including:

- development of open access transmission tariffs that require vertically integrated utilities to provide unbundled transmission service to third parties;

- "codes of conduct" for the transmission departments of the vertically integrated utilities, to insure that all transmission users receive impartial access to the grid;

- creation of independent system operators ("ISOs") and regional transmission groups ("RTGs") to further insure impartial transmission access and pricing; and

- divestiture of vertically integrated utilities.

Regardless of the specific measures implemented, the common characteristics of all open access regimes are impartial access for all producers and consumers, and a need for unbundled transmission prices. The latter point is especially critical, since transmission services will be paid for separately from energy production. Unbundled transmission prices are essential for competitive electric markets.

Perhaps the greatest challenge associated with introducing competition in generation is developing transmission prices that both (1) *promote efficient use and expansion of the grid* and (2) *facilitate effective competition in generation*. Literally volumes have been written on the theory and practice of transmission pricing to accomplish both of these goals, and we will not repeat those discussions. However, we will present some philosophical issues in transmission pricing for our readers' consideration (without presenting definitive answers to these questions):

- Most proposals for transmission pricing incorporate at least some degree of efficiency in transmission pricing. Efficient transmission prices attempt to assign the marginal costs imposed by particular uses of the transmission grid to the parties (energy buyers and sellers) using the grid in that manner[41]. The objective of these prices is to send the appropriate price signals to generators and consumers regarding their decentralized and competing decisions regarding energy production and consumption.

In particular, uses of the grid that cause or contribute to transmission constraints are typically assigned higher prices than are uses of the grid that avoid or relieve constraints. *Peak- load pricing* is an example of a transmission pricing methodology that attempts to assign greater transmission costs to producers and sellers that use the most constrained facilities.[42]

- While the broadest goal of efficient transmission pricing ("insure that generators and consumers pay the costs caused by their uses of the grid") is usually considered important, the results of such pricing may work against another goal for the transmission grid ("promote effective competition in the generation market"). By definition, efficient transmission pricing will mean that differently situated users (particularly generators) will pay different transmission prices for their uses of the grid. Thus, efficient transmission pricing will *necessarily* place some generators at a competitive disadvantage relative to other generators located in more favorable sites. "Efficient transmission pricing" and a "level playing field for all generators" (for a large enough electric market) are mutually exclusive.

This tension between economic efficiency in transmission and effective competition in generation is real, and should not be underestimated. We believe that this tension lies at the heart of many of the most controversial aspects of transmission pricing within regional ISOs in the United States. The art of designing transmission prices that promote both effective competition and efficient grid expansion is still being developed.

- In virtually all proposals for efficient transmission pricing, the cost of using the grid for certain transactions would increase as the transmission facilities required for those transactions became more constrained. Once the cost of using the existing grid for particular transactions exceeds the cost of expanding the grid to relieve the constraints, the competitive market participants in those transactions would (presumably) agree to financially support the grid enhancements. Perfectly efficient transmission prices should thus allow the transmission system to be developed in response to the commercial decisions of competitive market participants. The electric transmission system could thus be developed in a similar manner to generation ("build more capacity when the market demands it") and much as other transportation networks (especially gas pipelines) are developed. However, we note that the path from efficient transmission pricing to decentralized grid development has a crucial link still being developed and debated: enforceable property rights associated with electric transmission contracts. Other energy commodities (coal, oil and natural gas) and transportation networks (roads, railroads, waterways, gas pipelines, etc.) often face the same tension between effective competition in the underlying commodity market and efficient development of the transportation network. However, other transportation networks have the advantage of effective property rights associated with their transportation contracts. A coal company wishing to ship coal by barge, rail or truck can have an enforceable contract for the delivery of their product to the desired delivery point. Furthermore, the coal company can either (1) use their transportation right (and deliver the product), (2) sell the contract to another party (who might transport a different commodity) or (3) simply let the contract lapse. None of these options would affect other users of that transportation network who have similar contracts[43]. Thus, the benefits from transportation contracts can be fully internalized to the contracting parties.

- However, similar property rights for electric transmission are much harder to develop. The "contract path" method for purchasing transmission service attempts to assign contractual property rights to energy buyers and sellers for specific injection and withdrawal points, but the nature of interconnected electric grids produces unique challenges. We do not know of any commercial framework for allocating transportation property rights over conventional alternating current networks that has been widely accepted as both efficient and workable. However, we are confident that any workable proposal for fully decentralized planning of the transmission system *requires* such a system of property rights.

- Even if grid expansion is still centrally planned, transmission pricing methodologies with various degrees of efficiency[44] (and associated complexity) can be developed and used. However, since competing generators will have different transmission costs (due to their different locations), they will have different benefits from new transmission facilities, and will thus have different incentives for developing any specific project. Generators experiencing transmission constraints can be expected to lobby for new transmission facilities that might relieve their constraints, while generators closer the load centers will likely be less favorably inclined towards any new transmission facilities that would increase their competition. We believe that for effective development of the transmission network, the process of identifying and planning new transmission facilities must recognize that (1) competing generators will have different interests in such facilities and (2) increased efficiency in transmission pricing will magnify those differences. Any regional transmission planning body must have clear and transparent criteria for developing new transmission capacity, particularly if the cost of transmission constraints is allocated unevenly among competing generators. We are not certain whether the need for such criteria (or, in their absence, the need for enforceable and tradeable property rights associated with transmission) has been adequately addressed by the emerging ISOs in the United States.

Features of efficient transmission pricing methodologies. We will not delve into the details of methodologies for efficient transmission pricing, other than to note that the transmission sector exhibits the following features that make efficient and workable transmission pricing mechanisms particularly difficult (in addition to the tension noted earlier):

- Major transmission investments (such as new transmission lines) are "lumpy", with significant economies of scale. Often, the marginal cost of new transmission capacity is very high for the increment of capacity that triggers a major new project, and very low for larger versions of that project.

- Transmission systems often exhibit supply curves that are discontinuous in time as well as scale. For example, the cost of new rights-of-way for overhead transmission lines may be orders of magnitude more expensive than historical costs, and perhaps simply unavailable. In certain instances, underground transmission lines may be required for new transmission capacity, to address environmental concerns. However, such facilities are 2-3 times more expensive than overhead lines. The marginal cost of transmission capacity may be much greater than the historical cost, creating a problem of initially allocating transmission capacity.

- The short-run value of transmission capacity is often "all-or-nothing" and extremely volatile. Early works on the spot market pricing of electricity observed that the correct short-run marginal cost for transmission service between two physical locations could be defined as the difference in short-run marginal energy prices at those points. For an unconstrained transmission system, the short-run marginal value of additional transmission capacity is simply marginal losses. Typically, this value is far too low to justify new transmission investments. When the system becomes constrained, the value of additional transmission capacity increases, sometimes dramatically. In the extreme case, demand may be curtailed due to transmission constraints. The short-run value of additional transmission capacity in these instances is the average curtailed consumer's opportunity value for electric energy. In many competitive electric markets, this value is often above US $1000 per MWh. Therefore, the short-run value of transmission can be extremely volatile, making efficient pricing difficult to implement and presenting grid users with substantial uncertainty regarding their transmission costs.

- Finally, the causes and effects of transmission constraints are often difficult to establish, with a corresponding difficulty in allocating the cost of those constraints. Allocating cost responsibility for transmission congestion can be difficult, particularly given the non-linearities of a transmission system operating in real time. The question of who caused a transmission constraint when the lights go out (or local energy prices rise dramatically) can be difficult to answer.

These features of transmission systems challenge the development of transmission pricing methodologies that are both economically efficient and commercially workable. While no methodology for efficient transmission pricing has been universally (or even widely) accepted and implemented, most transmission pricing proposal attempt to send at least some marginal cost prices signals to transmission users. Common mechanisms for providing these signals include:

- Interconnection charges for generators that reflect at least some portion of the cost of the transmission upgrades caused by the generator's operation;

- Various "use of system" charges that attempt to allocate the actual use of the transmission systems (and the associated costs) to individual producers or consumers; and

- Location based spot market energy prices that explicitly reflect the local value of energy and (hence) the cost of any transmission constraints.

All of these mechanisms attempt to quantitatively recognize geographic variation in the value of energy in order to influence the decentralized commercial decisions of energy producers and consumers.

Regardless of the exact form of transmission pricing, a new investment planning process is required for transmission in competitive electric markets. Recall that for monopoly markets, transmission and generation are integrally planned, and that the monopoly utility's planned transmission investments are identified as part of its least cost resource plan. With the introduction of competition, this explicit link between generation and transmission is severed, and must be replaced with a new set of criteria for grid development. The next section proposes such a framework.

Transmission Planning in Competitive Electric Markets. Since all energy producers and consumers within a regional electric market use the regional transmission system, expansions of that system will affect all users, sometimes unequally. Producers in low price areas of the regional market typically benefit from greater access to consumers, while producers in high price areas may see reduced revenues[45].

As noted earlier, the process for planning and executing those expansions is often contentious given the competing interests in transmission expansions. Many regional transmission groups ("RTGs") in the United States are finding it difficult to establish a transparent and impartial process for identifying and commercially approving new transmission facilities.

A transmission planning process that recognizes that different projects will have different beneficiaries can facilitate transmission planning, while promoting a competitive energy market and retaining the benefits of coordinated network planning. Therefore, we suggest that new transmission enhancements in a competitive electric market be classified into two types of projects:

318   POWER SYSTEMS RESTRUCTURING

- Network transmission projects that mainly enhance the overall capabilities of a regional[46] transmission network, and that correspondingly benefit many (if not most or all) regional energy producers and consumers, and

- Dedicated transmission projects that mostly benefit specific and highly identifiable producers and consumers[47].

Examples of network transmission projects would include reconductoring of a backbone transmission lines serving the entire regional market, upgrades to high-voltage substations at major energy "crossroads", and other improvements to the integrated regional network that are not readily associated with specific users. Examples of dedicated transmission projects would include the following:

- new transmission required to a generating plants to the regional grid (generator leads),

- new direct current[48] transmission lines that connect separate electric markets("energy toll roads"), and

- "deep" network upgrades that are required to integrate a new generator into the regional grid (since those facilities are only required by the new generator, even if they are geographically remote from the generator).

In our proposed framework, network transmission projects would be centrally planned for collective benefit of all regional electric consumers, with limited competition to build, own and maintain the facilities. The annual costs for new network transmission projects would be supported in a similar fashion to the cost of the existing integrated grid. In contrast, we believe that competitive commercial arrangements and incentives can be used to develop dedicated transmission projects, and that no central planning process is required for these facilities. We will now explore the specific details of how these types of projects might be developed.

Proposal for centralized planning for network transmission projects. Within a competitive regional electric market, we believe that the regional transmission network[49] should be cooperatively planned and developed for the benefit of all regional consumers, while maximizing the net social welfare gain from the transmission system. Various methodologies for identifying and evaluating such projects can be used, including calculation of fully efficient "shadow" transmission prices (even if such prices are used only for system planning).

This regional transmission planning process should include a transparent and impartial process for analyzing, enhancing and financially supporting the regional transmission network. Elements of this planning process would include the following:

- Central planning through a regional network planning body, with defined voting and governance procedures;

- Transparent rules for transmission access and pricing to impartially allocate the overall costs and benefits of the regional transmission network, including the costs of any transmission constraints;

- Impartial economic evaluation of proposed network transmission projects to insure that the projected regional benefits (in the form of greater reliability and reduced regional congestion) are greater than the projected regional costs; and

- Explicit, transparent and impartial criteria to evaluate the need for new network transmission projects.

The output of this planning process would be a set of approved transmission investments and a clear process for allocating and supporting their costs. This coordinated planning process will promote effective competition within the regional electric market, while continuing to compare the costs and benefits of new transmission investments (although the benefits may be more difficult to forecast if energy market prices are set through competition).

While the regional transmission network should be planned in a coordinated manner, we believe that limited competition to construct, own and maintain specific network transmission assets is possible and desirable. By specifying the required network transmission facilities in appropriate detail, the right to build, own and maintain those facilities can be competitively bid. This bidding process would introduce competition in construction, ownership and maintenance of the regional transmission system and would help minimize the total cost of regional transmission service. Some competitive electric markets (such as the one in Victoria, Australia) have already implemented such a system of central transmission planning, and competitive bidding for the construction, ownership and maintenance of new network transmission projects.

Proposal for decentralized planning for dedicated transmission projects. In contrast to network transmission projects, we believe that *dedicated transmission projects* serving specific users can be effectively planned and developed through normal commercial incentives and decentralized investment processes. While relatively few transmission project may initially be considered as dedicated transmission projects, we believe that any project that could be financially supported solely through fully efficient transmission prices could be considered a dedicated transmission project.

Since dedicated transmission facilities will only be used by specific energy producers and consumers, those users should be solely responsible for planning, developing, and supporting the facilities as part of their energy transactions. These users will be solely responsible for obtaining the lowest costs and best terms for the transmission facilities. Hence, dedicated transmission facilities should not require commercial authorization by a regional network planning body, but should only be obliged to meet the technical requirements for interconnecting and operating the proposed facilities.

Dedicated transmission projects are suitable for competitive market pricing and independent project development. By allocating all of a dedicated transmission project's costs and benefits to the project's sponsors and users, this type of transmission investment can be placed on a fully commercial basis. The project sponsors will determine whether the competitive market value of the project justify the project's costs and risks. Since the project's sponsors and customers voluntarily assume all of the project's costs and risks, they should receive all of the project's benefits. Competitive market forces can thus determine the price of transmission service over dedicated transmission projects, while commercial investment criteria determine the level of investments in dedicated transmission projects. In short, we believe that dedicated transmission projects can be developed in the same manner as competitive generation projects.

Concluding caveats regarding proposed transmission development framework. The reader is reminded that this classification of transmission projects is not driven by fundamental technical differences that are always obvious for every transmission project. Rather, this dichotomy is rather more of a commercial convenience in considering transmission system planning. Most transmission projects will be grey, in that they will partially serve specific users and partially benefit the entire market. Also, the classification of similar projects may vary over time. Furthermore, different electric markets may consider very similar projects to be different types, depending on the degree of transmission pricing efficiency. For instance, the first direct current interconnection linking two heretofore unconnected electric markets is clearly a dedicated transmission project, since only specific, identifiable producers and consumers trading energy across the link will benefit from it. The fifth alternating current transmission line between the same markets more closely resembles a network transmission project, since the two electric markets would have probably been integrated into a single broader market, and the enhanced transmission capacity will be used by many producers and consumers in both regions.

Perhaps the most useful function of our proposed dichotomy is to help allocate costs rather than rigorously define projects. Very few transmission projects will qualify as either purely network transmission projects or purely dedicated transmission projects. A reasonable cost allocation for a given transmission project may assign part of the project's cost to all grid users (the "network transmission project" portion) and part of the cost to specific producers and/or consumers (the "dedicated transmission project" portion).

As more experience is gained with efficient transmission pricing methodologies and effective property rights for transmission, this artificial dichotomy might be complemented or supplanted by a more continuous cost allocation methodology. Mechanisms such as peak load pricing can more accurately allocate transmission use and costs. In essence, the "black-and-white" classification proposed above might be replaced with a single methodology that efficiently allocates *all* transmission costs (and, of course, simultaneously assigns effective property rights accompany such efficient pricing). However, the simple classification scheme proposed here may serve as a useful transitional measure during the initial years of newly competitive electric markets, or for electric markets in which fully efficient transmission pricing is not implemented.

Specific examples of a network transmission project and a dedicated transmission project. To better illustrate this proposed model for transmission expansions in competitive electric markets, we next consider two specific transmission projects. The first project is a transmission system enhancement under development at the end of 1997 that exemplifies the network transmission project discussed above. The second project is a recently completed high-voltage direct current (HVDC) transmission line that illustrates the dedicated transmission project model.

VPX Series Capacitor Project - A Network Transmission Project.

Background. Australia has been moving steadily to introduce competition into its electric market since the early 1990s, with the state of Victoria (on the southeastern coast of Australia) taking the lead. In Victoria, a fully competitive spot market for producers has been in effect for some time, with spot market energy prices set every half hour by the bid price of the last generator dispatched during that period. The Victorian Power Exchange (the "VPX") is in charge of both operating the Victorian transmission system (including the interconnections with neighboring states) and managing the physical energy market. In short, the VPX acts as the ISO, the RTG and the spot market administrator for Victoria.

In its role as transmission planner for Victoria, the VPX (1) identifies the need for transmission system enhancements, (2) compares the expected cost with the projected benefits from reduced congestion or increased reliability within the VPX market, and (3) implements cost-effective transmission enhancements through competitive bidding to build, own and maintain the desired facilities.

Another function of the VPX is to administer the transmission market. All energy producers and consumers operating in Victoria pay the VPX for use of the integrated Victorian transmission network. These payments are based on explicit "Use of System" formulae that allocate the total transmission costs among producers and consumers.

The VPX in turn pays all owners of Victorian transmission facilities ("Network Owners") for the use of their facilities. Total transmission payments from the Victorian producers and consumers equal the payments to the Network Owners, with a small deduction for the VPX's operating costs. The payments by the VPX to the Network Owners are collectively guaranteed by all of the producers and consumers participating in the VPX, and are made regardless of the actual day-to-day use of the facilities. Although the VPX reduces payments to the Network Owners in the event of the unavailability of their transmission facilities, the VPX (and through it, the Victorian generators and consumers) assume the "market risk" of the transmission network.

The transmission facilities existing in Victoria as of early 1997 are all leased to the VPX under a single long-term lease agreement with PowerNet Victoria (owner of those facilities). However, the VPX intends to acquire rights to new transmission facilities through competitive bidding for "build-own-maintain" proposals. Payments by the VPX for the use of these new facilities will be set by the results of the competitive bidding. That process was used to develop a significant upgrade to the existing transmission system near Melbourne, Victoria.

The project. In 1997, the VPX identified the need for additional transmission system development to maintain firm transmission capacity into Victoria. Without those facilities, voltage control considerations within the greater Melbourne metropolitan region could limit imports into Victoria from the neignboring state of New South Wales during the peak summer hours. The VPX identified that the addition of series capacitors to an existing 330 kV transmission line would economically relieve those constraints and maintain the full transmission capacity into Victoria during the peak summer hours.

This project was driven primarily by the possibility of not fully serving load in the Melbourne area during the peak summer months. Hence, the economic analysis for the project compared the project's expected annual costs to the expected increase in reliability (and the corresponding reduction in the expected unserved energy cost).

After this project was identified as economically feasible, the VPX developed a technical specification for prospective owners of this asset, and put the project out for competitive bidding. The technical specification described the electric environment in which the series capacitors would operate, and the gross rating requirements for the assets. Prospective bidders were also given the requirements for interfacing the project with the existing 330 kV transmission line (owned by PowerNet Victoria). Bidders were invited to submit proposals to build, own and maintain the series capacitors and associated equipment. The proposals submitted by the bidders included the annual payment that the bidder would require from the VPX to provide the requested transmission service for the duration of the 20 year agreement.

Under the terms of the bidding process, the winning bidder would be responsible for obtaining all necessary permits and licenses (including a "Transmission License" for operating as a transmission company within the state of Victoria), for financing, constructing and owning the project's facilities, and for maintaining the project in accordance with the VPX's instructions. In exchange, the winning bidder would be paid the annual tariff proposed in its bid response, net of any penalties for failure to achieve target availabilities for the equipment. Figure 8 illustrates the relationship among the winning bidder, the VPX, the Victorian producers and consumers, and PowerNet Victoria.

The VPX has logically extended competition in the electric supply industry from the generation sector to certain limited aspects of the transmission sector. Although competing transmission owners do not bear the market risk that competing generation owners assume (since the transmission payments by the VPX are independent of the market price of energy), the returns for the winning bidder are set by competition. In short, the cost to Victorian producers and consumers for the transmission capacity resulting from this network trans-

SYSTEM PLANNING UNDER COMPETITION   323

FIGURE 1
Project Diagram for VPX Series Capacitor Project

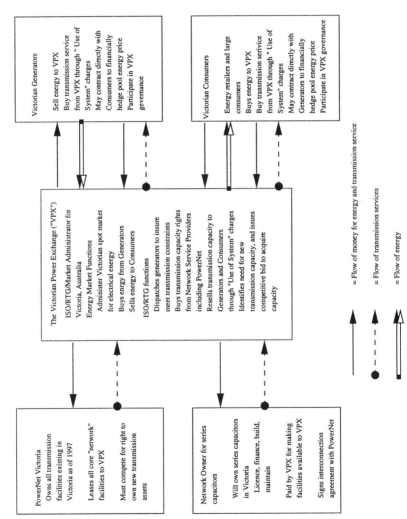

mission project will be set by competitive market forces, not by governmental regulation.[50]

The Hydro-Qubéc Phase II Interconnection - A Dedicated Transmission Project. We now consider an example of a specific dedicated transmission project between the Province of Qubéc in Canada, and the New England region of the United States[51]. This project (the "Hydro-Qubéc Phase II Interconnection"[52]) was developed during the 1980s, before significant restructuring of the New England electric markets was underway. Nevertheless, this project embodies most of the elements of the dedicated transmission project discussed above. Most saliently, the costs and benefits of this project are tightly coupled and assigned to the project's users. The rights to use this energy "toll road" between Qubéc and New England go only to those who pay the toll.

Background. After the oil price shocks of the 1970s, many New England utilities began searching for alternative, non-oil sources of energy. In addition to several nuclear and coal projects that were pursued in New England, the idea of an expanded transmission interconnection between New England and Qubéc was considered. Due to the extensive hydroelectric resources in Qubéc, New England utilities perceived an opportunity to purchase hydroelectric energy from Qubéc at prices that were more predictable and more stable than the cost of oil-fired generation.

A 10 year energy contract was negotiated between the provincial utility in Qubéc (Hydro-Qubéc) and a consortium of 47 New England utilities[53]. These New England utilities agreed to purchase an average of 7 TWh per year of energy from Hydro-Qubéc, for a total of 70 TWh over the 10 year period. This purchase was a significant fraction of the total annual energy consumption in New England at that time (approximately 100 TWh per year).

At the time of the project, the large Hydro-Qubéc transmission grid was not electrically connected to the large New England grid (that in turn is part of the even larger Eastern Interconnection grid of the United States) with alternating current transmission lines. Thus, the large Qubéc and New England grids could not be reliably connected through conventional alternating current transmission lines. To overcome this technical obstacle, a direct current transmission line was built between Qubéc and Massachusetts. This transmission interconnection (the "Hydro-Qubéc Phase II Interconnection" or simply "Phase II Interconnection") allows up to 2000 MW of energy to be exchanged between Qubéc and New England (although other considerations typically limit the effective transmission capacity to approximately 1400 MW).

The use of direct current technology in this project (and its predecessor) was a first for New England. However, perhaps the biggest innovation was the commercial arrangements that underpin this project. Unlike most transmission facilities, the Phase II Interconnection was specifically built to allow contractual energy sales from Hydro-Qubéc to specific New England utilities. These features of the Phase II Interconnection define it as a dedicated transmission project.

The project. Unlike the series capacitor project described above, the Phase II Interconnection was a very large project. The total capital cost of all Phase II facilities in the United States was over US $400 million, with somewhat greater costs in Qubéc. High voltage direct current (HVDC) converter stations were constructed (two in Qubéc, one in Massachusetts) to convert the alternating current produced in Qubéc into direct current and to convert the direct current received from Qubéc into alternating current suitable for injection into the New England grid. A 1500 km DC transmission line was also constructed to connect the converter stations. Finally, various ancillary facilities and reinforcements to the integrated New England alternating current grid were also required.

All of the HVDC transmission facilities in the United States were licensed, financed, built and owned by special purpose transmission companies established solely to execute the project. Unlike most other utilities in the United States, these special purpose companies have no generation assets or rights to any generation. They do not serve any load within New England, and they do not buy and sell any energy whatsoever. These transmission companies are simply the owners and operators of the energy "toll road" connecting Qubéc and New England.

The special purpose transmission companies entered into long- term transmission contracts with the New England utilities purchasing energy from Hydro-Qubéc. Under these contracts, the special purpose transmission companies provide dedicated transmission service over the Phase II Interconnection between the United States/Canadian border and the HVDC converter station in Massachusetts. *Only* the New England utilities that entered into those long-term transmission contracts received long-term rights to use the Phase II Interconnection facilities and were thus able to buy firm energy from Hydro-Qubéc.

In short, any New England utility that elected to not buy energy from Hydro-Qubéc did not require transmission capacity over the Phase II Interconnection and was not obligated to pay any portion of its costs[54]. Of course, that non-participating utility is not entitled to receive energy from Hydro-Qubéc over the Phase II Interconnection. Figure 8 shows the interrelationships of the 47 New England utilities, Hydro-Qubéc and the special purpose transmission companies.

The Phase II Interconnection is an early example of dedicated transmission project that connects two distinct regional electric markets. The direct current technology used for this project allows the energy flows over these transmission facilities to be directly and explicitly controlled, while the underlying energy contracts and transmission support agreements explicitly allocate the costs and benefits of the project to its users. Hence, the costs and benefits of the project are tightly coupled and fully aligned (the hallmark of a dedicated transmission project).

Other interconnection projects around the United States and throughout the world share these characteristics, and can be considered dedicated transmission market interconnections. As the trend to wider integration of regional energy markets continues, the first interconnections between different markets

FIGURE 2
Project Diagram for HQ Phase II Project

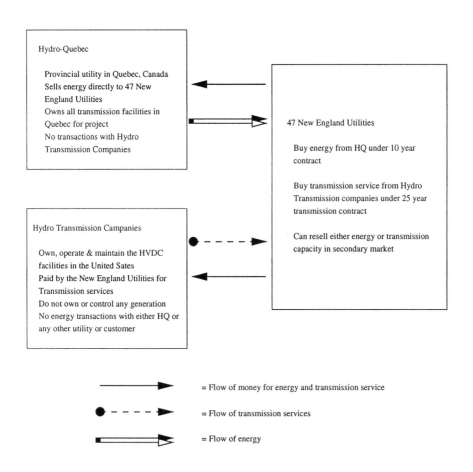

can often be considered a dedicated transmission project that benefits specific importing consumers and/or exporting generators (particularly for a highly controllable transmission line, such as an HVDC project). By commercially segregating the costs and benefits of such projects where possible and assigning them only to the users, these type of interconnections can be placed on a fully commercial footing, and developed solely in response to the market's needs.

Challenges in siting new electric projects in competitive electric markets. We close this chapter with a discussion of some of the social implications of siting new bulk power projects in competitive electric markets. Although much work has been accomplished in recasting the role of generation and transmission in competitive electric markets (particularly in efficient pricing for energy and transmission), some critical issues still require further work. In many newly or about-to-be competitive electric markets, the process of gaining social approval for new generation and transmission facilities is still based on approval processes used in monopoly electric markets. With the shift to competitive markets, new processes for siting and licensing generation and transmission facilities may be required. So far, the commercial changes in certain regions may be ahead of the legal and regulatory changes for those areas. This concluding section briefly considers the historical processes for siting and licensing new generation and transmission facilities, and possible changes to accommodate competition in generation.

In a monopoly utility market, the siting and licensing of new generation and transmission facilities are typically done by the monopoly utility, as part of its public service obligations. The commercial and environmental approvals of a new generation or transmission facility were often received simultaneously as part of a "certificate of public necessity and convenience" that the utility received for the project.

In the United States, the authority to site and construct new electric generation and transmission facilities is vested with the individual states and exercised by the monopoly utilities in that state. The utility's authority to exercise these powers is usually derived from its status as a public service provider. As a public service provider, the utility's approved activities (including the construction of approved new generation and transmission facilities) are deemed to be for the public good, and regulated by the state government.

Typically, a vertically integrated utility wishing to construct and operate new generation or transmission facilities first demonstrates that the proposed facilities are part of the "least cost" integrated resource plan (or some equivalent thereof) for the utility's consumers. The projects that are included in the least cost plan may then move to the permitting phase, in which the utility prepares a formal application to the relevant state siting authority for permission to construct and operate the proposed facilities.

Typically, the state's siting process requires that the utility again demonstrate that the proposed facilities are needed to meet the projected demand for energy at the least cost and with the lowest environmental impact. The state siting authority will then usually accept public comment or challenges to the proposed facilities, and often holds formal hearings on the project. Larger projects commonly draw more attention, comments and protests. If the project is judged by the state siting authority to be the least cost, lowest impact method of providing the necessary public service (perhaps with modifications proposed during the public hearing process), the state authority will approve the project, and issue a "certificate of public convenience and necessity" (or some similarly titled approval).

One key reason for this lengthy and legalistic process is that utility projects approved by the state siting authority often enjoy the police powers of the state to acquire and use privately owned land for those projects. After a project has been approved by the state siting authority, the utility can usually acquire property or easements (even from unwilling sellers) through the power of eminent domain in order to implement the project. Even if the utility never actually exercises any power of eminent domain, it can negotiate for the property or easements knowing that such legal powers are available to it. Other commercial projects (e.g., a new factory or a shopping mall) almost never have such powers available to the owners.

A second key reason for the lengthy siting process is the commercial endorsement implied by the certification process. Since the utility must usually demonstrate that the project is the least- cost method of meeting electric demand to receive siting authority, the granting of that authority may convey significant endorsement as a reasonable and prudent investment. Although further proceedings are usually required after the project enters service for the utility to recover the project cost in its rates, the original approval by the state siting authorities will often carry significant weight in determining the prudence of the investment. Hence, at least for rate- regulated projects, the siting authority often connotes commercial endorsement in addition to environmental approval.

This siting process must be revised for a competitive electric market. With the introduction of competition in generation, the least cost generation planning process of the vertically integrated utility is replaced by the decentralized decision-making processes of individual generation companies. In most competitive electric markets, no generation company is under any significant legal obligation to construct new generating plant. As discussed earlier, competitive generation companies rely more on energy price signals and commercial judgements, and less on reliability criteria and central planning processes, to determine the construction and operation of new generating facilities. Hence, the traditional process for approving new generating plants must be revised.

Many state siting authorities throughout the United States have begun to examine the need for changes in licensing new generating facilities. Typically, state authorities are considering a move to place more weight on the environmental and social impacts of a proposed project, and less emphasis on cost, or to even move away from explicit cost considerations. Since generation owners in a competitive electric market bear all of the financial risks and rewards of the project, the "least cost" test becomes less clearly defined and probably less meaningful. (The rationale for a competitive generation project receiving the powers of *eminent domain* is also less obvious.)

This revised paradigm for siting new electric facilities may also have to be extended to the transmission sector. With the unbundling of transmission from generation, the need for new transmission is now less tightly coupled to the need for new generation. As noted earlier, the responsibility for planning new transmission system enhancements may fall to the regional ISO rather than the local utility. The ISO may have broad geographic scope, but little contractual or siting authority, and uncertain abilities to promote specific transmission projects. Thus, the regional ISO may approve technical designs and commercial arrangements for new transmission facilities, but may find it difficult to license those facilities using the traditional process.

An open question for state siting authorities will thus be the appropriate siting criteria for network transmission projects that are primarily designed to enhance the competitiveness of the regional electric market, or for dedicated transmission projects that are not used by all generators and consumers within the local market. For many governmental authorities, the commercial changes in the electricity market have probably outstripped the regulatory changes that have been made to date. The challenge for siting authorities will be to stay relevant, without impeding the development of fully competitive electric markets.

CONCLUSIONS

This chapter has attempted to cover an enormous amount of ground. Rather than recapitulate the preceding pages, we simply close with a few observations regarding the introduction of competition into electric markets, and its effect on system planning:

- With the introduction of competition in generation, the planning and development of generating facilities undergoes a tremendous shift. Generation related reliability criteria are complemented, supplemented or even replaced by energy market prices and commercial performance incentives.

- Balancing effective competition in generation and efficient pricing in transmission is challenging many of the emerging electric markets in the United States. In particular, the development of workable property rights for transmission is still a work in progress.

- The introduction of competition in generation also changes the transmission planning process. We believe that fully decentralized transmission planning is only desirable only if effective property rights can be assigned to the decentralized projects. Otherwise, we believe that transmission projects should be planned and developed by a regional transmission planning body on behalf of all consumers within the regional electric market.

- The political and social processes for siting and licensing new generation and transmission facilities within competitive electric markets will probably need to be revised from the processes used for monopoly markets. In particular, the public service obligations and powers and obligations typically associated with electric projects developed by a monopoly utility may not be appropriate for fully commercial projects developed by competitive generation companies.

- Perhaps the most important point is that designing new commercial, regulatory and legal frameworks for competitive electric markets is a work in progress. While we may have discussed many of the features of the industry as of late 1997, we are certain that additional changes and refinements will be continuous and will modify some of our conclusions and proposals.

However, we are fairly certain that the trend towards increased competition into the electric supply industry will continue, and the new thinking will be required (by all parties) in this new world.

Notes

1. FIRST, A DISCLAIMER: Before we plunge into the detailed discussions, a strong disclaimer is in order. *Developing proven processes for system planning in fully competitive electric markets is a work in progress, not a finished masterpiece.*

2. Agora was the marketplace of ancient Greece at which buyers and sellers met to conduct their transactions.

3. For our purposes, the distinction between captive wholesale consumers and captive retail consumers is not meaningful. This discussion also applies to markets in which a monopoly utility controls the generation and transmission sectors, and sells energy to captive distribution companies at wholesale.

4. Many elementary regulatory economics textbook will discuss the concept of a natural monopoly, the loss of the social benefits under monopoly pricing, and the regulatory models used to approximate the socially optimal price of a monopoly product.

5. We will not discuss the definition of "prudent" or "reasonable" in this chapter, except to note that disagreements over "prudency" and "reasonableness" have historically constituted many of the rate case issues for electric utilities in the United States.

6. In defense of this practice, we note that relatively few consumers have a great interest in specific reliability criteria.

7. From *NPCC Document A-2*, "Basic Criteria for Design and Operation of Interconnected Power Systems", August 9, 1995, as adopted by the Members of the Northeast Power Coordinating Council. *Resources* include generation resources.

8. Although the Public Utility Regulatory Policy Act (PURPA) of 1978 introduced some degree of competition in the wholesale generation market in the United States, the planning process was not fundamentally changed, even if the outcomes were significantly different.

9. In the planning process discussed here, we only discuss generating resources. Many planning models used by vertically integrated utilities represent DSM measures as equivalent generation options. Hence, the process presented here can evaluate DSM alternatives.

10. Strictly speaking, the hourly load relative to the peak load may vary as a function of year, particularly if demand side management measures are introduced. Practically speaking, few planning processes explicitly model any shift in relative hourly loads over time.

11. Often, interconnections with other utilities are can reasonably be modeled as notional generating units.

12. As noted earlier, these alternatives can include DSM projects that are represented as generating units.

13. For new units, we use the subscript $N+m$ for convenience, as will be seen later.

14. Any introductory textbook on regulatory economics will discuss the details of rate making and financing for a regulated monopoly utility.

15. Under the criterion, discussed earlier, $LOLP_{crit} = 10\%$ per year.

16. For similar unit mixes, larger interconnected systems have higher reliability for a given ratio of generation capacity to peak load, due to the assumed independence of individual generating unit availabilities.

17. Typical installed capacity reserve margins used in utility planning range from 10-30%.

18. Strictly speaking, the value of RM should be recalculated as the system configuration changes, to reflect the addition of new generating units. However, many utility planning models and processes apply a single value of RM over almost all of the planning horizon.

19. A satisfactory generation plan is one that meets the annual reliability criterion.

20. In essence, the utility assumes that the existing generation is economical to operate until a certain defined date. The emergence of low cost combined cycle units has challenged this idea that existing generating units are always economical.

21. Typical planning horizons used in utility models are 30+ years. Competitive generation companies may use somewhat shorter investment horizons.

22. We do not present a quantitative representation of internal transmission constraints for the monopoly utility case.

23. Many utility financial models include detailed accounting mechanisms to properly allocate responsibility for income taxes. These mechanisms include the accrual and depreciation of various accounting funds that arise from the use of regulatory accounting. In the United States, vertically integrated utilities often keep three different sets of accounts (one for financial reporting purposes, one for income tax purposes and one for ratemaking purposes).

24. The term "bundled rates" refers to rates charged to a captive customer for the electric energy delivered to their facilities. The cost of generation and transmission is "bundled" into the single rate that the consumer pays.

25. This is an important issue in the United States, where individual utilities serving their own local franchise areas are connected, and thus share overall system reliability.

26. For example, the National Electric Reliability Council in the United States has an "honor roll" for control areas that minimize the unscheduled interchange of energy with their neighboring control areas. The direct financial penalty for failing to make the honor roll is zero.

27. Note that additional transmission interconnections to other regional electric markets can increase the availability of generating capacity within the regional market, and hence, could be considered as enhancements to generation-related reliability. Such transmission interconnections to other markets may thus directly compete with local generation.

28. Except for metering and/or load profiling considerations, we know of no technical barrier to allowing every retail consumer to choose their electric supplier.

29. We acknowledge that we have used qualitative words such as "sufficient", "reasonable", "adequate", "workable" and "few", without attached precise definitions to these terms. To our knowledge, no definitive and universally accepted definition of an adequately competitive electric market exists.

30. In other words, transmission is relatively unconstrained most of the time, such that few generators have market power within localized "load pockets", and no generator has extended market power for large numbers of hours. Alternately, the local ISO can have contractual arrangements with any must-run generators to mitigate the market power of such producers. Must-run generators are those plants whose operation is required for local reliability reasons, due to transmission constraints.

31. In fact, many of the transmission planning criteria used by a vertically integrated utility ("ensure access to multiple energy sources") are similar to the criteria that define a competitive regional electric market. In areas of the largest vertically integrated utilities, divestiture may be required to institute workable competition.

32. Since, at any instant in time, energy production and consumption are balanced, any shortfall in a given generator's output will have been made up by another generator. The spot market is the means of identifying and settling these instantaneous trades in electricity.

33. This simplifying assumption allows us to internalize the value of capacity into expected spot market energy prices, and thus consider only the market for energy.

34. This model for valuing generation reliability is the Australian model.

35. If the generator's actual output is less than its contracted output, the generator will purchase its shortfall from the spot market.

36. In contrast to the normally open planning process that is used by many utilities, competitive generation companies will closely guard their forecasted market prices and projected capital and operating costs as highly proprietary information.

37. As with the production cost problem, the expected production levels must be reflect the forced outage rate of the unit for_j.

38. The calculation of the income tax due for a specific project can be extremely complex, and will depend on the details of the specific project, and the financial structure of both the project and the company. For this example, we simply assume that the income tax due in year y can be computed directly from the expected revenues and costs.

39. Strictly speaking, the net present value of the capital investment stream should also be calculated, since the capital investment will be made over a period of time. Here, we assume that the project's capital cost is expended as a single lump sum just prior to the start of commercial operation.

40. We use the term "impartially" to denote that similar transactions will be treated similarly. We do not mean that all transactions will be treated identically.

41. The costs caused by different energy transactions are mostly sensitive to geographic location of the energy production and consumption for that transaction, since the geographic input and withdrawal points will determine the transmission costs. Hence, efficient transmission prices are often a function of "location, location, location" (to borrow a phrase from the real estate industry).

42. Lecing, B.S., Ilić, M.D., "Peak-load Pricing for Electric Power Transmission", Proc. of the Hawaii Conference on System Science, Jan. 1997, pp624-633.

43. Theoretically, one company's delivery of coal by rail may affect a competitor's coal delivery over the same railway system, to the extent that congestion on the railroad occurred due to the last transaction. However, the consequences of transportation constraints are much greater for electricity than for any other commodity.

44. Generally, the most efficient transmission pricing methods are the most complex, further hampering their acceptance and implementation. Part of the complexity is due to the highly dynamic nature of available transmission capacity. Unlike most other transportation networks, the maximum transportation capacity between two points on an electric grid is highly dependent on other uses of the interconnected grid.

45. The truth of this statement will depend on the exact mechanism for allocating transmission congestion costs among the regional producers and consumers.

46. In this chapter, we use "region" and "market" interchangeably when referring to electric systems.

47. A key characteristic of dedicated transmission projects is that the changes in regional energy flows due to the dedicated transmission project can readily be associated with specific energy producers and consumers.

48. Effective property rights can easily be assigned to direct current facilities, due to the high controllable nature of the energy flow over these facilities.

49. We have deliberately avoided quantifying the appropriate size for regional electric market or regional transmission network. To our knowledge, no technical or economic definition of an adequately competitive electric market has been widely accepted.

50. Many thanks to Tim George (formerly of the Victorian Power Exchange) for his kind permission to discuss this example as an illustration of a network transmission project, and to Tony Cook of PTI Australia for his help in confirming the project development process.

51. Thanks to Jeffrey Donahue of New England Hydro-Transmission for allowing this project to be discussed as an illustrative example of a dedicated transmission interconnection project.

52. The "Hydro-Qubéc Phase I Interconnection" project was a smaller (600 MW) interconnection developed earlier in the 1980s.

53. Although the energy purchased from Hydro-Qubéc is dispatched as part of the integrated New England power pool ("NEPOOL") and the energy dollars initially all flow through NEPOOL, the net effect of these arrangements is to allow the participating New England utilities to buy energy directly from Hydro-Qubéc.

54. In the end, most utilities within New England at the time signed both the energy contract and the transmission support contracts.

9 TRANSMISSION NETWORKS AND MARKET POWER

Ziad Younes and Marija Ilic

Department of Electrical Engineering
and Computer Science
Massachusetts Institute of Technology
Cambridge, MA 02139

Introduction

Combined cycle technologies and other electric power generation technologies have reduced the economies of and the critical size of a single generation unit, enabling the multiplication of generators and reducing their individual market power. This phenomenon is expected to facilitate the emergence of a competitive generation market and is one of the major arguments in favor of deregulating the electric power industry. Without transmission constraints and losses, the market for generation would have a unique equilibrium price and the features of a classical perfect or oligopolistic market, and a reduction in the critical size of a generator should definitely improve the competitiveness of the market. Nevertheless, because of loop flows, transmission constraints, and other network externalities, the electric power market is unique and must be studied carefully.

While, classically, the size of a market participant will certainly affect his market power, in electric power networks his geographic location will also play an important role because transmission constraints and loop flows could isolate him from the competition and protect his market share. This phenomenon can produce geographically localized zones where individual or small groups of suppliers can have significant "locational" market power independently of the market structure and regulations governing the short-term market.

336   POWER SYSTEMS RESTRUCTURING

This raises new challenges to the ongoing deregulation that the regulator cannot ignore when determining the long-term market rules and designing the expansion strategy for the transmission grid.

The first objective of this chapter is to present some of the potential conceptual impacts of transmission constraints and loop flow on market power, its locational dependence, and on the behavior of the market participants. The second objective is to propose solutions to reduce the locational market power. In the next section, the concepts of market power, residual demand, and relevant markets are discussed. In section 3, we derive the residual demand facing a generator or a group of generators in a network where one transmission line is constrained and show how this demand depends on the geographic location of the these generators independent of the market rules. To confirm these results, a numerical simulation of a 24-node network in the context of a nodal pricing scheme is presented in section 4, showing that one constrained transmission line out of the 38 can be enough to raise significantly the market power in some areas of the network. Section 5 discusses the nature of the oligopolistic competition that might take place in such areas and explains why tacit collusion is a plausible phenomenon of which regulators ought to be aware. Section 6 shows how transmission lines, even when they are not constrained, guarantee a minimal market share to suppliers that might lead them to operate a line at its limit in order to abuse this guaranteed demand. Section 7 discusses traditional solutions against market power and shows how, under certain circumstances, the expansion policy of the transmission network can be used to create credible threats against the use of locational market power. The last section concludes.

Some concepts

Market power   In a perfectly competitive market, all the market participants choose the level of their output such that their marginal cost is equal to the market price, which makes the market settle at an equilibrium that is optimal for the social welfare. Achieving, or at least coming close to this efficiency, which may be one of the goals so many governments are hoping to achieve by deregulating the power industry, relies on the assumption that the market participants do not have significant market power. In other words, they cannot raise the market price above the competitive level without losing so many sales so rapidly that the price increase is unprofitable (Landes et al. 81).

Residual demand Because of the detrimental effect of market power on social welfare, and, more precisely, on the prices seen by the consumers, the regulators are often eager to denounce and to try to reduce the market power wherever it can be found. There are many ways to assess the level of market power of a market participant or a group of market participants. One of the simplest is based on the concept of the residual demand, D_r, i.e., the market demand, D_m, minus the amount supplied by competing firms, S_c. As shown in (Landes et al 81), and used and developed in (Scheffman et al 87) and (Baker et al 88), a good indicator of the market power of a firm is the elasticity of the residual demand facing this firm, $\varepsilon_d = \dfrac{\partial D_r}{\partial P} \dfrac{P}{D_r}$, which measures the responsiveness of quantity demanded to a change in price. Whenever this elasticity is equal or less then 1, the firm has always an incentive to raise its price until the elasticity goes beyond 1 and its inverse reaches the "Lerner index"[1]. The concept of residual demand is interesting because its elasticity can easily be deduced from the elasticity of the market demand, ε_m, the elasticity of the competing suppliers, ε_s, and the market share, s.

$$\varepsilon_d = \varepsilon_m \frac{1}{s} + \varepsilon_s \frac{1-s}{s} \qquad (1)$$

The market share is an essential factor in determining the market power of a firm because the higher the market share, the lower the impact of the market demand elasticity, and the lower the impact of the competing suppliers. Nevertheless, market shares are not the only factor, as it is too often and too quickly assumed. A very elastic market demand would even deprive a monopolist of significant market power.

Calculating the elasticity of the residual demand can be useful in two situations. When a firm is suspected of exercising market power,

---

[1] The Lerner index (Lerner, 34) measures the used market power. It is equal to (P-mc)/P, the proportional deviation of price from the marginal cost.

the market power exercised can be deduced by the elasticity of its residual demand. Moreover, knowing this elasticity and its behavior as a function of the market price, we could evaluate the market power that a hypothetical group of producers would have if they acted as a monopoly.

Relevant markets A market participant can use its market power independently, and abuse it in order to raise its profits. The antitrust laws are intended to prevent an individual market participant from possessing enough market power to be able to disturb significantly the market equilibrium and reduce the social welfare. Moreover, market power, when possessed jointly by a group of producers, can be an incentive for these producers to engage in cartel-like behavior in order to exploit the market power that a hypothetical monopolist would have had. Because cartel-like behavior is more likely to be sustainable with a smaller number of participants (Stigler, 64), monitoring the concentration of such potential groups is important to prevent cartels and tacit collusion. This is one of the reasons why the US Department of Justice Merger Guideline defines a 'market' relevant for antitrust purposes as a geographical area and a group of products such that a hypothetical monopolist would have 'significant' market power. In the merger antitrust context, 'significant' market power means typically that a monopolist would impose a five percent price increase or more (Werden, 96). These 'relevant markets' are of course distinct from the classical economic definition of a market that is a geographical area and a group of products where prices are dependent variables linked by arbitrage opportunities. Relevant markets can be bigger or smaller than economic markets (Scheffmann et al 87).

In typical merger cases, the smallest relevant market is determined, the shares of the different competitors are assessed and the merger is then authorized or not. Our concern here is somehow different since in the regulated environment the relevant markets do not exist yet and the existing production units will determine the minimal size of a supplier in the deregulated environment. While the new combined cycle generation units can be small compared to the economic market, they might be too big compared to the sizes of some relevant markets that might emerge. Therefore, before deregulating the industry, the geographically relevant markets have to be determined and their sizes evaluated to verify that they are consistent with the emergence of a competitive market. We will show in the next sections how transmission constraints, because of their impact on the residual demand, can create relatively small relevant markets by raising the market power in specific locations.

The residual demand in electric power networks

The nodal pricing proposal advocated by Hogan (1992) has been frequently criticized because it ignores the potential market power that the market participants can have in such a framework. Singh et al. (1997) show how, with location-dependent nodal spot pricing and transmission constraints, a non-discriminating auction mechanism creates opportunities for strategic behavior. Nevertheless, the problem of market power lies beyond the auction mechanism or the features of the nodal pricing model. More generally, in any market structure, those generators in areas constrained by weak transmission lines should see their market power boosted because they are isolated, by the constraints, from the competition of other generators.

A simple example

To first give an intuitive sense of how transmission constraints affect the residual demand, we use a simple model where two generators G_1 and G_2 and an aggregated load L are isolated from the rest of the network by a transmission line whose maximum capacity is K. P and λ are respectively the electricity prices in the constrained and unconstrained areas and $d(P)$ is the demand of the load L.

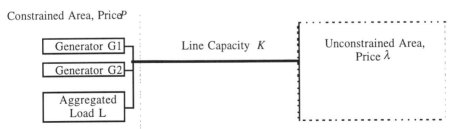

Figure 1. Market Power Due to Transmission Constraints

340 POWER SYSTEMS RESTRUCTURING

The market in the constrained area is similar to a duopoly with a competitive fringe where the fringe has a constant marginal cost λ and a maximum production capacity K. As long as the price P is above λ, the line is constrained and the generators G_1 and G_2 are facing a demand $d'(P)=d(P)-K$. They will act in this case as a perfect duopoly facing the new demand d', with the full market power that this situation gives them. When d' is sufficiently inelastic and K sufficiently small, the constrained area will be a relevant market where the generators could either engage in an oligopolistic behavior or even collude at the monopoly price. It is important to see that it is the transmission constraint that is enabling the market power to rise locally.

Mathematical derivation of residual demands

A constrained zone completely isolated by a constrained line is exceptional. More generally, the transmission lines are constrained because of loop flows and there might always be a path of non-constrained transmission lines between constrained and non-constrained generators. Unfortunately, and this is specific to power networks, loop flows prevent the capacity of the unconstrained lines from being fully used. The following mathematical derivation will show how the residual demands are affected by transmission constraints independently of market structure and how this can lead to geographical variations of market power.

We consider a network consisting of n nodes. At each node i, s_i and d_i are the power produced and consumed at this node and $q_i = s_i - d_i$ is the power injected from this node into the network. We will consider a simplified, linearized and per unit[2] model of the power system, ignoring losses and reactive power. Because there are no losses,

$$\Sigma q_i = 0 \qquad (2)$$

Every transmission line on this network linking nodes i and j is characterized by its admittance Y_{ij} and its thermal capacity K_{ij} with $Y_{ij} = 0$ when the line does not exist. Every node is characterized by the

---

[2] The voltage can be assumed to be equal to 1 at every node without loss of generality.

phase angle θ_i of the sinusoidal electrical waveform. The power flowing through the line from node i to node j is given by the linearized formula

$$q_{ij} = Y_{ij}(\theta_i - \theta_j) \qquad (3)$$

Of course, the power injected in the network at node i is equal to the total power flowing from node i through the transmission lines:

$$q_i = \sum_j q_{ij}, \quad i=1,\ldots,n \qquad (4)$$

or

$$q_i = \sum_j Y_{ij}(\theta_i - \theta_j), \quad i=1,\ldots,n \qquad (5)$$

Theorem 1. For any transmission line in the network joining two nodes that we can call 1 and n, we can choose the indexation of the other nodes such as:

$$q_{1n} = \sum_{i=1}^{n-1} \alpha_i q_i \quad \text{with} \quad 1 \geq \alpha_1 \geq \alpha_2 \geq \ldots \geq \alpha_{n-1} \geq \alpha_n = 0 \qquad (6)$$

where α_i, $i=1,\ldots n-1$, are coefficients that *depend only on the physical transmission network*.

Proof. Given in Appendix.

Theorem 2. For any set R included in [1,..n], when no line is constrained, the residual demand facing generators at the nodes included in R is given by (24). When line 1n is constrained from n to 1, the residual demand is given by (25).

$$\sum_{i \in R} s_i = \sum d_i - \sum_{i \notin R} s_i \qquad (24)$$

$$\sum_{i \in R} s_i = \sum_{i \in R} d_i - \left(K_{n1} - \sum_{i \in R}(1-\alpha_i)q_i + \sum_{i \notin R}\alpha_i q_i \right) \qquad (25)$$

Proof. The first assertion is a straightforward consequence of (2) and the definition of q_i.

Using (6) and the definition of q_i, we have

342 POWER SYSTEMS RESTRUCTURING

$$\sum_{i \in R} s_i = \sum_{l \in R} d_i - \left(q_{n1} - \sum_{l \in R}(1-\alpha_i)q_i + \sum_{i \notin R}\alpha_i q_i \right) \qquad (26)$$

When only line 1n is congested from n to 1, a constraint is added on the feasible set of injections in the network, and qn1=Kn1, which gives (25)

When the transmission network is not constrained, the residual demand facing the generators in R, given by (24), is, classically, equal to the total market demand minus the supply of other generators. The market power that the generators on nodes 1 to n would have will only depend on the elasticities of the market demand and the competing suppliers, and on their market share. Their geographical position on the network is not critical.

When one transmission line is constrained, the situation is different. In order to be able to exploit this equation fully and to calculate the elasticity of the residual demand, and therefore the market power of the generators on nodes 1 to p, one must know the relationship between the prices at the different nodes of the network. This relationship will depend on the chosen market rules and structure. Nevertheless, and independently of the market, we can draw several conclusions from (25).

⟨ For R={p}, the residual demand facing the generators on node p is

$$s_p = d_p - \frac{1}{\alpha_p}\left(K_{n1} + \sum_{i \notin p}\alpha_i q_i \right) \qquad (27)$$

The elasticity of the residual demand, and therefore the market power, will depend on the elasticities of the market demand and the competing suppliers. It will also depend heavily on the coefficients α_i, and therefore on the network topology and the geographic position of the generator p on the network. Moreover, all other things remaining constant, the smaller p is, the larger the market power of the generator on node p because a larger α_p implies a smaller perceived competing supply $\frac{1}{\alpha_p}\left(K_{n1} + \sum_{i \notin p}\alpha_i q_i \right)$.

Therefore, a priori, and independently of the market structure and regulation, when a line linking two nodes is congested, it affects the market power on all the nodes of the network, raising the most the market power on the node congested in generation and raising the least (or decreasing the most) the market power of the generators on the node congested in demand.

⟨ For R = {1,...,p},

The set of generators on the p nodes that would benefit from the larger market power when line n1 is congested from n to 1 is a natural candidate to be a relevant market. The residual demand facing these generators is

$$\sum_{i=1}^{p} s_i = \sum_{l=1}^{p} d_i - \left(K_{n1} - \sum_{l=1}^{p}(1-\alpha_i)q_i + \sum_{i=p+1}^{n-1}\alpha_i q_i \right) \quad (28)$$

Once again, the topology of the network will have a big influence on the market power of this group of suppliers. All other things remaining constant, the market power of these generators will depend on the size of the perceived competing supply $\left(K_{n1} - \sum_{l=1}^{p}(1-\alpha_i)q_i + \sum_{i=p+1}^{n-1}\alpha_i q_i \right)$. The closer $\alpha_1,..., \alpha_p$ and $\alpha_{p+1}, ..., \alpha_{n-1}$ are to 1 and 0 respectively, the bigger the market power of the studied group of suppliers and the more likely that nodes 1 to p will constitute a relevant market.

It is noteworthy that when the nodes 1 to p are linked to nodes p+1 to n by only one transmission line (1n), then $\alpha_1=...=\alpha_p=1$ and $\alpha_{p+1}= ...= \alpha_{n-1}=0$. The residual demand facing generators in nodes 1 to p would then be

$$\sum_{i=1}^{p} s_i = \sum_{l=1}^{p} d_i - K_{n1} \quad (29)$$

(29) is the residual demand facing a monopoly with a competitive fringe producing K_{n1}. When K_{n1} is sufficiently small, it is very likely that nodes 1-p will constitute a relevant market, especially if the

market rules do not encourage the demand to be elastic through appropriate pricing.

Simulation of a 24 node network

In the last section, we showed how the residual demands depend on the topology of the network and the geographical location of the associated nodes. However, to illustrate this result numerically, we have to choose the market rules explicitly and set the way the prices at the nodes are derived. We have chosen the context of a nodal pricing scheme à la Schweppe/Hogan because their pricing proposal is the most widely discussed. The conclusions that are drawn from this simulation should be valid in other contexts, since the physical topology of the network is the primary source of locational market power, not the market rules.

We have simulated the realistic IEEE reliability system (IEEE, 1979) using a modified version of software developed by Macan (1997). This system consists of 24 nodes, 38 transmission lines, 19 inelastic loads and 14 generators with quadratic cost functions and finite generation constraints located at 10 nodes. We assume that the generators are communicating their cost functions[3] to a central Independent System Operator (ISO) that determines the nodal prices and the physical dispatch. The software calculates this socially optimal dispatch, the nodal prices at the generation nodes, and the profits of every generator. It is straightforward to conclude that in a competitive market the generators will maximize their profits by communicating their true cost functions. Nevertheless, the number of generators is finite and every one of them has the power to influence the nodal price and sometimes to raise its profits by cheating on its cost function. The goal of our simulation is to examine how transmission constraints can influence this power and whether they can raise it locally, transforming small groups of nodes into relevant markets.

Five transmission lines (lines 7 and 14 through 17), separating the network in two distinct sub-networks, can potentially be constrained while all other lines are operating far from their limits. We consider three different levels of demand. At the first level, only

---

[3] The cost function coefficients may be estimated on the base of price bids.

line 7 is constrained; at the second, lines 7 and 16 are constrained. At the third demand level, lines 7 and 17 are constrained while line 14 is operating so close to its limit that some generators can cause it to be constrained by changing their bids. For every demand level, we have raised by 10% the cost function processed by the ISO of every one of the fourteen generators, leaving the costs of the other thirteen unchanged, and observed which nodal prices were sensitive to which costs. This experiment has suggested that nodes 1, 2 and 7, where generators 2, 3, 11, 12, and 4 are located, could constitute a potentially relevant market.

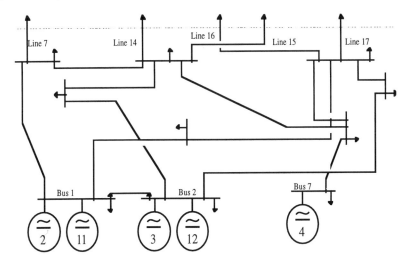

Figure 2. A Relevant Sub-Market of the IEEE 24-Bus Reliability System.

To confirm this suggestion, we have raised by 10% the bids (the cost function communicated to the ISO) of these five generators together as a monopoly could do, with (Table 1) and without (Table 2) transmission limits on the five critical transmission lines. The following two tables summarize the *increases* in prices and profits that this 10% increase in bidding has produced in the two cases.

Demand level	Congested lines	PRICES				Profits at nodes 1, 2, 7
		Node 1	Node 2	Node 7	Other nodes	
1	7	+5.7%	+5.4%	+4.9%	-1.3% to +3%	+17%
2	7,16	+9.8%	+9.7%	+10.3%	-0.6% to +1.4%	+26.3%
3	7,17,(14)[4]	+9.2%	+9.3%	+10%	-1.4% to +0.2%	+18.5%

Table 1. Changes in Prices and Profits **With** Transmission Constraints

Demand level	Congested lines	PRICES				Profits at nodes 1, 2, 7
		Node 1	Node 2	Node 7	Other nodes	
1	none	+1.3%	+0.7%	+1.1%	+0.5% to +1%	+1%
2	none	+1.3%	+1.2%	+1.3%	+0.8% to +0.9%	+2%
3	none	+3.1%	+3.2%	+2.9%	+2.3% to +2.6%	+6.9%

Table 2. Changes in Prices and Profits **Without** Transmission Constraints

It is clear from these results that the generators at nodes 1, 2, and 7 have significant market power for the three levels of demand and that they constitute very probably a relevant market[5], especially for levels 2 and 3. Moreover, it appears that it is the transmission constraints that are raising this market power in the first and second levels of demand, since removing the transmission constraints dilutes the market power of these generators by putting them

---

[4] Line 14 becomes constrained only after the generators raise their price.
[5] Formally delineating the relevant markets would necessitate determining the price increase that would maximize the profit of a monopolist. This increase in price is not, *a priori*, 10%, but is expected to be higher than 10% in the constrained case and lower in the unconstrained one. Nevertheless, our objective is not to determine explicitly the relevant markets but rather to see how the constraints could affect their delineation.

directly in competition with the other generators on the network. We will further note that their size contributes to their market power in the third level of demand, where they have significant market power in the absence of transmission constraints, because the cheap generators that are competing with them are already working at their full capacity. In all three levels, locational market power is created by the location of the sub-market. In the third level of demand the generation capacity of the generators raises their market power even in the absence of transmission constraints because too many of their competitors are operating at full production capacity and their supply is no more elastic to price. However, it is of primary importance that, in the first and second levels of demand, only one and two lines, respectively, are constrained, out of the five that separate the sub-market from the rest of the network, and that these constraints raise the increase in profits by an order of magnitude. This indicates that lines constrained by loop flows can produce sub-markets that are relevant markets from an antitrust perspective while these sub-markets are still connected by many other non-constrained transmission lines to the rest of the network. Locational market power created by transmission constraints, a classical problem in almost all network economies, is thus increased dramatically by loop flows, the specificity of electric power networks. Finally, it is interesting that the level of demand affects the market power in such sub-markets; markets that are not relevant in off-peak periods could become relevant in peak periods and we therefore need to add the temporal scope to the geographical one when searching for relevant markets.

Oligopolistic competition

As we have seen in the last section, locational market power will exist because of loop flows and transmission constraints that produce geographically and temporally localized relevant markets. Moreover, global market power can also exist, transmission constraints set aside, because of the relative size of a competitor, as shown by the experience of deregulation in the British electric power market (Wolfram, 1995). Therefore, one should not assume, *a priori*, that the market is perfect, but rather take the potential existence of market power into consideration in any form of proposed deregulation and try to limit this power. To do this, we will have to replace our traditional assumptions of a perfect market with a more realistic oligopolistic model.

We give in the Appendix a short tutorial on a game theoretical approach to oligopolistic competition and present the concepts of static and inter-temporal Nash equilibria.

Modeling the oligopolistic competition

The choice of competition model adapted to study an oligopolistic market for electric power depends on the relevant framework of interest, and, therefore, assumptions have to be made. Before choosing the appropriate model, and deriving explicitly the market behavior, the institutional framework (centralized pool or bilateral transaction, bundling or unbundling of generation and transmission cost, etc.) has first to be set. The time frame of interest will also be critical to determine if generation capacities are fixed (short-term analysis) or endogenous to the model (long term analysis). When this framework is set, a careful analysis must be done to choose the type of oligopolistic interaction (price, quantity or supply function competition) that is most appropriate.

For the long term analysis, the most suitable model seems to be the classical Cournot model where the firms compete by choosing the quantity they want to put on the market (their generation capacity) and an independent auctioneer (the market) sets the clearing price. Well adapted to the study of long-term competition where the generation capacities are the main variables (Borenstein and Bushnell, 1997), Cournot competition models are also useful in scenarios where the firms first commit themselves to a production capacity and compete next by choosing prices in a second period This is because competition by prices (Bertrand model) in the second period yields the same results given by a one-phase game where the strategic variable chosen by the firms is their output (Kreps and Schneikman, 1983). Therefore, and as used in (Wei and Smeers, 1996), the Cournot competition model might be adapted to examine generation competition in a long term strategic interaction framework where the generators have to choose their generation capacity *à la* Cournot before competing *à la* Bertrand every day.

For short-term analysis of the market, the institutional framework is critical. In a centralized market, where a Power Exchange (PX) takes the bids of the generators and loads and determines the physical dispatch, the type of competition is exogenous and depends on the bidding procedure. When the generators are bidding prices and the PX determining quantities, the short-term competition will be a Bertrand competition (applied by Hobbs (1986) to electric power markets). When the generators are to bid their production levels as a function of the different prices they might receive, the short term equilibrium prices will then be given by the "supply function equilibrium" developed in (Klemperer and Meyer, 1989) and applied to the British power market in (Green and Newberry, 1992). The equilibrium price will be higher than those yielded by a Bertrand competition and lower than those given by a Cournot competition in the unlikely case where the generators are only bidding quantities.

In a decentralized market where the transactions are settled in bilateral and multilateral markets, the type of short-term competition is endogenous and probably not unique[6]. However, besides some specific cases (the oil market in certain periods of its history, for example), competition by quantities is fairly unrealistic to analyze short-term competition. This is especially true in decentralized multilateral and bilateral electric power markets where firms rarely bid for just quantities and where the central auctioneer that sets the price does not exist. One could argue in favor of using Cournot competition to determine what happens on a daily basis because of the equivalence proposed by Kreps and Schneikman (see *supra*). Nevertheless, this reasoning makes two very strong and unrealistic assumptions: (1) the demand characteristics should be the same in the "second period" as those expected when the capacity choices were made (first period); (2) the demand should also be fairly stable in the very short term, and very close to its average, to enable the long-term expected outcome (the outcome of the Cournot competition) to be used to interpret what happens on an hourly basis. Wei and Smeers (1996) show how the equivalence between the two-stage game and the Cournot competition might not be valid when the

---

[6] However, it might be affected by the rules adopted by the PX to have the players respect the transmission constraints: Curtailment of output, surcharges for using constrained lines, etc.

demand varies with time. Furthermore, when transmission lines are constrained and generators are competing for their use, it is unrealistic to consider that when taking its strategic decisions, a generator will not expect his competitors to react immediately to any change in his own output. For all these reasons, this model seems unable to give valuable insights on short-term (hourly, daily) competition in the generation market.

An alternative is the classical Bertrand oligopolistic competition model where the strategic variables are the prices that each competing firm chooses to maximize its profit, considering as fixed the prices of its competitors. Under this model, the equilibrium price will be the marginal cost of production when the products are undifferentiated, the firms can serve all the demand they face at a constant marginal price, and the players bid once. This result, combined with the observation that the prices on the British electric power market are above marginal prices has sometimes been used to reject this competition model in favor of a Cournot competition (Oren, 1997). Nevertheless, Edgeworth (1897) has shown that, if no single firm can serve all the demand, the output of a Bertrand competition with production capacity constraints is no longer competitive and the equilibrium price can exceed the marginal cost. As reported by Tirole (1988), this result is valid in the more general context of price competition between firms with increasing marginal costs. The price competition model can give interesting results if used in the context of generation capacity constraints and increasing marginal costs, but these results must be interpreted carefully, since in this model every player assumes that the price decisions of his competitors are weakly linked to his own decision.

The supply function model, where generators bid functions linking the price to their output, can also constitute a credible alternative and a good compromise between Cournot and Bertrand competition in a highly decentralized market, where the prices of the transactions are not public, where large transactions are likely to be made at different prices than smaller ones and where, as reported by Green (1996), negotiation is more prevalent than the 'take it or leave it' offers of price or quantities implied by Bertrand or Cournot competition. Therefore, in a decentralized market, we are likely to observe a combination of Bertrand competition and supply function competition in different geographically and temporally localized submarkets.

Finally, the way the transmission network is modeled will determine how realistic, and complex, the oligopolistic modeling will be. The simplest assumption is that all market participants are at one node (Green and Newberry, 1992). The most sophisticated models acknowledge the second electric law of Kirchoff and loop flow (Cardell et al., 1996; Smeers and Wei, 1997).

However, none of the models we discussed takes into consideration the repeated nature of the interactions between the players. When this interaction is periodic, and especially in a centralized scheme where the prices are public, inter-temporal Nash strategies in which tacit collusion could be enforced by retaliation threats is a credible alternative that must be investigated. Oren (1997) suggests that such behavior can be sustainable in specific situations. The next section discusses, more generally, why tacit collusion might occur because of the specificities of electric power markets.

Tacit Collusion

In the geographically localized relevant markets created by transmission constraints, it will be likely that oligopolistic competition will take place, raise the market price or prices, and lower the social welfare. However, as was said before, one of the main reasons why antitrust regulators are interested in relevant markets is because these markets, when too concentrated, because they are too small for example, can encourage cartel-like behavior that seriously threaten the social welfare. One such behavior is tacit collusion.

Chamberlain (1929) suggested that within a framework of oligopolistic competition and homogenous good, market participants, because of the threat of price war, could sustain a monopoly price without explicit collusion. Friedman's Folk Theorem (1971) illustrates how tacit collusive behavior can appear in the context of an infinity of repeated basic games with price competition (Bertrand supergame). This theorem states that any average payoff vector that is better for all players than the Nash equilibrium payoff vector of

352 POWER SYSTEMS RESTRUCTURING

the basic game can be sustained as the outcome of a perfect equilibrium. Under certain conditions, betting for example the monopoly price as long as all other players do the same, and coming back indefinitely to the Nash bet of the single stage game after any deviation, can be a Nash strategy in the framework of the inter-temporal infinitely repeated game.

Tacit collusion can appear in the framework of a Bertrand or a supply function competition (or even a Cournot competition) and its real danger lies in the Nash quality of its equilibrium that confers its stability, unlike classical cartels where the incentives to cheat can be important. Furthermore, tacit collusion does not necessitate the communication channels that cartelisation needs and will therefore occur more frequently and be more difficult for regulators to prove and monitor. The electric power market seems, unfortunately, to constitute a credible candidate for tacit collusion.

A simple example.

To illustrate how and why tacit collusion can occur, we consider here a simple model similar to that of Brock and Schneikman (1985) where n generators, with constant marginal costs c and individual production capacities k, are facing a demand $q = a - p$, where p is the price and a is a coefficient larger than c^7. The generators are competing by prices in an infinitely repeated game. In a one-stage game, the equilibrium would be a Bertrand-Nash equilibrium (BNE) of pure or mixed strategies[8]. In the repeated game, S^* is a set of strategies $[S^*_1,..S^*_n]$ for players 1 to n that consist of betting the monopoly price as long as all other players do the same, and sticking to the Nash bet of the single stage game after observing any deviation. If S^* is an inter-temporal Nash equilibrium as defined in the annex, we say that the players can tacitly collude at the monopoly price. At every stage of the repeated game, if generator i thinks that the other generators are playing the strategies S^*_j, $j \neq i$, he will have to choose between colluding by bidding the monopoly price and sharing the demand with the others, defecting by bidding a price that is slightly

---

[7] For a more in-depth analysis of this model and its implications, see (Younes and Ilic, 1997a)
[8] For $(a-c)/(N+1)<k<(a-c)/(N-1)$ there exists no pure strategy, *i.e.*, the equilibrium bidding prices are random variables.

under the monopoly price and producing at full capacity, or playing the one-stage Bertrand Nash strategy if he is expecting other generators to do the same. When the profits from defecting in one period are smaller than the discounted future profits that a generator will lose by defecting, he will choose to collude if no one has defected until now but he will bet the BNE bets if anyone has defected because he knows that his competitors will follow their strategies S^*_j, $j \neq i$, and do the same. Therefore, in this case, the strategy S^*_i is his optimal strategy and S^* is a Nash inter-temporal equilibrium that satisfies (15). When r is the annual discount rate, T the period between two interactions, C the profit from colluding, D the profit from defection and B the profits at the BNE, tacit collusion is an inter-temporal Nash equilibrium if:

$$D(n,k) - C(n,k) \leq \frac{C(n,k) - B(n,k)}{e^{rT} - 1} \qquad (30)$$

It is clear from (30) that the period T between two interactions will determine whether the strategies S^* are Nash strategies and whether we are likely to observe tacit collusion at the monopoly price.

Tacit collusion and spot pricing

Tacit collusion is an unusual phenomenon, but a centralized framework where daily or hourly posted bids are managed publicly by a central ISO seems to constitute a perfect framework for it to occur, the major reason being that *electricity is hard to store*.

Since every market participant evaluates a tradeoff between the immediate benefit from a free riding behavior and the future losses from the competitive equilibrium, his decision will be a function of his discount rate between two consecutive interactions. In classical spot markets, the buyers will come to the market whenever their stock need to be refilled. Because electric power is hard to store, buyers of electric power will come to the market every time they need power. Therefore the interaction is quasi-continuous and the discount rate will be extremely low in a framework where power companies are bidding against each other every day (every half hour in certain propositions).

Exceptional demand size encourages occasional free riding and hinders collusion. Because electric power can hardly be stored, no exceptional quantities can be traded on a short-term market.

Detection lags of price changes usually play an active role in hindering tacit collusion by increasing the benefits from defecting and reducing the incentives to collude because collusion is based on the threat of retaliation and retaliation cannot occur before the detection of a deviation[9]. With posted bids, the prices and quantities supplied are known immediately and retaliation can take place with no delay.

Forgiving trigger strategies might encourage free riding. The time lag between two interactions is so small that even if renegotiations are possible to restore a collusion after an occurrence of free riding, the renegotiation will not take place before the once-free-riding company gets hurt by the competitive behavior it caused and therefore no temporal defection followed by re-renegotiation will take place.

Collusive behavior is usually very difficult to sustain in large systems because the very high number of market participants with different characteristics increases, for some of them, the incentives for defection[10] and any instability or irrational behavior spreads quickly to the whole system by making collusive behavior sub-optimal. In the market for electric power, transmission constraints create geographically and temporally relevant sub-markets with reduced numbers of generators, sub-markets that are separated from each other's direct influence and protected from any price collapse in another sub-market.

---

[9] With a fair number of competitors, keeping the bids hidden and releasing minimal information about prices may encourage 'free riding' among generators and thus discourage collusion.

10 See (Brock and Schneikman, 1985) for more details on the effect of the number of participants in Bertrand repeated game with capacity limits.

Gaming

Gaming opportunities because of finite transmission capacity

We have seen in previous sections how, when a transmission constraint is active, a reduced number of players interacting in a small market can raise the price above the competitive levels and therefore reduce the social welfare.

Equation (25) gives a mathematical illustration of how a transmission line operating at its thermal limit can influence the residual demands facing a generator or a group of generators and therefore their market power. However, even when no transmission constraint is active, from (26) we get

$$\sum_{i \in R} s_i \geq \sum_{l \in R} d_i - \left(K_{nl} - \sum_{l \in R}(1-\alpha_i)q_i + \sum_{i \notin R} \alpha_i q_i \right) \quad (31)$$

This inequality shows how, even when the transmission capacity is 'large', the mere existence of a finite limit of this capacity guarantees to a generator or a group of generators a minimal residual demand, a guarantee that these profit maximizers might want to exploit.

A simple example

The goal of this sub-section is to give a sense of the type of strategic behavior that could take place, through a series of very simple simulations of three-bus power systems.

First, to develop an intuitive sense of what could happen, we consider the following simplistic network with 2 identical generators G_1 and G_2 at nodes 1 and 2, and an aggregated load L at node 3.

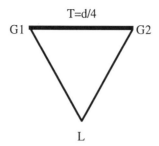

Figure 3

The three lines of this network have the same impedance and the line joining the two generators has a thermal constraint of d/4, where d is the inelastic demand of load L.

For the couple of node 1 -2, q_i being the quantity injected at node i, equation (6) becomes

$$q_{12} = \frac{2}{3}q_1 + \frac{1}{3}q_3 \qquad (32)$$

From (31) and $q_3=-d$,

$$q_1 \geq \frac{d}{8} \qquad (33)$$

We can see how, because of the constraint, the generator G_1 has a protected output equal to d/8 and can therefore decide to raise its price and its profits as much as it wants. Generator G_2 can do the same. The mere existence of a transmission constraint transforms a duopoly that would have respected this constraint at its equilibrium[11] into a quasi-monopoly.

Transmission constraints and Nash equilibrium.

Using the same simple example, we now consider the case where the load is elastic and examine the results given by different topologies and transmission network constraints.

The two generators G1 and G2 have quadratic cost functions:

$$C_i = b_i \cdot q_i + a_i \cdot q_i^2, \quad i = 1,2 \qquad (34)$$

---

[11] Because the network is symmetric.

The two generators are a duopoly and are assumed to compete by providing linear supply functions:

$$p_i = x_i \cdot \frac{\partial C_i}{\partial q_i}, \quad i=1,2 \text{ or,} \tag{35}$$

$$p_i = x_i \cdot (b_i + 2a_i \cdot q_i), \quad i=1,2 \tag{36}$$

where x_i is a strategic variable set by player i.

The load is elastic and has a utility function:

$$U = b_L \cdot (q_1 + q_2) - a_L \cdot (q_1 + q_2)^2 \tag{37}$$

The load is assumed to be an aggregation of smaller loads without market power. Its demand function will therefore be:

$$p_L = b_L - 2 \cdot a_L \cdot (q_1 + q_2) \tag{38}$$

In a market without network externalities, the unique market price that makes supply match demand maximizes the (apparent) social welfare, total utility of the consumers minus the apparent total cost of production: $U - x_1 \cdot C_1 - x_2 \cdot C_2$ or $\int_0^{q_1+q_2} p_L(q) dq - \int_0^{q_1} p_1(q) dq - \int_0^{q_2} p_2(q) dq$. The presence of transmission constraints may not allow such a unique price to exist and the maximum welfare is attained for different prices at different nodes through a coordinating PX that takes the supply function bids of the generators and determines the optimal outputs, or through an appropriate set of trading rules which would make the system attain this optimal scheme through decentralized transactions. The way the outputs are determined is irrelevant to the simulation as long as the outputs and prices are the same. Moreover, whether the lines to be constrained and the outputs of the generators are formally set by a coordinating power exchange (PX) organism or indirectly reached through an appropriate set of rules for decentralized bilateral or multilateral trading, it will always be the *choice of x_1 and x_2 by the generators that will effectively determine which lines are constrained* as well as the optimal outputs $q_1(x_1,x_2)$ and $q_2(x_1,x_2)$.

Knowing these functions, the generators deduce their profits as a function of x_1 and x_2. They can then calculate their reaction functions, which give the optimal choice of x_i for every given x_j. More specifically, depending on what constraints are active, the analytical formulation of the output functions q_1 and q_2 could change. The players will calculate primary reaction functions associated with every type of constrained or unconstrained dispatch. Every player will then build his global reaction function by comparing the profits yielded by his primary reaction functions. These global reaction functions will reflect when a player will choose to constrain the network and when he chooses to leave it unconstrained.

These reactions functions, whose intersections constitute a Nash equilibrium in pure strategies, will show us when a generator chooses to leave the network unconstrained, selling to the residual demand in (24) and when he constrains the network abusing the minimal guaranteed residual demand (31).

In what follows, these reaction functions are presented in different scenarios. In all graphs, the gray line separates the area where the bids (x_1, x_2) will lead the critical transmission line to be operating at its thermal limit (constrained area) from the area where it would be operating below this limit (unconstrained area). The solid lines are the reaction functions $R1$ and $R2$ and the dotted lines are the reaction function as they would have been if the line had no thermal constraint.

Topology 1: no loop flows

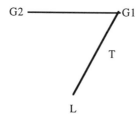

Figure 4

The line linking G1 and the load has a thermal constraint T and therefore,

$$q_1 + q_2 \leq T \quad (39)$$

When the constraint is relatively low (Figure 5), the Nash equilibrium is the same as would be obtained without constraints. When the constraint is tightened (Figure 6), the Nash equilibrium is obtained in the constrained area and for higher prices because the generators, having market power, can exploit the constraint more effectively than a price-taker load.

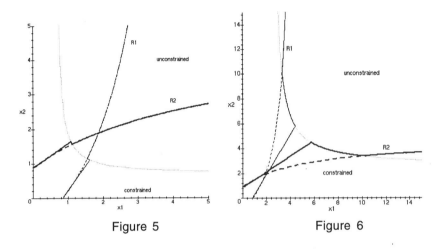

Figure 5   Figure 6

The last case, shown in Figure 7, is an intermediate case where the unconstrained equilibrium would lie in the constrained area but the prices are not high enough to be in the situation of Figure 6. We observe a peculiar situation where a continuum of Nash equilibria seems to be sustainable on the boundary separating the constrained from the unconstrained area. This situation is similar to that observed in the model developed by Wei and Smeers in the framework of a Cournot competition (1997). As in the simulation discussed here, their model deals with transmission constraints but not loop flow.

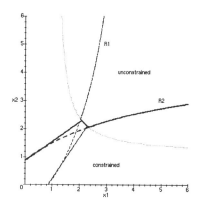

Figure 7

Topology 2: Introduction of loop flow. We add in this example an unconstrained line between G2 and the load, giving to G2 a competitive advantage on G1 because of loop flows.

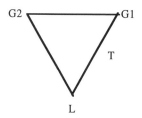

Figure 8

To respect the constraint we must have:

$$\alpha_1 \cdot q_1 + \alpha_2 \cdot q_2 \leq T \qquad (40)$$

with

$$0 < \alpha_2 < \alpha_1 < 1 \qquad (41)$$

where α and β reflect the physical characteristics of the network.

In the first simulation, (Figure 9) the first generator had more power to affect the state of the system, constrained or unconstrained, than G2 because of loop flow. In this example, because the transmission constraint is not very tight, the unconstrained equilibrium is however sustainable.

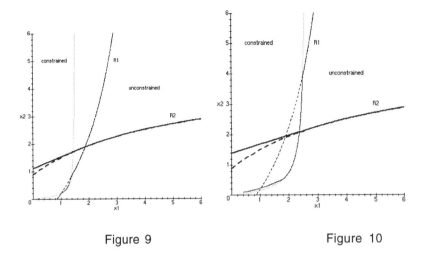

Figure 9   Figure 10

A higher constraint on the network (Figure 10) moves the Nash equilibrium on the boundary separating the constrained from the unconstrained areas. When we raise the inelasticity of the load[12], (Figure 11) we observe how *R2* becomes discontinuous and no longer intersects *R1*, preventing a static Nash equilibrium from existing. This discontinuity is due to the fact that generator G2, because it has a better strategic location, will prefer not to follow the first generators when the bids of the latter are too low, but rather concentrate on raising the price and the profits from the market share that the constraint is protecting for him. It is mostly notable that in this example, an unconstrained equilibrium exists in the unconstrained area and that it would be a feasible Nash equilibrium if the players did not know of the existence of the constraints. However, because G2 knows that the constraint is protecting a portion of his market share, he will deviate from this equilibrium and try to raise its profit further.

---

[12] To raise the inelasticity, we have changed the coefficients of the utility function to make it steeper, under the condition that the outputs at the competitive equilibrium (when the generators are bidding the true cost functions) stay unchanged.

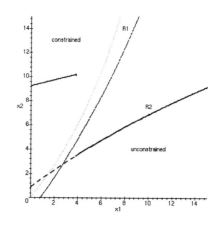

Figure 11

Topology 3: Two-way constraints. In this example, it is the line linking the two generators that has a thermal limit T imposing two constraints on the system:

$$\alpha.q_1 - \beta.q_2 \leq T, \quad (42)$$
$$\beta.q_2 - \alpha.q_1 \leq T \quad (43)$$

with

$$0 < \beta < 1, \ 0 < \alpha < 1 \quad (44)$$

where α and β reflect the physical characteristics of the network.

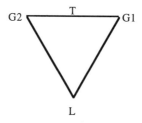

Figure 12

When the line linking the two generators is weak, it can be constrained in its two directions, which creates two constrained areas: "Constrained area 1" where (42) is an equality and "Constrained area 2" where (43) is an equality (Figure 13).

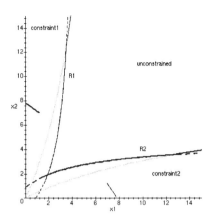

Figure 13

The generators now have new opportunities for strategic bidding: in the last example, only one generator had a market share protected by loop flow and constraints. Here, both do; and that is why we observe discontinuities in both reaction functions in Figure 13. Nevertheless, the constraint is not very restrictive in this example and the reaction curves intersect in the unconstrained area at the unconstrained Nash equilibrium.

However, with a tighter transmission constraint, as in Figure 14, or a more inelastic load, as in Figure 15, the reaction functions may no longer intersect, even if, once again, the unconstrained equilibrium (where the market would settle if the generators did not know the existence of the constraint) lies in the feasible unconstrained area.

Interpreting what would happen in reality when the reaction functions do not intersect and where there is no static Nash equilibrium is not easy. A first basic interpretation might be that every generator will react myopically to the bids of his competitors, observing these bids, reacting accordingly to its reaction function, and leading the market to an unstable situation. In this case, the examples developed above seem to indicate that the bids will always remain above the bids of the unconstrained Nash equilibrium, independent of the physical feasibility of this equilibrium. This is due to the fact that in the cases where the reaction functions did not intersect, at least one of them lies completely above the bet of the unconstrained equilibrium, the other having either the same property or being monotonic.

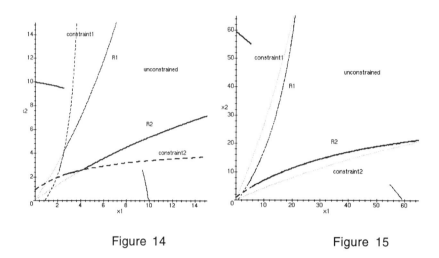

Figure 14   Figure 15

A more sophisticated interpretation is that the generators will play mixed strategies, i.e., that their bets will follow a random process maximizing their expected profits, and also leading to instability. Nevertheless, myopic instability or mixed strategy equilibria may not be sustainable in the long term in a repeated game where continuous interactions (and maybe the fear that a high instability will lead the regulator to investigate and change the rules) could make a more sophisticated inter-temporal Nash equilibria sustainable. In all these scenarios, the load seems to be worse off because the generators have more market power and more opportunities for strategic bidding, which are created by thermal limits on some of the transmission lines.

Solutions

Traditional Solutions

Market power in general, and tacit collusive behavior in particular, seriously threaten the social welfare in the future deregulated power industry and must be treated as such when designing the regulatory framework of this deregulation. Reducing general and locational market power and hindering collusion must be part of the goals of the federal regulator. As reported by Joskow (1996), two methods are usually proposed to reduce market power. They consist of imposing price caps or multiplying the ownership of the generators.

However, imposing price caps does not eliminate the market power but reduces its expression. When significant market power does exist, the prices will remain at the maximum authorized by the caps and the regulator will be virtually setting the transaction prices in peak periods by setting the maximum price cap. Such a practice can lead us closer to mandatory pricing than to true competition during these periods and gives a very high discretionary power to the regulator. Moreover, high prices during peak periods can be a useful and legitimate economic signal to build the marginal generation capacity needed for reliability. It will be very difficult for the regulator to differentiate between the abuse of locational market power in peak period and very high but legitimate prices, useful to reimburse the capital cost of the marginal production units.

On the other hand, multiplying ownership is a useful way to reduce the global market power that utilities might have because of their total size, independent of transmission constraints. Nevertheless, it might not help reduce locational market power created by transmission constraints since some of the local and temporal relevant markets may not be sufficiently large to enable this multiplication. Artificially multiplying the number of firms can be detrimental to the social welfare as shown by Baumol (1982) in his theory of contestable markets. Furthermore, as shown by Brock and Shneikman (1985), on a market where market players with production capacity constraints are playing an infinitely repeated price competition game (or Bertrand supergames), increasing the number of firms has a non monotonic and thus ambiguous effect on collusion because raising the number of firms first encourages collusion by raising the losses incurred during a price war before hindering collusion by raising the relative profits of a defector.

Long-term markets can hinder collusion

Another way to hinder collusion is to change the discount factor between two interactions by changing the frequency of these interactions. As reported by Tirole (1988) and confirmed by Brock and Shneikman (1985) in the case production capacity constraints exist, the discount rate between two interactions is an essential stability factor of collusive behavior because every participant evaluates a tradeoff between immediate benefits and the future cost of non-cooperation, and a larger discount rate always diminishes the value of future retaliation. Making the interaction less frequent raises this rate and could prevent collusion. It seems reasonable that while bidding every half hour might discourage free-riding and make collusion sustainable, long term contracts imply a very large discount rate that might discourage, in turn, collusive behavior.

To illustrate this claim, we consider again the infinitely repeated game between n generators presented previously. It is shown in (Younes and Ilic, 1997) that tacit collusion is not sustainable for any values of k and n if T satisfies:

$$T \geq \frac{\ln(2)}{r} \qquad (45)$$

In other words, for any number of generators of any size facing any linear demand and for a discount rate of 10%, contracts for delivery of electric power during 7 years or more should always help hinder collusive behavior. Of course this result gives only an order of magnitude, and more realistic data should be collected to analyze this phenomenon and determine *Tmin* with non-linear demand, increasing or asymmetric marginal cost functions, and a potential "supply function" competition scheme. Furthermore, the nature of the interactions between a long-term market and a spot market is critical because uncertainty makes a long-term market alone not feasible.

Grid expansion strategies can reduce locational market power

Hindering collusion is not sufficient to ensure a competitive market. As we have seen previously, even a simple Bertrand competition is likely to give-non competitive outcomes in the relevant sub-markets. A more global strategy is needed to eliminate, or at least decrease, the locational market power created by constraints. Price caps and the multiplication of ownership might certainly be useful but still have the side effects and limitations discussed earlier. A more natural strategy would be to eliminate the cause of locational market power, i.e., transmission congestion, via a network expansion, or, even better, via the threat of network expansion. An expansion policy for the grid that takes into consideration the *used* market power in relevant sub-markets (e.g., via the prices that generators request) could constitute a credible threat against the use of market power. The credibility of the threat seems to be linked to the use of ex-ante characteristics of the market to remunerate the investors. For example, an expansion strategy that would yield the appropriate financial revenues to link two nodes whose electricity prices are sufficiently different *when* the investment is decided, and not *after* the line is built, could discourage generators at the expensive node from abusing their market power and raising the price too much, in order to stay separated from the unconstrained area and benefit from the modest rents that this strategy can give them[13].

To illustrate how a policy for grid expansion that determines lines to be expanded as a function of ex-ante conditions on the market can reduce the market power of the market participants, we have made a second simulation on the IEEE reliability system. Using the same software as previously (Macan 97), we are able to estimate the optimal expansion for the lines 7 and 14 through 17 under the peak load pricing method (Lecinq and Ilic, 1997; Macan, 1997)[14]. This capacity is a function of the costs of expansion, the demand functions, and, most importantly, the bids that the generators are making. It is noteworthy that, in this model, the generators are giving their bids

---

[13] Nevertheless, because of environmental considerations, new lines cannot always be built and this impossibility reduces the credibility of the threat. Investment in Flexible AC Transmission Systems (FACTS) devices could help reestablish this credibility.

[14] In the peak load pricing method, the expansion of the transmission lines and the nodal prices are calculated to maximize the total social welfare, taking into consideration the generation price and the consumer surplus (as in Hogan's nodal pricing approach (1992)) as well as the capital cost of the grid expansion (as distinct from the Hogan approach). It is interesting that the nodal prices given by the peak load pricing are the same as those the Hogan nodal pricing calculations would give when the line capacities are set at the optimal values given by the peak load pricing method.

368 POWER SYSTEMS RESTRUCTURING

before the expansion of the line is decided; i.e., the expansion policy of the grid depends on the ex-ante behavior of the market participants and characteristics of the market. To analyze how the expansion can affect the market power, we have added a fourth demand level, greater than the former three, and set the original capacities of the five critical lines to their optimal values determined by the peak load pricing method when all the generators are bidding their real cost function in the four demand levels. We then consider two cases: (1) the transmission capacity is fixed, and (2) an expansion of lines 7 and 14 through 17 can take place to reach the 'optimal' capacity given by the peak load method for the present bids. The two scenarios would give the same result when the generators communicate their true cost functions and no further expansion would occur in either case. We have then made the generators that are at nodes 1, 2 and 7 (that we have shown to possess significant locational market power) raise all together their bids by 10% in the four demand levels. The following two tables summarize the increases in prices and profits that this 10% increase in bidding has produced in these two cases.

Demand	PRICES at nodes				Profits at nodes 1, 2, and 7	
level	1	2	7	Other nodes	Range	Aggregated
1	1.4%	0.9%	1.6%	1.2% to 0.6%	0% to 2.9%	
2	3.3%	3.4%	2.6%	-0.2% to 1.7%	0% to 8.4%	+ 21%
3	10.2%	10.0%	10.1%	-0.9% to 0.1%	24.5% to 45.1%	
4	10.8%	10.6%	9.5%	-1.3% to 1%	19% to 32.3%	

Table 3. Changes in prices and profits with fixed transmission capacity

Demand	PRICES		at	nodes	Profits 1, 2, and 7 at nodes	
level	1	2	7	Other nodes	Range	Aggregated
1	1.4%	1.1%	1.8%	0.5% to 1.8%	0% to 3.6%	
2	1.4%	1.5%	1.2%	0.5% to 1.2%	0% to 2.6%	+1%
3	3.3%	3.1%	3.1%	3.3% to 3.7%	6.3% to 10.7%	
4	-0.4%	-0.5%	0.3%	1.2% to 3.2%	-3.2% to -0.3%	

Table 4: Changes in prices and profits with 'optimal' expansion

When colluding together and raising their bids by 10%, and when the transmission capacity is fixed to its 'optimal' value, the five generators at nodes 1, 2, and 7 are able to raise their cumulative profits by 21%. If we allow the transmission lines to expand accordingly to the bids of these generators, the same increase in bidding raises the cumulative profits by less than 1%, some of the generators being worse off. The locational market power in the studied sub-market is seriously weakened and cartelisation is not sustainable. It is noteworthy that, in these conditions, non-competitive bids are therefore unlikely to be observed as is an *effective* expansion of the grid. The *threat* of a transmission grid expansion is sufficient to remove the market power of the generators and alter the relevant character of this sub-market.

Implementation

The practical implementation of a long-term market seems feasible. A long-term market, competing with the spot market, already exists in the UK and in the majority of the other deregulated electric power markets. If it is true that such a market is excluded from the purest approach to nodal pricing, however, even the market that will be the closest in its approach to the nodal pricing, California, is very likely to allow long term contracts between producers and consumers[15].

The implementation of an expansion strategy for the transmission grid that would yield credible threats to those abusing their market power could be, in some situations, more difficult. The ultimate degree of deregulation would be achieved when the investments in the transmission network are done in an uncoordinated way by independent investors owning the lines they build and earning benefits from operating these lines. In that case, the remuneration of this investor depends only on the ex-post congestion of the network (e.g., Transmission Congestion Contracts). An investment that would reduce congestion and eliminate the abuse of market power might not yield enough revenues to the investors and the threat of such an investment would not be credible. Nevertheless, there are very few proposed methods that could send appropriate signals for investment in transmission and there are even fewer methods that can do it for decentralized investors. The most elaborated one, proposed by Bushnell and Stoft (1996), rely on Transmission Congestion Contracts that would be distributed to the investors following to a feasibility rule. This proposal would yield the correct incentives for the investors under restrictive conditions. However, when these conditions are not met, market participants might find it profitable to make alterations in the transmission grid that are detrimental to the system.

Most likely, the expansion of the transmission grid will be undertaken by an independent, but regulated, centralized operator or by those that are benefiting directly from the transmission network: Utilities and distribution companies. For example, in England and Wales the transmission is controlled, managed, and its expansions are planned by the National Grid Company (NGC). The NGC is jointly owned by the regional distribution companies, though recently these companies have considered divesting their shares in NGC and making it an independent company. Nevertheless, NGC remains tightly regulated and it is required to facilitate competition and meet users' reasonable requests for connection to the system (Green, 1997). Implementing an expansion strategy similar to that recommended in

---

[15] As confirmed by several Utilities delegates at the 18th IAEE conference (Sep. 97, San Francisco).

this report is relatively easy and could be done through the regulations binding this company. Specifying, through these regulations, that the NGC have to build new lines when locational market power is abused would give the needed credibility to the threat of expansion. In the future deregulated California market, the transmission lines are owned by the Utilities and the ISO may have no financial interest in any transmission facility. Under a proposal submitted by California's three largest investor-owned electric utilities to the Federal Energy Regulatory Commission, the transmission owners would have an obligation to build all transmission facilities the ISO decides are needed for economy or reliability (Burkhart, 1997). Implementing the expansion strategy proposed here in the California market will only require the ISO to have a transparent policy for expansion of the network that would specify that such an expansion will take place when locational market power is abused. Finally, New Zealand is a very interesting case. The New Zealand system is very close to the spot pricing method and Trans Power, a state owned utility, is responsible for operating the transmission system and for planning expansions. System expansions are justified if the difference in prices with and without a line equals the cost of the line (Green, 1997). This system might very well yield appropriate and credible threats against abuses of market power that would raise some nodal prices and justify new investments. Nevertheless, the market-based rule used by the industry is that investment should only be done if a coalition of users is willing to pay for it. The problems of free riding could alter the credibility of the threat.

Conclusion

Transmission constraints and loop flow create a double threat to the emergence of a competitive electricity market. (1) When these constraints are too tight, they separate sub-markets from the rest of the network and raise the locational market power of generators located there, even if these markets are still linked by other unconstrained lines to the rest of the network. The reduction in the number of players in interaction leads to higher prices as usually recognized, and sometimes to tacit collusion due to repeated periodic interactions. (2) A Transmission line can also lead to inefficient behavior even when it has a capacity large enough to allow normal operations of the network without being constrained. It could become a source of inefficiencies and higher market prices if the generators realize that they can make profits by strategically constraining and abusing their protected market share with very high bids. This behavior would raise the expected prices and could create market instabilities.

The problems of global and locational market power have frequently been evoked and sometimes used to offer counter-examples in the debate over the deregulation of the electric power industry. However, they are rarely directly taken into consideration in the different general proposals that have been made so far. Multiplication of ownership and regulatory price caps are the most frequently advocated solutions against market power; they are classical methods that can be used with any deregulatory scheme but do not deal explicitly with the specificities of this market.

It is our opinion that the problems raised by locational market power and its corollaries, tacit collusion and gaming, cannot be resolved using artificial tools that ignore the specificities of this market. The solution should rather be an integral part of any proposal that claims to produce an efficient market. We have investigated in this chapter two of the features that an efficient deregulatory scheme might or should have to reduce the effect of market power. First, long term contracts could help hinder tacit collusive behavior by encouraging free riding. More generally, locational market power can be decreased through a transmission grid expansion policy that depends on the *ex-ante* behavior of the market participants and other characteristics of the market and that would expand the weak links when locational market power is abused.

We have shown that such an expansion policy can produce the appropriate threats against the market participants, drastically decreasing their market power without necessarily producing any real expansion of the transmission grid.

Explicitly defining such an expansion policy must be the next objective of further research. This policy must meet two goals. (1) It should first induce, *ex-ante*, credible threats against the abuses of market power. (2) It must also produce, *ex-post*, a transmission network that has the appropriate dimensions. Moreover, it must also acknowledge the technical, environmental, and financial constraints.

In order to achieve the second goal, the threats of over-building the network must be dissuasive enough that it has only to be rarely executed. However, due to technical constraints, the time needed to build a new line might be significant enough to make it worthwhile for a market participant to raise the market prices even if a new line will be built, thus reducing the dissuasive power of the threat. Moreover, environmental constraints make it unlikely that new transmission lines will be built on certain locations, reducing the credibility of the threat of expansion.

It seems that FACTS devices might be used to restore the credibility and dissuasive power of the threat, because their installation can be less time and capital consuming and more environmentally friendly than building physical transmission lines.

Funding the expansion strategy can also be a serious concern since no threat is credible, and no needed transmission capacity can be built, when funds are not available. The transmission costs paid by the market participants, or at least their expected value, must be sufficient to cover the capital cost of the grid and to remunerate the investors. If there are, as many authors agree, economies of scale in the transmission of electric power, pricing at the marginal cost will not yield enough revenues to cover the costs and some additional charges must be paid in order to equilibrate the budget. These charges ought to be chosen in a way that minimizes economic distortions. A good candidate would be a method similar to the voluntary price discrimination used by insurance companies. Different combinations of fixed and variable charges, from which market participants would have to choose, could be proposed.

Finally, once the problems of credibility and funding resolved, we will still have to determine how the investment decisions will be taken, who will execute them, how would he be paid for them (as distinct from how to get the money to pay him), and who will own the physical lines. When the regulator decides that a centralized and regulated body (e.g., an ISO) will own and develop the network, designing an expansion strategy that has the characteristics described above can be (relatively) easy since the regulations that bind this body will give it the power to make his threats credible. Nevertheless, if the regulator chooses to go a step further and allow decentralized investments in the transmission network, an independent and unregulated investor would not execute a project unless he is expecting benefits. If the remuneration of this investor depends only on the ex-post congestion of the network, the threat of investment will not be credible. Future research should try to find a mechanism that would create a credible threat of individual investors over-building the network when market power is abused, while producing, *ex-post*, a well designed but competitively built network. This is definitely one of the most challenging goals to be achieved in this field.

Appendix: Oligopolistic competition and game theory

In a dynamic game with discrete stages, a number of players are making a strategic decision at each stage. The combined decisions of all players determine, through a middleman or in a decentralized market setting, the quantities of good exchanged between the players as well as the prices at which they are exchanged, and, therefore, the profits of the players for the stage.

Let n be the number of players

- k the stage index
- x_i the strategic variable whose value player i chooses in the set E_i
- $x_i(k)$ the strategic decision of player i at stage k

π_i the one-stage profit function of player i,
$\pi_i(x_1(k),..,x_n(k))$ is its profit at period k

r the discount rate between two consecutive stages

X the vector $[x_1,...,x_n]$

X_{-i} the vector $[x_1,..., x_{i-1}, x_{i+1},...,x_n]$

One-stage competition

In a one-stage competition framework, player i assumes that all other players are taking fixed decisions Xo_{-i}. He will give to his own decision variable x_i the value xo_i that maximizes his profit.

$$\pi_i(xo_i, Xo_{-i}) = \underset{\alpha \in E_i}{Max} \pi_i(\alpha, Xo_{-i}) \qquad (A1)$$

Or, when x_i takes real values and π_i is differentiable,

$$\frac{\partial \pi_i}{\partial x_i}(xo_i, Xo_{-i}) = 0 \qquad (A2)$$

When π_i has continuous second derivatives, the implicit function theorem (Ramis, Odoux and Deschamps, 1988) ensures locally the existence of a reaction function R_i, whose partial derivatives exist and are continuous, so that

$$\frac{\partial \pi_i(R_i(X_{-i}), X_{-i})}{\partial x_i} = 0 \qquad (A3)$$

where,

$$R_i(X_{-i}) = x_i \qquad (A4)$$

gives the optimal choice of player *i* for X_{-i} fixed.

Definition 1: We say that X^* is a static Nash equilibrium if

$$\forall i, \frac{\partial \pi_i(X^*)}{\partial x_i} = 0 \qquad (A5)$$

At a static Nash equilibrium, every strategic variable x_i^* is the optimal choice of player i for X_{-i}^* fixed. X^* is also an intersection point of all reaction functions:

$$\forall i, \; x_i^* = R_i(X_{-i}^*) \tag{A6}$$

In an oligopolistic situation with only one stage, the static Nash equilibrium is likely to be obtained if all players can observe the decisions of all other players and change their own decisions in real time. If the bids are revealed only after the market clears, attaining this equilibrium is unlikely and depends on the information that every player has since he will make his decision as a function of what he *expects* the other decisions to be.

Dynamic process and stability of static Nash equilibria

In a dynamic framework, at every stage the players learn more about the decisions of the others and the strategic variables may converge to the static Nash equilibrium even if the bids are not revealed before the market clears at any given stage. Assuming that a player knows the bets of other players in the *previous* stage, not knowing what they will be at the current stage, the simplest strategy would be for him to assume that the bets are staying as they were in the previous period and to maximize his profit accordingly. In this case,

$$\forall i, \; \forall k, \; x_i(k+1) = R_i(X_{-i}(k)) \tag{A7}$$

This set of equations describes dynamically how the bets at the current stage depend on the bets at the previous stage. It is clear that if the process defined in (A7) converges, its limit is a static Nash equilibrium. The stability of this equilibrium is an important question. Studying the stability of an equilibrium requires determining whether the system can converge to this equilibrium independently of the initial conditions (general stability), and/or whether it can return to it after a small perturbation (local stability).

When X^* is a static-Nash equilibrium,

$$\forall i,\ x_i(k+1) - x_i^* = R_i(X_{-i}(k)) - R_i(X_{-i}^*) \tag{A8}$$

$$\forall i,\ \forall k,\ x_i(k+1) = x_i^* + \sum_{j \neq i}\left(\frac{\partial R_i}{\partial x_j}(X_{-i}^*)\right)\cdot(x_j(k) - x_j^*) +$$

$$O\|X_{-i}(k) - X_{-i}^*\|^2 \tag{A9}$$

From (A3) we get:

$$\forall i,\ \sum_{j \neq i}(\frac{\partial^2 \pi_i}{\partial x_i x_j}dx_j) + \frac{\partial^2 \pi_i}{\partial x_i^2}\cdot \sum_{j \neq i}\frac{\partial R_i}{\partial x_j}dx_j = 0 \tag{A10}$$

When $\frac{\partial^2 \pi_i}{\partial x_i^2}$ is different from 0 for all i, equation (A9) can be rewritten as:

$$\forall i,\ \forall k,\ (x_i(k+1) - x_i^*) = \sum_{j \neq i} -\left(\frac{\frac{\partial^2 \pi_i}{\partial x_i x_j}}{\frac{\partial^2 \pi_i}{\partial x_i^2}}(x_{-i}^*)\right)\cdot (x_j(k) - x_j^*)$$

$$+ O\|X_{-i}(k) - X_{-i}^*\|^2 \tag{A11}$$

Linearizing equation (A11) gives:

$$(X(k) - X^*) = A^k \cdot (X_0 - X^*) \tag{A12}$$

where the elements of matrix A are,

$$a_{ij} = -\left(\frac{\frac{\partial^2 \pi_i}{\partial x_i x_j}}{\frac{\partial^2 \pi_i}{\partial x_i^2}}\right),\ i \neq j\ \text{and}\ a_{ij} = 0,\ i = j \tag{A13}$$

Theorem: The stability of the linearized system, and, therefore, the local stability of the static Nash equilibrium, is ensured when the eigenvalues of matrix A are less than 1 in module. Furthermore, when equation (A11) is linear, the same condition implies general stability.

This theorem is an immediate application of (Aoki, 1976).

Inter-temporal Nash strategies

Nash static equilibrium, although useful for analyzing many situations, has a major weakness in multistage games since it does not recognize that the decision of a player at one stage might affect the decisions of other players at further stages and that this player might take his decision accordingly. In order to recognize this interdependence in multistage games, we can define Nash equilibria in terms of strategies rather then in terms of strategic variables, where a strategy chosen by a player determines his decision at a stage as a function of what the other players did at past stages of the game.

We say that the function S_i is the inter-temporal *strategy* of player i when the decision of player i at the stage k is given by

$$x_i(k) = S_i(k, X(0), \ldots X(k-1)) \qquad (A14)$$

If $S = [S_1, \ldots S_n]$ is the set of strategies of all players, their strategic decisions at every stage will be defined recursively by the strategies S and the vector X at stage k will be $X(k, S)$.

Definition 2: We say that $S^* = [S^*_1, \ldots, S^*_n]$ is an inter-temporal Nash equilibrium if every strategy S^*_i maximizes the discounted sum of the profits of player i given the other strategies $S^*_{-i} = [S^*_1, . S^*_{i-1}, S^*_{i+1}.., S^*_n]$ as fixed. Or,

$$\forall i, \ \sum_k \frac{\pi_i(X(k, S^*))}{(1+r)^k} = \underset{S_i}{Max} \sum_k \frac{\pi_i(X(k, S_i, S^*_{-i}))}{(1+r)^k} \qquad (A15)$$

When X^* is a *static Nash equilibrium*, staying at X^* at every period (strategies S^* where $X(k, S^*) = X^*$) is also an *inter-temporal Nash equilibrium*. However, it might not be the unique or even a Pareto-optimal[16] inter-temporal Nash equilibrium.

---

[16] An equilibrium is Pareto-optimal if there is no other *equilibrium* that makes *all* the players better off or indifferent.

Cournot competition, Bertrand competition and perfect markets

A Bertrand competition is an oligopolistic framework where the strategic variables $(x_i)_{i \leq n}$ are the prices $(p_i)_{i \leq n}$ of the goods produced by the players, a static Nash equilibrium is then called a Bertrand-Nash equilibrium. In a Cournot competition, the strategic variables $(x_i)_{i \leq n}$ are the quantities $(q_i)_{i \leq n}$ of goods produced by the players, and an equilibrium would be a Cournot-Nash equilibrium. A perfect market is an extreme case of Cournot competition where the number of players is infinite and the price of the market p is an exogenous constant rather than a function of the quantities. In a perfect market, equality (A5) shows how every player will produce the quantity that equalizes its marginal cost with the market price.

Appendix 2: Proof of theorem 1

Rewriting equation (5) in a matrix form gives us:

$$\begin{bmatrix} q_1 \\ q_2 \\ \cdot \\ q_n \end{bmatrix} = \begin{bmatrix} \sum_{i \neq 1} Y_{1i} & -Y_{12} & \cdot & -Y_{1n} \\ -Y_{21} & \sum_{i \neq 2} Y_{2i} & \cdot & \cdot \\ \cdot & \cdot & \cdot & -Y_{n-1n} \\ -Y_{n1} & \cdot & -Y_{nn-1} & \sum_{i \neq n} Y_{ni} \end{bmatrix} \begin{bmatrix} \theta_1 \\ \theta_2 \\ \cdot \\ \theta_n \end{bmatrix} \quad (7)$$

We can eliminate the last equation, to get a system of independent linear equations and take the last column out of the matrix to have a square matrix:

$$\begin{bmatrix} q_1 \\ q_2 \\ \cdot \\ q_{n-1} \end{bmatrix} = \begin{bmatrix} \sum_{i \neq 1} Y_{1i} & -Y_{12} & \cdot & -Y_{1n-1} \\ -Y_{21} & \sum_{i \neq 2} Y_{2i} & \cdot & \cdot \\ \cdot & \cdot & \cdot & -Y_{n-2n-1} \\ -Y_{n-11} & \cdot & -Y_{n-1n-2} & \sum_{i \neq n} Y_{n-1i} \end{bmatrix} \begin{bmatrix} \theta_1 \\ \theta_2 \\ \cdot \\ \theta_{n-1} \end{bmatrix} - \theta_n \begin{bmatrix} Y_{1m} \\ Y_{2n} \\ \cdot \\ Y_{n-1n} \end{bmatrix} \quad (8)$$

Since,

$$\begin{bmatrix} \sum_{i \neq 1} Y_{1i} & -Y_{12} & . & -Y_{1n-1} \\ -Y_{21} & \sum_{i \neq 2} Y_{2i} & . & . \\ . & . & . & -Y_{n-2n-1} \\ -Y_{n-11} & . & -Y_{n-1n-2} & \sum_{i \neq n} Y_{n-1i} \end{bmatrix} \begin{bmatrix} \theta_n \\ \theta_n \\ . \\ \theta_n \end{bmatrix} = \theta_n \begin{bmatrix} Y_{1m} \\ Y_{2n} \\ . \\ Y_{n-1n} \end{bmatrix} \quad (9)$$

using (8) and (9)

$$\begin{bmatrix} q_1 \\ q_2 \\ . \\ q_{n-1} \end{bmatrix} = \begin{bmatrix} \sum_{i \neq 1} Y_{1i} & -Y_{12} & . & -Y_{1n-1} \\ -Y_{21} & \sum_{i \neq 2} Y_{2i} & . & . \\ . & . & . & -Y_{n-2n-1} \\ -Y_{n-11} & . & -Y_{n-1n-2} & \sum_{i \neq n} Y_{n-1i} \end{bmatrix} \begin{bmatrix} \theta_1 - \theta_n \\ \theta_2 - \theta_n \\ . \\ \theta_{n-1} - \theta_n \end{bmatrix} \quad (10)$$

Let this matrix be Y and its inverse $Z=[z_{ij}]$, $i,j=1,...,n-1$.

$$\begin{bmatrix} \theta_1 - \theta_n \\ \theta_2 - \theta_n \\ . \\ \theta_{n-1} - \theta_n \end{bmatrix} = Z. \begin{bmatrix} q_1 \\ q_2 \\ . \\ q_{n-1} \end{bmatrix} \quad (11)$$

Since

$$q_{1n} = Y_{1n}(\theta_1 - \theta_n) \quad (12)$$

then (11) and (12) give

$$q_{1n} = \sum_{i=1}^{n} \alpha_i q_i \text{ with } \alpha_i = Y_{1n} z_{1i}, \ i=1,...n-1 \text{ and } \alpha_n = 0 \quad (14)$$

This equation shows how the injections in the network are linked to the flow through line 1n. More information regarding the coefficients α_i of this equation is now useful.

Y being a matrix with a dominant diagonal and negative values of this diagonal,

$$z_{ij} \geq 0, \quad i,j=1,...,n-1 \quad (15)$$

Moreover,

$$Y.\underline{1} = \begin{bmatrix} Y_{1n} \\ Y_{2n} \\ \cdot \\ Y_{n-1\,n} \end{bmatrix} \text{ or } \underline{1} = Z. \begin{bmatrix} Y_{1n} \\ Y_{2n} \\ \cdot \\ Y_{n-1\,n} \end{bmatrix} \qquad (16)$$

Z being symmetric because Y is symmetric, $z_{1i}=z_{i1}$, $i=1,..n-1$, and

$$\alpha_i = 1 - \sum_{j=2}^{n-1} z_{ij} Y_{jn} \leq 1 \qquad (17)$$

$$\alpha_i = z_{1i} Y_{1n} \geq 0 \qquad (18)$$

Using a similar calculus we get

$$q_{n1} = \sum_{i=1}^{n} \beta_i q_i \qquad (19)$$

with $\beta_i \in [0,1]$ and $\beta_1 = 0$

using equation (14),

$$0 = \sum_{i=1}^{n} (\beta_i + \alpha_i) q_i \qquad (20)$$

and, using (2), we get

$$0 = \sum_{i=2}^{n} (\beta_i + \alpha_i - \alpha_1) q_i \qquad (21)$$

The n-1 variables $q_2,...,q_n$ are independent when they take values small enough so that no constraint on the network is active. Therefore, the coefficients of this equation must all be equal to zero and

$$\alpha_1 - \alpha_i = \beta_i \geq 0, \quad i=1,..,n \qquad (22)$$

or

$$\alpha_1 \geq \alpha_i, \quad i=1,..,n \qquad (23)$$

which proves theorem 1.

References

Aoki, M (1976). "Optimal Control and System Theory in Dynamic Economic Analysis." North-Holland: 123.

Baker, B. and T Bersnahan. (1988). "Estimating the Residual Demand Curve Facing a Single Firm." International Journal of Industrial Organization 6:283-300.

Baumol, W. (1982). "Contestable Markets: An Uprising in the Theory of Industry Structure." American Economic Review 72(1): 1-15.

Baumol, W., P. Joskow and A. Kahn. (1994) "The Challenge for Federal and State Regulators: Transition from Regulation to Efficient Competition in Electric Power" Edison Electric Institute, Dec. 9

Borenstein, S. and J. Bushnell (1996) "An Empirical Analysis of Market Power in a Deregulated California Electricity Market" University of California Energy Institute. Berkeley.

Brock, W. and J. Schneikman (1895). "Price Setting Supergames with Capacity Constraints." Review of Economic Studies 11: 371-382.

Burkhart, L (1997). "ISO/PX Plan Goes to FERC; BPA Unhappy." Public Utilities Fortnightly May 15: 46-47.

Bushnell, J. and S. Stoft (1996). "Electric Grid Investment Under Contract Network Regime." Journal of Regulatory Economics 10: 61-79

Cardell, J., C. Hitt and W. Hogan (1996) "Market Power and Strategic Interaction in Electricity Networks". Technical Report. Harvard Electricity Policy Group.

Chamberlain, E. (1929). "Duopoly, Value Where Sellers Are Few." Quarterly Journal of Economics 43:63-100.

Edgeworth, F. (1897). "La Teoria Pura del Monopolio." Giornale degli Economisti 40:13-3 or (1925). "The Pure Theory of Monopoly." Papers Relating to Political Economy vol.1, ed. F. Edgeworth (London: Macmillan).

Friedman, J. (1971). "A Non-Cooperative Equilibrium for Supergames." Review of Economic Studies 28:1-12.

Green, R. (1997). "Electricity Transmission Pricing: An International Comparison." Presented at the 15th Energy Modeling Forum At Stanford University, September.

Green, R. and D. Newberry (1992). "Competition in the British Electric Spot Market." Journal of Political Economy 100: 929-953.

Hobbs, B. (1986). "Network Models of Spatial Oligopoly with an Application to Deregulated Electricity Generation" Operations Research 34(3) 395-409.

Hogan, W. (1992). "Contract Networks for Electric Power Transmission." Journal of Regulatory Economics. 4: 211-242.

IEEE (1979). "Reliability Test System." IEEE Transactions on Power Apparatus and Systems PAS-98(6).

Ilic, M., E.H. Allen and Z. Younes (1997), "Transmission Scarcity: Who Pays?" The Electricity Journal July:38-49.

Joskow, P. (1996). "Restructuring, Competition and Regulatory Reform in the US Electricity Sector." MIT Department of Economics WP. December.

Klemperer, P.D. and M.A. Meyer (1989). "Supply Function Equilibria in Oligopoly Under Uncertainty." Econometrica 57:1243-1277.

Kreps, D. and J. Schneikman (1983). "Quantity Pre-commitment and Bertrand Competition Yield Cournot Outcomes." Bell Journal of Economics 14: 326-337.

Landes, W and R. Posner (1981). "Market Power in Antitrust Cases." Harvard Law Review 94:937-996.

Lecinq, B. and M. Ilic (1997). "Peak Load Pricing for Electric Power Transmission." Proc. of the Hawaii International Conference on Systems Sciences January.

Lerner. (1934). "The Concept of Monopoly and the Measurement of Monopoly Power." Review of Economic Studies 157.

Macan, E. (1997). "Peak Load Pricing for the IEEE Reliability Test System." MIT LEES TR 97-004.

Oren, S. (1997). "Economic Inefficiency of Passive Transmission Rights in Congested Electricity Systems with Competitive Generation." The Energy Journal 18(1): 63- 83.

Ramis, E., Cl. Deschamps, J. Odoux (1988). "Cours de Mathématiques Spéciales." Masson, Paris.

Scheffman, D and P. Spiller (1987). "Geographic Market Delineation Under the US Department of Justice Merger Guidelines." Journal of Law and Economics 30:123-147.

Schweppe, F., M. Caramanis, R. Tabors and R. Bohn (1988). "Spot Pricing of Electricity". Kluwer Academic Publisher.

Singh, H., S. Hao and A. Papalexopoulos (1997). "Power Auctions and Network Constraints." IEEE 608-614.

Smeers, Y. and J.-Y. Wei (1997). "Spatially Oligopolistic Models with Opportunity Cost Pricing for Transmission Capacity Reservations. A Variational Approach." CORE Discussion 9717, Université Catholique de Louvain, Louvain-la-Neuve.

Stigler, G. (1964). "A Theory of Oligopoly." Journal of Political Economy 72:44-61.

Tirole, J. (1988). "The Theory of Industrial Organization." The MIT Press.

Wei, J.-Y. and Y. Smeers.(1996). "Procedures for Computing Equilibria of the Oligopoly Expansion Game Under Predetermined Multi-period Pricing, Part 2: Second Lowest Marginal Cost Pricing." CORE Discussion 9622, Université Catholique de Louvain, Louvain-la-Neuve.

Wei, J.-Y. and Y. Smeers (1996). "Spatial Oligopolistic Electricity Models with Cournot Generators and Regulated Transmission Prices." Forthcoming in Operation Research.

Werden, G.J. (1996). "Identifying Market Power in the Electric Generation" Public Utilities Fortnightly 134(2): 16-21.

Wolfram, C.D. (1995). "Measuring Duopoly Power in the British Electricity Market." MIT Department of Economics, WP, November.

Wu, F., and P. Varaiya (1995). "Coordinated Multilateral Trades for Electric Power Networks: Theory and implementation." University of California Energy Institute, PWP-031

Younes, Z. and M. Ilic (1997a). "Transmission System Constraints in Non-Perfect Electricity Market." Proc. of 18th Annual North American Conference USAEE/IAEE 256-265.

Younes, Z. and M. Ilic (1997b). "Generation Strategies for Gaming Transmission Constraints. Will the Deregulated Electric Power Industry be an Oligopoly?" to appear in proc. of the Hawaii International Conference on System Sciences, January 6-9, 1977, Kona, Hawaii

10 COMPETITIVE ELECTRIC SERVICES AND EFFICIENCY

Stephen R. Connors

Director, Electric Utility Program
Energy Laboratory
Massachusetts Institute of Technology
Cambridge, MA

Introduction

While FERC restructures the wholesale market for electrical energy, and the interstate network of transmission lines that supports it, state agencies and electric companies are rapidly redefining themselves to compete in the retail market. How vibrant the retail market for electrical and energy services becomes will either shut the door or open the floodgates for end-use technologies. Customer choice is required at some level if new technologies beyond central station generation, transmission and distribution are to enter the market at accelerated rates. Sophisticated load management in response to spot market prices is not enough. Fuel diversity, protection from shifting environmental regulations, rate stability, environmental performance, localized reliability and power quality are all "products" that certain customers are willing to pay for. Integration of these functions, technologies, products and markets is key.

State regulators have recognized this problem, and have come up with some monopoly-competitive market hybrid proposals which not only enable companies to offer such services to customers who want them, but also require that each electric customer support some energy efficiency and renewable energy initiatives. Systems benefits charges, portfolio standards and green pricing are just several of these. Such initiatives *intend* to promote broadly defined efficiency improvements, ranging from more energy efficient electricity consuming devices, optimized operation and coordination of devices, more sophisticated portfolio management, and ultimately the development and deployment of innovative new electric consuming technologies and processes—the so-called "electrotechnologies." Initiatives by environmental regulators may also have a beneficial long-term impact, as multiple emissions constraints provide a set of industry-wide performance targets requiring longer-term strategies.

Simultaneously, businesses are recognizing the synergies between technological innovation and the ability to attract and retain customers. Key to the development of such efficient strategies are the kinds of market signals and constraints that will be put into place via various restructuring rules. What will their effect be on the introduction of energy efficiency and other "customer-based" technologies? This chapter reviews the technological trends, corporate strategies and regulatory policy initiatives associated with the development of the retail market, and its prospects for the introduction of more efficient end-use technologies.

The Competitive Environment

It is recognized that competition at the wholesale level requires that generators compete against one another on price, energy, environmental performance, availability, and other such factors. They may have multiple customers for their electrical energy; distribution companies, cooperatives, municipals, the power exchange, brokers, marketers, and ultimately individual consumers. Based upon their location and the responsiveness of their machines, they may also be selling capacity, frequency and voltage support, and other such services in the ancillary services market. Superior environmental characteristics may also allow them to participate in the growing markets for tradable emissions permits such as sulfur dioxide, nitrogen oxides, and ultimately carbon dioxide. These secondary markets, and the retail market, are essential in order to differentiate products beyond price, and create niches for new technologies.

In the United Kingdom and elsewhere, where a mandatory pool (PoolCo) structure was initially set up, primary customers have been distribution utilities who purchase electricity though a central clearinghouse—the power exchange—on behalf of their native load customers. This was essentially a pure commodity market where electrical energy was differentiated only on price, and customers had no leverage to select certain classes of generators, let alone individual power producers. In California, which has lead the charge towards a competitive market in the United States, a voluntary PoolCo arrangement has been set up, whereby the power exchange still handles the majority of the electricity purchases, but side-deals–or bilateral contracts–between specific generators and customers are permitted. This provides customers greater leverage to choose their generator, but is still essentially a price-dominated commodity market. Then there are the proposals which emphasize retail over wholesale competition. These are aimed at providing consumers greater choice over who is generating on their behalf. From the power system's viewpoint, a primarily bilateral system–by allowing greater customer choice–takes the fundamental generator dispatching function away from the system operator, allowing consumers' collective preferences for power dictate unit commitment.[1] However, this bilateral or "direct access" approach still requires a spot market or power exchange to manage variations due to temperature sensitive loads and myriad operational dynamics such as generator outages and transmission system constraints. Each architecture therefore is comprised of both a firm and a spot market for electrical energy, the primary difference being their relative size.[2]

One of the fundamental aspects of competitive markets is the ability to incorporate new technologies, new approaches to business operations, and new products more responsive to customer needs and desires. This was certainly a driver in the restructuring of the telecommunications industry, and is represented by a shift from undifferentiated cost-based monopoly rate-making to a hybrid consumer market exemplified by commodity priced long-distance rates, value-based services–such as call waiting and caller ID, and re-bundled services such as flat fee rate packages. Such opportunities exist in the electric and energy sectors as well, and competition at the retail level is a prerequisite for unleashing such competitive and technological dynamics at the customer level. In the

---

[1] As discussed in previous chapters, each of these competitive architectures offers their own set of operational, planning and business challenges to generators and transmission and distribution system companies.

[2] It may be argued that for operational reasons a voluntary PoolCo arrangement is put in place at the onset of competition, but the share of the energy market executed by bilateral contracts is allowed to increase with time as transmission system operators develop tools to "herd" generators and manage the ancillary services market. Furthermore, it allows customers time to become familiar with prices, price volatility and generator performance characteristics.

discussions of the technological, business and regulatory aspects of competitive electric services and efficiency below, the fundamental aspects of customer access versus customer choice are explored, with a emphasis on technology development and deployment, long-term market stability, and environmental performance. Special attention is given to the role performance based and environmental regulation plays in promoting more efficient and innovative uses of electricity.

Energy Services and Products

Competition at the retail level is necessary if the technological and operational dynamics of a competitive industry are to reach all the way down to the consumer. If a more efficient market, including the utilization of electrical energy, is one of the primary goals of introducing competition to the industry, then "efficiency" becomes essentially synonymous with "productivity." Productive and efficient use implies much more than just more efficient energy consuming devices such as light bulbs and motors. It is this broader definition of efficiency that his chapter emphasizes.

Historically, end-use activities pursued by electric utilities and others have encompassed a broad range of technology development, deployment and utilization activities. Through the late 1980s and early 1990s many electric utilities—usually at the insistence of their regulators—implemented large-scale demand-side management (DSM) programs aimed at increasing the efficiency of electricity use at their, then-captive, customers' facilities. Commonly encompassing the entire range of customers (often as an equity issue), these programs went well beyond the installation more efficient devices. Initially referred to as C&LM (conservation and load management), these initiatives included operationally oriented programs in order to shift or reduce peak loads. Approaches such as subsidizing cool storage technologies, direct load control of pool heaters or central air conditioning units, the institution of time-or-use rates, as well as interruptible rates for large industrial consumers to manage sever capacity constraints were all pursued, particularly where generation margins were tight. The motivations for such programs included not only the need to ride out capacity shortages, but to avoid the cost or siting problems associated with new, expensive generating capacity. Environmental and local employment benefits of such programs soon became a significant factor in regulators', politicians' and consumer advocates' calls for utility integrated resource planning (IRP).

COMPETITIVE ELECTRIC SERVICES AND EFFICIENCY 389

While such regulatory initiatives were new to the utilities, and often pursued under a light bulb versus power plant mentality, they built upon the experiences of an already existing business environment which had grown in the aftermath of the OPEC oil shocks to provide energy management services to large industrial and commercial customers, as well as governmental facilities such as cities, towns, and defense installations. Companies such as Johnson Controls and Honeywell were major players in this early 1980s industry, and while they shifted emphasis to implement utility bankrolled programs in the integrated resource planning area, they survive as the major source of technical expertise in the design and implementation of facilities-specific energy efficient technology and management projects. Just as we are seeing technology developers, fuels companies, and generation utilities merge to gain competencies and leverage in the generation market, we are seeing end-use technology companies merge and acquire DSM deployment corporations to serve the retail market. Companies such as Xenergy and The Demand-Management Company, which sprang up or grew prodigiously via the implementation utility DSM programs are being acquired by both technology developers, such as those mentioned above, and utilities' newly formed energy service subsidiaries. Another mergers and acquisition trend focuses on the growing partnerships between energy service companies and information technology companies. Companies such as Cell Net Data Systems and Asea Brown Boveri are battling for the high ground in the smart metering market, as the ability to communicate, control and therefore manage customers' energy requirements is seen as key position in the competitive end-use service landscape.

This brief review of the history of DSM serves to delineate the vast array of business and technological approaches to providing customers with innovative energy services.[3] But there are still a few new spins to the energy management game that are being introduced via competition in the electric sector. These include smart equipment such as direct operational monitoring and control and total building systems integration as well as energy service portfolio management. New technological applications that are increasingly becoming available via the competitive retail market

---

[3] Products actually being offered by retail "energy service" companies go well beyond diversified energy services such as electricity and natural gas, and include such things as telephone, cable television, internet access, and home security. For the purposes of this chapter, however, we limit ourselves to electrical energy services.

begin with more efficient-end use equipment, but are quickly followed by the deployment of integrated systems management tools. These may optimize the internal operational performance of a building's heating, ventilation and air conditioning (HVAC) systems, but are rapidly being adapted to provide information pertaining to reliability centered maintenance (RCM), combined comfort, indoor air quality and cost-control, including increased responsiveness to real-time electric rates. Such applications are particularly attractive to both buyers and sellers of enhanced energy services since instrumentation retrofits can be installed quickly and with little disruption to recipients than most major direct energy efficiency improvements. Additionally, for some customers, electronics manufacturers and corporate data centers for example, high power quality and uninterruptible supply issues are exceedingly important "products."

On a longer-term and facilities-wide basis, reducing energy costs—including their month-to-month fluctuations—is becoming increasingly important. Fuel and emissions diversity to stabilize "effective rates" in a more uncertain and cost volatile competitive electric market should not be overlooked. Although "market hedges" such as electricity futures may be available to large consumers, the market is likely to be immature or at least relatively homogeneous early in the transition, requiring a more direct approach to managing energy costs such as diversified firm contracts for supply, and on-site management of energy consumption. In the tradition of outsourcing services not directly in line with a company's core businesses, energy service companies have a potentially large commercial and industrial market. The lessons learned in the long-distance telephone market, regarding customer retention and market share, have not been lost on electric utilities seeking to retain their once-captive customers, and new competitors attempting to lure customers with new products and services.

Behind the scenes, with respect to all these energy service enhancements, is the need to increase the real-time intelligence of the wires businesses. Time and space differentiated transmission (and distribution) pricing, two-way communication, and smart metering enable energy service providers to be more responsive to market conditions, beyond just the cost of wholesale electrical energy. Broader issues of demand forecasting and load profiling also become more important as energy service providers attempt to deliver lower-cost energy services to small residential and commercial customers via load aggregation. Energy service providers are simultaneously buyers and sellers of products, and the managers and maintainers of end-use technologies.

Building upon these layers of more efficient devices, systems integration, and energy-portfolio management, is the design and implementation of even newer electrical processes, products and services. "Electrotechnologies"–as they are generally referred to–look beyond the optimization of today's electrical uses, to devise and implement new technological processes that reduce costs and process steps in other areas. Substituting electricity for other energy sources, electrotechnologies offer significant productivity enhancements, usually in terms of fewer and shorter manufacturing steps, reduced product and process wastes, often with significant cost and environmental benefits.

Everyday examples of electrotechnologies include fax machines and electronic mail, whereby physical conveyance of written materials is substituted by near-instantaneous electronic transmission of documents. Industrial process examples include electrostatic coating of machine components and plasma processing of hazardous wastes. Electrostatic coating employs a powder-based paint that is attracted to the piece via a differential electric charge. Cured via infrared lamps or some similar process, the finish is durable enough to be applied prior to final machining, eliminating the need to mask the surfaces of the piece, further reducing time, process steps, and opportunities for wastage. The substitution of radiate curing for chemical evaporation reduces air emissions, and the total volume of "paint" is reduced as powder which is not attracted to the piece can be recirculated and reapplied. The thermal and electrical decomposition of hazardous wastes by allowing the waste to flow through an electric arc plasma offers significant advantages over chemical processing, since it reduces the need for other, possibly hazardous treatment chemicals and their by-products. If cheap clean sources of electricity can be found to drive this process, so much the better. Treatment of contaminated solid wastes via a hot plasma process yields an inert glassy slag which is easier–and cheaper–to dispose of than conventionally treated solid wastes, with a greater flexibility in the types of waste streams handled. Similar, but less dramatic electrotechnologies have been developed–technologically–for residential and commercial applications. Ground source heat pumps for heating and cooling, and electric vehicles for transportation, provide a shift in energy source but with improved life-cycle efficiencies, and therefore environmental performance. Commercial food processing applications, such as flash lamp cooking, reduce preparation times and kitchen thermal loads improving both service and HVAC system demands. In hospitals, use of ultraviolet lamps to kill airborne germs in waiting rooms reduces levels of infection and their associated human and health care impacts. Full scale retail competition will be required in order to capitalize on these innovations sooner rather than later.

The Corporate Dimension

Discussion of electrotechnologies, along with more conventional energy services, illustrates the potential role of customer level competition as a driver for innovative technological change and improved overall productivity. Businesses however are the mode by which such services are identified and implemented. Implicit in the above discussions are companies which have designed themselves to develop and deliver different technologies, products and services. Honeywell and Johnson Controls consider themselves primarily technology companies. Xenergy–now a subsidiary of New York State Electric and Gas, and The Demand-Management Company–now a subsidiary of Honeywell, help specify and install more efficient energy utilization systems. There are many other such mergers occurring as companies position themselves to compete in the new energy market. In the future, as in the past, they may make their money, in part, by taking a percentage of the cost savings realized by their efforts. Such "shared savings" projects can be expanded to include operation and maintenance services once new systems are in place. These technological service companies are also being extended to become purchasing agents on the customers' behalves in the competitive electricity and fuels markets. Of course some "energy service" companies will focus exclusively on being sophisticated purchasers for their customers, whether it be from individual generators, diversified generation companies, the power exchange, load management companies, or a mix thereof. Other companies will develop the technological components or sophisticated software applications necessary for these customers to provide these services. There are layers upon layers of "service providers" in the competitive retail market, just as there are with any other dynamic and diversified market.

Who you are, or who you want to be, in the competitive electric/energy marketplace is a central question for any company. It defines who your customers are, what suppliers you must develop relationships with, as well as the range of core competencies and products you need to develop. Fortunately, as the above discussion indicates, the essential technical elements of such an industry exist.

However, successful scale-up, to where such companies, manage and deliver *all* electrical energy services is more problematic, and depends upon lawmakers and regulators putting into place a set of fair and balanced competitive rules. Because energy services often involve a capital investment in both enhanced metering and energy efficient hardware, issues regarding customer turnover, customer retention, contract length, and exit charges emerge. These transition and fairness issues are just several of the key elements industry regulators must consider when crafting new market rules.

The Regulatory Dimension

Ensuring that end-use products and services are not left out of the competitive electric sector marketplace is best served by allowing, if not requiring, customer choice. Whether customer *access*–the ability to choose among different providers, or customer *choice*– where consumers actually have a diverse range of suppliers to choose among occurs depends very much on how regulators set up the initial competitive market.

Since the potential benefits in terms of improved consumption and utilization efficiencies are so great, several states are considering various ways to ensure that a diversity of resources, both supply and demand, are included in the resource mix during the transition to retail competition. These are many of the options, such as DSM, renewable energy sources, small cogeneration and others, that were promoted via integrated resource planning initiatives. Many draft rules on electric industry restructuring, at both the state and federal levels, have included policy mechanisms that are designed to jump start the retail market by ensuring that a diversity of supply and end-use opportunities exist. There are four basic mechanisms being considered. First is the extension of DSM and other programs, agreed to by utilities and regulators in previous regulatory proceedings. Second are "system benefits charges" (SBCs) which collect money via a fee on the transmission and distribution of electrical energy, which is then used to support specific initiatives. Third are "portfolio standards" which set a minimum level of resource diversity in a market, often accompanied by a tradable permit system for the reallocation of funds, and subsidization of DSM and renewable energy projects. The fourth, hardly a mechanism at all, is "green pricing" whereby regulators allow, and set up verification processes for, the sale of cleaner, premium (or value) priced electric services.

The extension of IRP initiatives essentially ensures that distribution utilities continue to offer DSM services, previously agreed to, to local customers as the industry transitions to retail choice. The second and third mechanisms mandate niche markets for DSM and/or renewables. Systems benefits charges have been designed to support "public benefits" activities such as low-income rates and basic research, as well as DSM and renewables. Many of these SBCs are intended to replace activities once performed by vertically integrated utilities, that more narrowly focused businesses are unlikely to pursue. In support of DSM and renewables, several states have designated that SBC proceeds be used to support innovative, emerging, or "pre-competitive" technologies. In this form the SBC promotes the development and deployment of new technologies and energy resources, as opposed to existing ones. This is not the case with the third mechanism, the "portfolio standard." As designed, portfolio standards require that some percentage of electricity sales, or total demand for electric service, be met with renewable energy sources and/or end-use energy efficiency. In the case of the "renewable portfolio standard," old as well new renewable resources such as hydropower, geothermal, wind, biomass and solar generation are eligible technologies. In some states waste-to-energy, and new fossil sources such as natural gas fueled fuel cells are also eligible. While customers can purchase electricity directly from such providers to meet their portfolio requirements, others can continue to purchase electricity from the power exchange, or traditional fossil and nuclear generation. A tradable allowance system, much like the sulfur dioxide emissions allowance system, called for by the 1990 Clean Air Act Amendments (CAAA), ensures that the target is met and that least-cost renewables are used to make up the balance of the renewable energy mix. The transfer of allowance proceeds to renewable generators enables them to price their output more competitively in the firm and spot energy markets.

Continuation of IRP programs, systems benefit charges, portfolio standards, and green pricing can all exist simultaneously. Continuation of IRP programs ensure that current service providers continue to have a market in the early stages of competition. Systems benefit charges promote the early introduction of new technologies, hopefully "buying down" the price of such systems so that they reach competitive price/performance levels earlier. Portfolio standards support both new and existing projects, and provide a more stable market environment for their continued existence. Green pricing benefits, since a continued presence of qualified and cost-competitive resource providers improves the credibility of, and therefore willingness to buy, value priced services.

Arguably, these mechanisms increase the administrative burden of a competitive industry, and may detract from the anticipated cost-savings customers may realize from competition in the electric sector. Customers, regulators and service providers alike are concerned that multiple system benefits charges for R&D, low income support, DSM, and renewables, when coupled with the collection of stranded assets, will eliminate anticipated reductions in energy costs. However, one of the factors that leads regulators and others to suggest such mechanisms, is a concern that the long-term technological balance of the industry will not be maintained if a purely cost dominated, commodity-like system is put in place.

Utility regulators themselves are facing a fundamental change in their function, and related job skills. As the industry transitions towards full scale competition, in the energy business at least, they will no longer be "rate regulators." The public utility commission of the future will look much more like the Securities Exchange Commission in function, than it does an adjudicatory agency today. Fairness, not cost, will be the criteria for review. Portfolio managers will sign up customers based upon their "prospectus" of energy services; cost, cost stability, energy mix, reliability, power quality, integrated building management, etc. Regulators will evaluate them on how well they adhere to their advertised services. These shifts have already begun, promoted by the activities of some market participants in pilot programs. In New Hampshire, for example, some marketers were advertising "green" resource portfolios, which upon closer review included pumped storage generation, essentially nuclear and coal-fired power once removed. Performance-based regulation of transmission and distribution utilities offers its own challenges, particularly at the onset. What will be the implicit rate of return that regulators will allow "wires" companies, and if there are significant cost savings to be achieved, how will regulators allocate the savings between customer and company? How can market sophistication in terms of transmission pricing, power exchange operation and performance monitoring aid overall industry balance and long-term cost-effectiveness?

The Environmental Dimension

Long-term cost-effective provision of energy services is not the only performance goal the electric sector must face. Long-term environmental issues also play a large role. Unlike the deregulation of the telecommunication industry, the electric sector has a large environmental dimension, rivaled only by the transportation sector. Yet, unlike transportation, the electric sector has been able to manage its collective environmental performance via historically long-term planning performed by regional, vertically integrated electric utilities. Obviously, this central management function is going away. Yet the industry, and its collective environmental burden remains, just as environmental regulators step up their implementation of the 1990 CAAA, and identify new emissions requiring monitoring and control. Sulfur dioxide (SO_2), nitrogen oxides (NO_x) and volatile organic compounds (VOC_s) are the primary emissions that have been dealt with by the U.S. Environmental Protection Agency (EPA) so far under its implementation of the 1990 CAAA. The control of toxic emissions is also called for under the CAAA, and it can be expected that mercury and other emissions will come under control in the future. In addition to these efforts, state and federal environmental regulators are aggressively pursuing the additional control of currently regulated and unregulated emissions. In the Fall of 1996, promoted by several lawsuits claiming they were not living up to their legislative obligations, the EPA announced its intent to further control ground-level ozone (smog) related emissions, and expand its regulatory authority over the emission of fine and ultrafine particulates. Despite a considerable degree of scientific uncertainty, the EPA has decided to pursue these efforts, just as the federal government is conducting international negotiations which may restrict the emission of greenhouse gases, of which carbon dioxide (CO_2) is most prominent. These of course are all air emissions, and are matched equally in terms of land-use, solid waste, and water-use issues on the regional and local level.

While increased control of emissions such as SO_2, NO_x, particulates, mercury, and CO_2 offer considerable challenges to the disaggregated, competitive electric industry, they may offer the ultimate industry-wide performance constraint, whereby long-term balance between energy sources can be restored. As mentioned before, the CAAA called for the reduction of sulfur emissions, via a tradable emissions allowance system. Such "market based regulation," as opposed to the more traditional "command and control" approach, provides much greater flexibility and cost-effectiveness. When the Clean Air Act was reauthorized in 1990, economists predicted the cost of reducing sulfur dioxide emissions by one ton (synonymous with the cost of an SO_2 allowance) would be around $750. The actual cost has been dramatically lower, around $100, due to unanticipated drops in the cost of rail transport, increased productivity in mining, (reducing the delivered cost of low sulfur coal), the availability of offsetting natural gas fired generation, and the ability to reduce scrubber costs (due to decreased design redundancy).

Such market-based approaches to environmental regulation have been instituted in Southern California for smog, and are also envisioned for summertime NO_x and VOC emissions in the northeastern United States. Such "cap and trade" approaches, as they are also called, are being promoted by the administration for the long-term, global reductions in greenhouse gas emissions. In some states' restructuring legislation, emissions accounting mechanisms have also been proposed in order to identify and manage the environmental burden of the spot market's emissions and those of various brokers' resource portfolios.

With the uncertainties that such generators, energy brokers and power marketers face in the newly formed competitive industry, it is not hard to see that one way to manage one's diverse environmental emissions, is to reduce one's overall environmental burden. Most SO_2, NO_x, CO_2, particulates and air toxic emissions–such as mercury, come from the combustion of fossil fuels. One way to minimize the administrative, as well as environmental, burden of managing all these emissions is to reduce the overall amount of fossil generation one uses. As fossil fuel costs are the major source of quarter-to-quarter cost variability, reducing a customer's reliance on fossil generation also stabilizes a their energy costs. End-use efficiency (DSM), renewables, and nuclear generation all assist in this regard, however with different cost and performance characteristics, as well as hidden cost dimensions, such as decommissioning. Increasing the efficiency at which fossil resources are utilized is also an effective management strategy, and reflect the current popularity of new, highly efficient natural gas-fired combined-cycle generation.[4] Other types of efficiency improvements include cogeneration, and moving generation closer to the customer. On average, seven percent of all electrical energy generated is lost in transmission and distribution between the central station generator and the ultimate customer. Small incremental efficiency increases in the transmission and distribution of electricity therefore have large aggregate impacts on fuel consumption, fuel costs, and environmental emissions. Measurement and tracking of T&D system performance by independent system operators can therefore aid overall industry-wide cost and environmental performance.

Thus, market-based regulation of multiple environmental emissions, with its implicit monitoring and tracking of those emissions, can provide a valuable industry-wide performance constraint, to both the day-to-day operation of the competitive system, and perhaps more importantly, to the long-term performance of the industry. Coupled with portfolio standards and other policy instruments to ensure the availability of efficient, cost-effective and clean electric service options, market-based environmental regulation can help signal technology companies and project developers regarding the types and amounts of resources needed in future years.

The Ultimate in Electric Service Integration – The Distributed Utility

The above sections have covered a broad range of technical and policy issues related to retail competition. The integration of energy efficient equipment with sophisticated monitoring and operations is a key element of ongoing technological shifts. Thinking "out of the box" to identify process enhancements using electrotechnologies extends this integration at the customer level, linking not only energy but total corporate cost, business and environmental performance. Links between customers, their electricity suppliers

---

[4] Although gas-fired generation has been touted as clean way to generate electricity–and it is–this is not necessarily true with respect to CO_2 emissions. Natural gas-fired generation can only reduce CO_2 emission when it is used to displace higher CO_2 emitting resources. When used to meet additional demand, or as replacement power for out-of-service nuclear units, despite its lower carbon content and greater efficiency, natural gas-fired units are still significant emitters of CO_2.

and the market-at-large extend this integration to respond to changing market conditions, be they short or long-term. Integrated portfolio approaches to manage multiple uncertainties and risks, and to adapt to shifting environmental regulations, completes the litany of activities that customers must incorporate into their thinking.

For other market participants integration is also the key. Optimized performance of generation units, regional transmission systems, local distribution feeders, monitors and meters, will ultimately work in intimate coordination to deliver electrical energy and related energy services. Appropriate regulatory and market signals will facilitate their development and deployment. Well crafted market-based environmental regulations will work to integrate cost, reliability and environmental performance. This ultimate in integration, where customer level integration receives as much attention as central station coordination is commonly referred to as the "distributed utility" concept.

The figure below outlines the breadth of the distributed utility's ultimate architecture, and some of its technological components. As can be seen the distributed utility concept exploits technological integration at all levels. Technological applications have been grouped into four system levels. As mentioned previously, transmission and distribution losses account for approximately seven percent of all central station, or "busbar" generation. Getting generation closer to the customer, either at the "T&D Interface" or "at the Meter" reduce these losses. Generation from remote sources like mountain hydropower or wind, or extra-regional fossil incurs additional line losses, or increased costs via the installation of DC lines to offset those losses.

Technological integration has temporal as well as spatial implications. Roof-mounted or building integrated photovoltaic systems (PVs) are likely to become more cost-effective for various niches in a competitive market. Although still more expensive on average, the ability to compare the cost of PV generation versus the cost of peak load delivered power, with its peak generation price, congested T&D system delivery charges, and avoided peak load line losses and summertime NO_x emission costs, will provide the purchaser a better understanding of the true price premium being paid. The same cost calculation holds true for all efficiency improvements and other types of end-use control and on-site generation.

Technological Components of the Distributed Utility

"System Level"	Representative Technological Applications			
"Remote"	Extra-Regional Imports (Hydro, Nuclear, Etc.)	Remote Solar, Wind	DC Power Transmission	
"Busbar"	Central Nuclear Station	Regional Hydro	New and Existing Oil/Coal/Gas	Hi-Tech T&D (FACTS, HTS)
The T&D "Interface"	Strategically Placed Gas Turbines	Local Biomass Generation	Local Wind & Hydro	
at the "Meter"	Cogeneration/ District Energy "Golden Carrot" Refrigerators	On-Site Gen. (Fuel Cells) Ground Source Heat Pumps	Building Integrated PVs Electric Trans- portation	Spot- Pricing of Electricity Energy Efficient Housing

Time and space differentiated transmission prices, and locationally directed ancillary service provision will introduce this same level of sophistication in the ultimate siting of generators, and enhancements to transmission and distribution systems. One of the challenges will be to get these competitive markets, and their associated market signals, in place in a timely and consistent fashion. Large wire charges to collect stranded costs or support multiple social goals may serve to jump start additional customer level services, as many so-called "nonbypassable" charges can only be collected on power delivered over the grid.[5]

Conclusions

In real estate, the axiom for success is "location, location, location." In the competitive electric industry it may well be

---

[5]Some customer-level activities such as distributed generation may not benefit however, as many states are putting provisions into their restructuring proposals that may require self-generators above a certain size/output to pay an "exit fee" to cover both stranded costs and T&D operators backup system costs.

"integration, integration, integration." Integration in all aspects, hardware, utilization software, two-way communications, portfolio management, etc. will all play their role. Businesses picking their spot in the market, and developing their products, suppliers and customers to match is its own form of integration. Integration of utility and environmental regulations will also be key, as will developing an adequately sophisticated set of market institutions, including ISOs, power exchanges, futures markets, emissions trading forums, and so on. Development of such forums is essential, as advanced technological applications require such environments to act in, and respond to. Not only must a successful competitive market integrate such functions on the day-to-day, week-to-week and month-to-month basis, but it should be able to signal developers of new technologies, what to build and develop over the very long-term, whether they be generation or end-use technologies.

Competition down to the customer level is an essential element of this competitive integration. Failure to institute true competition, providing merely customer assess as opposed to real customer choice, will ultimately lead to an anemic industry, whereby opportunities for efficiency and productivity improvements along the entire supply chain will be missed.

V POWER SYSTEMS CONTROL IN THE NEW INDUSTRY

11 NEW CONTROL PARADIGMS FOR DEREGULATION

Lester H. Fink

Introduction

Restructuring of the electric energy utility industry is well underway, even as wide-ranging debate continues about desirable and achievable nature and goals. At this stage of the process, it would be a mistake to assume that any of the experiments in restructuring that are now under way necessarily represent the final patterns that will develop over the next decades. It is important to remember that the traditional industry structural and operating paradigms evolved together, in symbiosis. Energy management in general, and automatic generation control in particular, have been part of that evolution, which has reflected a generally stable pattern of vertically integrated, regulated corporations. The industry's management and control structures have been effective, in large part because they reflected the existing industry structure. If the industry structure had evolved differently, operating practices no doubt would be quite different. It is to be expected, then, that a radically restructured industry will need, and will develop, new control structures adapted to the new circumstances. One should consider *what operating practices, irrespective of past practice, would be most congruent to the new, disaggregated structure?*

Modern technology in the service of modern control theory is more than adequate to provide effective controls to meet the needs of any possible new structure or structures. Hence, at this juncture, the major difficulty in considering what management and control structures will be required within the industry-to-come is not technological. Rather, it lies in attempting to foresee *where the possibilities that have opened up, and, most importantly, where the incentives that are inherent or legislated, will lead?*

This discussion attempts to survey the current reconceptualization of the industry in light of these two fundamental technical questions. It is not taken for granted that the future will always remain like the past - although it is recognized that it may well be very like the past for some time to come. It is suggested that contrived new corporate and operational patterns that do not reflect congruence of prescribed operating practices with both corporate structure and profit based incentives will not survive. In the first major section, following, we attempt to sketch those aspects of the emerging new structure that impact potential system control requirements. Some extended discussion of the concept ot ancillary services is necessary, since the allocation of responsibility for these services directly affects control structures and objectives. In the second major section, we discuss control problems that may emerge as the new industry structure materializes.

A NEW ENVIRONMENT FOR SYSTEM CONTROL

Industry Restructuring

As a global phenomenon, restructuring of the electric utiilty industry has been driven by a variety of motives, and has taken a variety of paths. Broadly speaking, in the United Kingdom and Latin America, privatization has offered a means of attracting funds from the private sector to relieve the burden of heavy public subsidies. In Central and Eastern Europe, it follows the general trend away from centralized public control and toward a market economy, also providing a vehicle to attract needed foreign capital. In the United States and other countries where the industry has long been privately owned, the trend is toward increased competition and decreased regulation. Nearly universally, proponents argue that deregulation and competition will result in lower energy prices for the end consumer, and lead to more efficient and environmentally sound utilization of resources. All this remains to be proved [1].

In the United States, restructuring of the industry has come to be understood as involving (*de facto*) *deintegration* of vertically integrated utilities, *unbundling* of various, conceivably separable, aspects of electric energy service, and *deregulation* of wholesale bulk power sales. These are negative terms representing a *de-structuring* of the traditional industry. Positively, once the industry has been (conceptually) taken apart, *restructuring* must be considered as it affects both the *institutional* (*ownership, business*) and the *technical* (*operating and control*) aspects of the industry. Unfortunately, to date, most of the attention and ongoing debate has focussed on the former, to the neglect of the latter aspect. In this discussion, we are concerned specifically with this latter, neglected, aspect. But, before turning our attention to that, we must put the discussion in the context of the overall restructuring question.

Corporate De-integration Within the United States, the Federal Energy Regulatory Commission (FERC), in its prodding and guidance of the industry, has been constrained by its statutory authority. Even though it may desire a radically reinvented industry structure, it cannot require of existing corporations the divestiture of any of their assets. Therefore, it has perforce been constrained to requiring *functional* separation of generation and transmission, and has encouraged *operational* separation. The broader discussion has debated, along with these, the relative effectiveness also, of *structural* and *corporate* separation of generation, transmission, distribution, and market operation. Broadly speaking, only the last requires change in ownership of assets. *Functional* separation requires separation of generation, transmission, and distribution functions into three distinct organizations within the same corporation, each with their own costs and employees. *Structural* separation involves the establishment of separate corporate subsidiaries for generation, transmission, and distribution assets and functions under a holding company. *Operational* separation (which involves some institutional restructuring in that it adds a new corporate entity) focuses on control of the transmission system by means of establishing an *Independent System Operator* (ISO), with individual utilities still retaining ownership of their transmission assets. Finally, *corporate* separation requires at least divestiture of most or all of a utility's generation assets; thorough-going corporate separation would entail separate, independent generation, transmission, and distribution companies.

Service Unbundling In the interests of opening the sale of electric energy to competition, it has been observed that traditionally such sale has involved not just energy per se, but also a wide variety of ancillary services that have been bundled into the product. The inference drawn from this observation is that establishment of a competitive market requires that such ancillary services be unbundled, priced, and marketed separately, each to only those end users who need a subset of them. From the perspective of this discussion, it follows in turn that the nature, sources, and consumers of these services must be considered in the quest for new and appropriate control regimes.

Deregulation Ostensibly, deregulation of bulk power sales has been the prime objective of the restructuring process. Letting the market, instead of a *pas de deux* between utilities and regulators, set prices is the necessary crux of the matter, and before this can be granted, conditions must be established to ensure that no party has sufficient market power, either globally or locally, to foreclose potential competition. This *sine qua non* leads to the requirement for equal access to transmission, i.e., that each utility shuld open up its transmission grid to any and all comers (competitors) absolutely on a par with their own access in all respects. Means of achieving this objective have dominated most of the restructuring debate.

It is important to note, however, that this argument has implicitly been focused on equal access by generating entities, whether other utilities, non-utility generators, or power marketers (brokers). The momentum of the restructuring process, however, quickly moved beyond equal access by a utility's peers to retail access, i.e., enabling end users to select their supplier(s) of choice. Unfortunately, the parameters of the debate have not changed to accommodate the interests of such end users. The implications of this disjuncture between the parameters of the debate and the set of interested parties have been unfortunate [2]. Access rights are invariably discussed as pertaining to suppliers, rather than consumers. This involves provision of complex procedures for acquiring and trading such rights. The tie between access rights and individual transactions will inflate the demand (rights must be inplace before credible offers of evergy supply can be made). Such such activities will encourage gaming to gain market power in the energy market. Suppliers pay for such rights, but in the end they are paid for (plus overhead) by consumers, the sole source of revenue. Most of these problems and undesirable side effects disappear if access rights are assigned to consumers (represented by load aggregators or load serving entities). The existing transmission grid has been designed and built to ensure adequate access to all consumers. If all consumers are assigned their proportionate share of rights, including their share of limited access across critical cut sets, they can deal with whatever suppliers they wish, so long as they do not exceed their allotted capacity. Their access would be paid for as in the past, to the transmission entity in whose territory they reside. Conversely, assigning rights to suppliers will reintroduce via a back door the market-empowering tie between generation and transmission, the elimination of which was the major objective of this whole deregulation process.

Influence of Incentives In order to assess what patterns might emerge after the present experiments in restructuring have run their course, it is prudent to consider what parochial interests will have been unleashed, and what responsibilities may be identified with those interests. Long term implications of current trends in the industry could result in a very different control structure and control objectives. Anderson [3] has stressed the importance of getting the incentives right, and Hill [1], from a different perspective, has questioned the wisdom of ignoring economic incentives as the resructuring of the industry proceeds. The following inferences are based on consideration of the implicit *operating-related incentives* - and disincentives - that may be built into new structures.

Traditional industry structure. Under the traditional vertically integrated industry structure, corporate financial objectives were such as to favor highly reliable service achieved by high quality equipment with redundant capacity, both for generation and transmission. Some possible gains in efficiency that were not achievable by capital expenditures often were neglected, because, fuel costs being passed through to customers, the costs required to achieve such efficiency gains could not be recovered. Company interconnections, and later regional interconnections, were undertaken in the interests of increased reliability. Interconnections provided improved quality of performance: more stable frequency resulted from increased system inertia and from the tighter regulation required to control tie flows; emergency assist power became widely available, along with economy purchases of energy. Operating objectives included satisfaction of load demand, maintenance of system security, minimization of production cost, environmental protection, and fuel conservation. The resulting control tasks involved control of generation and storage (e.g., pumped hydro), load ("demand side management"), power flow, and system status. Thus, in the Normal State of operation the general system control regime was that of controlling generation in response to changing load demand while minimizing production costs and maintaining an adequate level of security.

However, even within this mature structure, which evolved to satisfy industry objectives that reflected the implicit incentives inherent within the regulated industry structure and environment, human nature sometimes acted counterproductively. Instances such as the following will hereafter be referred to as *operator parochialism*. Plant operators (and plant superintendents) within vertically integrated utilities resisted and even obstructed regulation (e.g. by placing units on "load limit," or opening up governor deadbands). This was a major factor in poor system performance for many years. Some utilities were known to minimize their contribution to regulation by maintaining wide governor deadbands. Again, it was noted in a recent paper [4] that "it is not uncommon to operate a 75 psig H_2 machine at 60 psig H_2. Such operation reduces the windage losses in the synchronous generator... There is also less hydrogen consumption at the lower H_2 pressure... Lower hydrogen consumption results in the purchase of less hydrogen which reduces the operating cost of the synchronous generator." The significance of this lies in the fact that a machine's reactive capability, which is extremely important to voltage security, is a direct function of hydrogen pressure, and is significantly reduced by such operating practice.

We may conclude that operating practices reflect incentives and disincentives, not just as perceived at the corporate level and incorporated in company policy, but also as perceived at the operating level.

A restructured environment. With such operator parochialism having flourished in a conventional vertically integrated utility structure, under unitary management, it is difficult to imagine that similar motives will not influence the manner in which systems will operate in the future, as generating plants become profit centers, not to mention when they are owned by independent production companies! We are told that, with independently owned plants, no such problems will arise - the market will supply whatever there may be needed - but operator parochialism argues otherwise. Management of the restructured industry will not be primarily by engineers as in the past, but largely by entrepreneurs with an eye to balance sheets and near term profits. Possible implications of incentives to minimize both capital investment and operating costs will be discussed below under *System Management: Generation Control.*

Unbundling of Services

In the long run candidates for ancillary services will be judged in terms of their respective characteristics and requirements. Most importantly, it is necessary to ask: *Who or what gives rise to the need for a particular service? For whose benefit is it provided? Can responsibility for use of the service always be traced to the responsible party?* For those services that are system requirements (as opposed to end-user requirements), additional questions must be addressed. *How can costs be assigned equitably?*[1] Then, because of the disparate nature of the four categories we have defined, *the manner in which each should be procured must be considered.*

In Order 888, FERC proffered a candidate list of such services. The North Amercan Reliability Council (NERC) subsequently developed a revised list. Without discussing these lists in detail, it is important to note that certain principles must be observed in defining and assigning responsibility for, on the one hand, providing, and on the other hand, paying for such services. Candidate services should be distinguished from two perspectives. First, a distinction should be made between modes of energy supply other than constant power base load, on the one hand, and on the other hand, services per se. The former comprise energy in various guises, or specific purposes for which which energy might be used; the latter comprise functional responsibilities such as coordination, security enhancement, etc. Second, a distinction should be made between those services necessary to the functioning of the transmission system, system specific services, and those required by individual users. In this case, since the former, literally, are *sine qua non* to the functioning of the network, no user of the system can opt to forego them. In regard to the latter, it is contended that they are not the responsibility of the network, which therefore should not be required to procure or provide them.

Accordingly, we distinguish four distinct categories: active energy (frequency control, which implies <u>default</u> regulation and loss compensation), reactive energy (voltage support; load compensation), capacity (reserves), system management (security and coordination). Each of these, *but only insofar as it relates to the network*, provides *indivisible* support of network functioning. The term "indivisible" is used *vis-a-vis* individual energy market transactions. In addition, because of the current popularity of the poolco and independent system operator concepts, which imply continued centralized control of all generation, we discuss (under system management) the concept of system control as a candidate ancillary service.

Active Energy: Default Regulation The spectral density of loads has a very wide bandwidth [2] Various types of generating units have significantly different response rate capabilities. Whether explicitly or tacitly, it has been found necessary to structure the control of generation in such ways as to accommodate load bandwidth and unit response capabilities. Some load changes (and much "noise" which does not reflect actual load changes) are *de facto* uncontrollable, and should be (and increasingly have been) filtered out of signals that control generation. In mid-century, explicit distinction was made between "sustained" and "fringe" load [6]. Fringe (frequency-regulating) response was recognized as being disruptive to economic dispatch. Some AGC algorithms have accommodated this insight by developing separate economic and regulating signals for application to assignable groups of generators.

Within a vertically integrated utility, the line between economic and regulating generation may be soft. The distinction between the two is usually between MW bands within which units assigned to regulation are constrained to generate. If a negative fluctuation in load drives a regulating unit below its lower (regulating) limit, the only penalty suffered is a relatively small one for a short period of time. Within a disaggregated industry, however, with unbundled modular classes of energy, the distinction cannot be so soft, since different suppliers may well be involved. In such a case, if a lower regulating limit were violated, the fringe supplier presumably would be unable, or at least contractually unobligated, to reduce his output below zero for a given contract. In such a case, then, the only recourse would be for the transmission entity (TCo) to absorb the difference by reduction of the generation it has under contract to supply its own needs.

On a tightly interconnected system, this function may be provided by (adequately responsive) generation almost anywhere on the system. Ostensibly, then, it is suitable for being made available as a distinct "service" product. However, a primary question is: (1) *Who is responsible for causing the need for fringe control, and therefore, who should pay for it* (see note 1 above)? The need for fringe control arises from the fact that generation cannot respond instantaneously to rapid changes in load [7]. Accordingly, generation movement, in general, lags load movement, creating a fluctuating imbalance between the two. Obviously, then, the responsibility for this need lies formally on the user, since he creates demands that physically cannot be met.

It should be noted immediately that not all end users are equally chargeable. Some users may present loads that are essentially constant, or very slowly changing, over relatively long periods of time; such users should not be charged with, and could be expected to resist paying for fringe control "services" they do not require. Other users present loads that are qualitatively worse than most (e.g. arc furnaces); such users require costly customized treatment, the charges for which should not, in equity, be spread across other users.

Two additional questions, then, are: (2) *How might users be buffered from undesirable side-effects of the loads of other users?* (3) *How may costs for fringe control be assigned to those users who bear the responsibility?* In a sense, fringe control requirements imposed by end-user non-constant loads are a form of capacity (demand) requirement. It should not be infeasible to treat distribution companies, or even retail brokers, as small control areas, tracking the margin between their contracted load trajectories and their actual demand, and billing them for capacity needed to cover their fringe load.

A fourth question is: (4) *How should fringe control be provided?* Economic dispatch of regulation is problematic for at least two reasons. First, effective regulation requires parallel operation of multiple units in order that their combined (additive) rates of response be adequate to follow imposed load fluctuations. Second, non-monotonic, or even monotone-negative incremental costs are not uncommon - e.g., combustion turbines may have cost characteristics of the latter type. In such case, economic dispatch can be achieved only under dynamic optimization, which requires load prediction [8], which in turn is infeasible for fringe energy. In short, fringe control is disruptive of cost minimization on most thermal generating units to which it might be applied [6,9]. For this reason, wholesale generators are likely to wish to insulate their units from such duty, making fringe control energy likely to be more expensive than load following (in the conventional sense) energy. *Thus it may be attractive for end users, or distribution companies or retail brokers, to control their own fringe demands by local small, dedicated generation and/or load management techniques.*

A first finding is that only *default regulation*, necessary to maintain system frequency within acceptable bounds, passes muster as a *system* ancillary service.

Active Energy: System Loss Compensation Losses are an inherent, unavoidable consequence of delivering energy. The energy is "lost" by the transmission entity (TCo), but since it is an inescapable "overhead" cost of doing business, it will be paid for by the end user. It is relevant to note, also, that the effect of uncompensated losses is to depress system frequency, since they represent a mismatch between generation and total effective load. This in itself means that loss compensation is an essential, system-specific service.

One's first thought (and it is seriously proposed by some) is that the generator (GCo) should deliver more energy into the transmission system than the customer receives; the transmission provider (TCo) should not be concerned. However, this approach suffers from the very real difficulty that, since transmission losses are loading dependent, they will be greater, for a given transaction, during periods of heavy network loading than during periods of light loading. This means that it would be inordinately difficult for any GCo to determine by what magnitude of losses he would have to over-generate to meet at any point in time. (It has been proposed [10] that losses ascribable to individual transactions can indeed be calculated. The accuracy of these calculations, however, will be open to question on the basis of the assumptions, and limited data, on which they would be based, and default correction would be necessary by the TCo in any event.)

An alternative might be to assign responsibility to the end user: the GCo could generate to schedule, and the end user would make up what was lost in transmission through his "load following" contract(s). This suffers from the fatal defect that there would be no way for the user to detect and track how much of his energy was being lost in transit, and how much therefore should be charged to his auxiliary contracts.

For these reasons, then, it seems rational to assign to the TCo responsibility for maintaining, under contract, for the common benefit of all users, sufficient generating capacity to cover total transmission losses, and to bill load serving entities for these losses as a function of their loading of the system. Loss compensation is inherently related to overall functioning of the system, formulae and algorithms for assigning "responsibility" to individual transactions notwithstanding. Thus, a second finding is that *loss compensation qualifies as a system ancillary service.*

An open question is, *should load and generation be indexed to scheduled frequency?*

Reactive Energy: Support of System Voltage Profile
Voltage support pertains directly to the maintenance of system integrity. Hence, this "service" is as integral to the transmission function as is its hardware "plant" (transmission towers, conductors, transformers, etc.). Reactive power injections at major system buses (both generating buses and load buses) are necessary to maintain a viable system voltage profile. This is as true of an unloaded transmission system as of a heavily loaded one, but the magnitudes of reactive requirements are dependent on system loading. The effectiveness of reactive injections is generally local in nature, so that the sources of the required injections must be sited according to network topology.

In the early days of the industry, systems were small and isolated, and reactive power was provided naturally and conveniently by synchronous generating units. As companies began to interconnect, and larger interconnections developed, circumstances changed: the need for reactive support outgrew the ability of generators to provide it, and shunt capacitors were installed to provide support at critical locations on the system. For many years, most transmission systems were generally overbuilt by today's standards. Since reactive transmission losses are a quadratic function of current, lightly loaded transmission lines do not give rise to excessive voltage drops. This encouraged specification of new generators with higher power factors (meaning reduced reactive capability) to hold down their cost.

On today's systems, reactive support is provided by a combination of generator capability and reactive compensation, including not only conventional shunt reactors, but, increasingly, static VAr compensators (SVC), and prospectively, new power electronic devices [11]. Such network compensation is becoming increasingly important because, especially with heavily loaded systems, the effectiveness of support is site specific. The amount of reactive required to support system voltage also increases quadratically with system loading, and the transmission of reactive from remote sources to support affected buses in itself further increases that loading (this is one major factor in voltage collapse failures). Hence the greater the need for reactive support, the more important it is that it be supplied where needed, not remotely.

Increasingly, generators are not sited with regard to the need for reactive support.[3] Moreover, even to the extent of a machine's reactive capability, available reactive energy usually decreases with increased MW loading, and limiting MW output in order to maintain reactive availability can be very costly.

A third finding, then, is that *reactive support of the transmission system voltage profile* is a quintessential *system* ancillary service. It is an inherent need of the network (towers support the conductors, reactive injections support the voltage profile), and reactive compensation (primarily in the form of static VAr compensation) will have to be provided where needed, as an inherent aspect of network design. Its capital cost necessarily should be an embedded cost of the network physical plant (no production costs are involved).

Reactive Energy: Support of Customer Loads A second need for reactive power is quite distinct from the first: the need that results from "lagging" loads, notably induction motors. This need is implicit in the nature of a load, and hence is a need of the consumer. This reactive power cannot be provided efficiently by the generating entity supplying that consumer's (active) load, by virtue of the electromagnetic phenomena that govern functioning of the electric energy system.[4]

The upshot is that consumer reactive requirements should either be supplied by equipment purchased by the customer himself (power factor correction), or provided locally by his retail provider. It should be noted that, even at the load site, compensation is more effective at the low side of a load transformer than at the high side, because of reactive losses within the transformer itself. Vertically integrated utilities themselves have understood the desirability of distinguishing between transmission voltage support and load compensation, and the economic benefits of handling the latter at the load site [12,13]. This responsibility of power factor correction at the distribution level should emerge naturally from the greater costs involved in drawing reactive power from the bulk power transmission network, as well as from the fact that the requirement is end user load specific.

Thus, reactive load requirements should be the responsibility of the end user, and obtained through its load serving entity, whether distribution company (DCo), retail broker (RBr), or other load aggregator, so that only active (unity power factor) load is supplied over the network. This conclusion parallels that reached earlier with regard to regulating energy. The general principle that emerges is that bulk power transmission is too valuable for enabling economy energy transfers to be pre-empted by transferring "ancillary" modes of energy that usually can be provided, effectively and economically, on site. A fourth finding, then, is that *reactive support of customer loads* does not qualify as a *system* ancillary service.

Capacity: Operating Reserves Who should be responsible for covering for the loss of a major unit? Two responsibilities present themselves: that of the GCo for fulfilling his contracts to provide power on schedule, and that of the TCo for maintaining the security of the system. Requiring all GCos to provide their own spinning reserves would seem equitable from a free market standpoint, but would suffer from obvious disadvantages. First, if every GCo carried enough reserve to cover his own exposure (e.g. loss of his largest unit), the total spinning reserve on the system would be uneconomically large. Second, effectiveness of spinning reserve requires that it be spread over a large number of units, since if they are confined to one or a few units, the effective rate of response may be inadequate for response to a sudden loss of generation.

Requiring the TCo to provide spinning reserve presents different advantages and disadvantages. Providing reserves on a system wide basis would be more economic than doing so on an individual provider basis, since, for instance, reserve to cover the traditional loss of the largest unit or largest import would presumably be adequate system wide. In this case, costs for maintaining this reserve spinning capacity would be spread across all suppliers, off system as well as native, in the transmission tariff. In this case, it would be necessary to require that the reserve contracts be spread across a large number of native suppliers, and/or that reserve capacity be provided by partially loaded units, in order to avoid the problem of excessive generation due to minimum load requirements on otherwise unloaded units.

It should be observed that the willingness of a customer to purchase non-firm energy (subject to arbitrary interruption) does not remove the necessity of providing spinning reserve support for the generation in question. This is because, even if all such customers were willing to suffer interruption of service, loss of a large unit dedicated to such service could still have potentially catastrophic consequences for the entire system. This does mean that the "spinning reserve" for such loads could be constituted by the loads themselves being "transfer tripped" upon loss of the supplying generators. Apart from this, however, provision of spinning reserve by individual GCos, either by spare synchronized capacity or by requiring immediate dumping of contracted load, is obviously infeasible. Thus, a fifth finding is that, by default, *system wide spinning reserve* qualifies as a *system* ancillary service.

Procurement of spinning reserve capacity, as in the case of procurement of frequency regulation and loss compensation energy, could be on a periodic bid basis, with provision for after-the-fact compensation of energy costs for reserve capacity actually used.

System Management: System Security System security control may be viewed as the "covering" ancillary service (provision for operation of the network), the most basic, comprehensive, or even the ultimate one, to which the others are necessary but not sufficient contributors. Security is the resiliency of a system *vis-a-vis* potential (defined) significant disturbances. Formally it may be defined as a joint function of the contingent probability of violation of critical operating constraints and of the probability of disturbances [14,15]. Pragmatically, it may be seen as a matter of adequate reserve margins in critical system variables [16]. A basic requirement for effective system security control is the ability to conduct continual security assessment, which in turn requires access to system-wide data. The mechanism of security control is that by which the findings of security assessment are brought to bear so as to ensure ongoing normal operation of the system. A direct consequence of any invoking of this function is an increase in system marginal cost, due to operating within constraints that are calculated to contain the system's exposure to major disruption of its operation.

Of the many approaches that might be taken to maintaining system security within a restructured industry, two of those being discussed may be suggested as compatible with the framework being developed herein. One is that of direct curtailment by the TCo of access to the network that would impair security margins viewed as hard constraints [10]. The other is that of indirect control of access by escalation of transmission tariffs as a function of diminishing margins viewed as soft constraints [17]. Each of these is subject to apriori objection. Depending on whatever specific criteria might be developed, the former could be viewed as arbitrary in curtailing access on fortuitous grounds, while the latter could be viewed as arbitrary in assigning prohibitive prices bearing no relation to actual transmission operating cost.[5] The latter would, however, have the advantage of discouraging use of the network by less "valuable" transactions.[6] In either case, justification would have to be provided by regulatory approval of judiciously developed criteria for curtailment of transmission access when necessary to maintain security of system operation.

In any event, the finding is that *system security* is a *system ancillary service*.

System Management: Centralized Control of Generation It has been suggested, and widely accepted, that system control might be considered as a separable (ancillary) function, implicitly invested with the responsibility (and authority) to control generation and transmission facilities owned by others; such control is essential to the poolco concept. Another suggestion has been that the transmission owner might retain direct control over plants within its territories. These suggestions seem ultimately improbable. There are a number of reasons, even apart from operator parochialism, for departing from the current consensus that centralized control of generation will continue to be a system (ancillary) service. These reasons involve containment of both investment costs and operating costs within a competitive environment.

With proliferation of independent power producers, those able to shave costs the most will have a distinct competitive advantage, especially during periods of low demand. Producers who are aware of this will endeavor to minimize their capital as well as their operating costs. Barring strong incentives to do otherwise, they could very well prefer *high power factor* units designed for *base load* use, in preference to more expensive low power factor units. High power factor means very little reactive capability, reinforcing our contention that reactive support will become an intrinsic component of the transmission system. Base load use means very constrained response capability, reinforcing our contention that fringe control will not be available from bulk power suppliers. In any case, generators are unlikely willingly to forego control of their own production costs. In England, newly "privatized" and "de-integrated" generating companies saw a sellers market: they considered NGC as a captive customer, and overcharged significantly for providing regulating power and reactive support.

Furthermore, unless generators would lease their plants to the transmission entity - which in effect would be a reversion to the vertically integrated structure - it seems unlikely that the plant owners would be willing to permit others to exercise direct control without requiring such "hold harmless" and indemnity provisions as would be unacceptable to the candidate controlling entity. In the context of operation of a huge, complex system that is always vulnerable to widespread disturbances, owners of large generating units will be reluctant to cede active control of those units, which will represent not inconsiderable capital investment, to a third party.[7] Conversely, transmission entities will be correspondingly reluctant to assume responsibility for operation of such units when the consequences of operator error could be considerable.

In addition to such reasons, a clinching consideration is that, from a control theoretic standpoint, there will be no necessity for traditional, centralized control of units; necessary coordination can be provided by properly structured availability of information. Coordination need not imply control.

<u>Our finding, then, is that *centralized control of generation* not only does not qualify as a system-specific ancillary service, but that it is not a necessary function</u>, and is likely to fade away, like the proverbial old soldier.

We have concluded that the system specific ancillary services that impact system control requirements include (i) frequency control, comprising <u>default</u> regulation and loss compensation, (ii) voltage suppport, (iii) capacity reserves, and (iv) security control. This means that these system support requirements should be provided or procured by the transmission entity, with costs folded into load-sensitive tariffs. It further implies that, to the degree possible, end users (or their immediate suppliers) should provide their own ancillary requirements (including more or less rapid regulation of fluctuating loads, and inductive load compensation), since they are end-user specific, and, importantly, they can be met most efficiently locally rather than via the bulk power network.

SYSTEM CONTROL WITHIN THE NEW ENVIRONMENT

Assumptions

Based in large part on the foregoing discussions, the ruling assumptions of this major section are as follows:

- We are considering the end result of the current wave of restructuring, wherein generation, transmission, and distribution will be independent entities, along with marketers and brokers.
- Energy will be traded in an open market, involving sellers (generators, marketers, brokers) and buyers (end users, brokers, distributors, or other aggregator).
- Transmission and distribution will remain as natural, regulated monopolies.
- Functional responsibilities will follow ownership; wherever applicable, authority, responsibility, and liability will not be diffused.
- Given a proper incentive structure, cooperative decentralized control behavior can be elicited effectively from independent agents (generators).

Energy will be obtained by various classes of users by a mix of contracts reflecting diverse rates of response and terms of delivery, and that in each case the cost will reflect the rate of response.

It is assumed also that the following *Transmission Principles* pertain:

TCos should be responsible for security of the grid, not economics of generation, and should not participate in the functioning of the market.

TCos should not be required to provide energy, active or reactive, to end users, other than as a general network-specific service.

"Costs should be recovered in the rates of those customers who utilize the facilities and thus cause the costs to be incurred." (FERC's "fundamental theory of Commission ratemaking" [5]).

Primary consideration in policies and procedures should be objectivity and avoidance of ambiguity.

Conceptual Control Structures

The picture that has emerged is that of a large electrically unified system comprising a multiplicity of agents independently controlling energy injections at a multiplicity of nodes within a common network. The motivating principle behind this picture is that of control following ownership: generation control devolving to generation owners, and distribution control and load management to distribution companies and/or other retail load aggregators, while network control remains with an independent network owner. The latter is responsible for overall network security. With limited (hopefully necessary and sufficient) control means at his disposal, he will provide system-specific services to ensure system security and promote optimal use of the network. (It would be consistent with this picture for network reinforcement to be the responsibility of the network, with generation expansion planning and maintenance scheduling devolving to generation owners, perhaps in consultation with the transmission entity.)

Control Horizons Given a proper incentive structure, there seems to be no reason why cooperative decentralized control behavior could not be elicited effectively from independent generation owners. However, to the extent this behavior is volitional, i.e. under the direct conscious control of human agents, it cannot be effective over the full spectrum of control actions required for effective power system operation. One over-riding reason for this is that maintenance of system security may require drastic arbitrary actions without regard to economics.[8] Another reason, which should not be ignored, is that (as discussed above) regulation *per se* is distinct from, disruptive of, and cannot successfully be subsumed within, economic dispatch [6,9]. Accordingly, it seems necessary to define some *lead time threshold*, outside of which slower control actions will be taken by independent agents on their respectively owned generating equipment, in accordance with market related decisions, but within which higher frequency default fringe and emergency control will be effected automatically by centralized control acting on identified control means (including active and reactive power sources). Once the need for this *lead time threshold* is accepted, its definition can be effective in accommodating a number of considerations.

Within the traditional utility centralized control structure (which was not affected by the functioning or requirements of any external market), the effective delineation has been between the threshold within which no effective control can be exerted (which will continue to delimit control within a market structured system), and the threshold within which regulation (fringe control) is required, and beyond which economic dispatch can be effective. Under this traditional structure, the AGC function observes system frequency, net tie flow, and total generation. On the basis of these data and forecasts of changes in system load, it dispatches economy (including base load and peaking) generation, and controls regulating generation. Within this traditional structure, economic dispatch is (or can be) as automatic and centralized as regulation, so the threshold is determined by the minimal horizon over which economic dispatch can be meaningful. Under these circumstances, this threshold should certainly not be any shorter than five minutes. (A convenient rule of thumb has been to ignore apparent load fluctuations beyond 0.01 Hz, to regulate for load changes between 0.001 and 0.01 Hz, and to apply economic dispatch to load changes slower than 0.001 Hz.)

Within a market driven industry, other considerations may come into play at the discretion of individual agents (generators), and it may be desirable to define individual thresholds for each agent in the market. In this case, thresholds could range from five minutes, for agents with responsive generation and effective load-forecasting-based dynamic control, out to an hour for agents with less responsive generation. In either case, users would be responsible, utilizing contractual or other mechanisms, to procure energy responsive to their changing needs; such energy could be procured by a mix of means, including contracts for regulation, self-owned regulating generation, self-imposed load control, or other. In either case, higher-frequency (more responsive) energy to satisfy requirements *within the threshold*, would be provided by default by the central (transmission) control agent, and be charged to the generators or users according to their departure from schedule.

Games and Information Structures One would expect that the fluid situation provided by the changing structure of the utility industry would provoke considerable analysis by the game-theoretic community. This discussion does not seem to have developed however; the author is aware of only two papers [18.19] that explicitly address the topic,[9] although undoubtedly there are numerous others.

From a game theoretic perspective, the potential new structure of the industry seems be heterogeneous. On one hand, deregulated independent power producers seem to be faced with a zero sum game - a quasi fixed market of which each producer is trying to gain a larger share at the expense of the others. On the other hand, these same entities are mutually interested in enjoying maximally efficient operation of transmission grid - which implies a cooperative game.

The incentive for the transmission entity, under a regulated monopoly structure, would be to maximize use of the transmission grid, and to ensure that adequate transmission capacity is available to satisfy all potential transactions (sales/purchases). These two incentives are not equivalent. With a guaranteed rate-of-return on capital investment, the incentive would be to overbuild transmission; within a given physical plant, however, and with a given approved tariff, the incentive would be to maximize use of existing capacity within constraints set by system security. The incentive for generating entities in either case would be for each to maximize its sales within its available capacity, whenever necessary at the expense of the sales of competitors, in other words, obtain as large a portion of the market as possible within governing constraints.

Threats to security, for which the transmission entity would be responsible, include exhaustion of various categories of reserve capacities relative to contingent disturbances. These may include, inter alia, overloading of lines, exhaustion of reactive reserves, depletion of active (spinning) reserves, etc. These threats to security are inter-related, since, for instance, overloading of lines (i) results in disproportionate increases in reactive losses or (ii) may impair steady state stability; level of spinning reserves may affect transient stability following loss of equipment, etc. Thus, the transmission entity will wish to maximize system loading to the extent possible without incurring any of these dangers. Individual generators would prefer not to impair system security, but will wish to utilize as much of the available transmission capacity as needed for their own transactions, if necessary at the expense of use by competing generators. This would seem to imply a Nash strategy, although in an antagonistic competitive situation a min-max strategy might evolve [20].

Control and information structures are tightly interrelated, to the extent of implying each other. The mutual implication is not always explicit, or obvious, but the treatment of system control options would not be complete without some consideration of relevant information structures.

If the situation were considered as a Stackelberg leader-follower structure, with the TCo as leader and the GCos as multiple independent followers [21], the information structure that is adopted would have a critical influence on the optimality of the solution that is achieved [22]. (The information structure denotes the precise information that is available to each player at every stage of the game.) Thus the current discussion that FERC has initiated on a "real-time information network" (RIN) [23] would have a significant impact on the optimality of the operation of the restructured industry.

Control Areas NERC encompasses more than 150 control areas that range in size from less than 100 MW to greater than 10,000 MW - roughly two orders of magnitude. This diversity reflects the manner in which the industry has grown - historical happenstance as well as economic rationality. With restructuring of the industry, and the trend toward independent TCos, it seems likely that motivations that have led smaller companies to prefer independence to acquisition by larger companies will change. That incentive may well continue for GCos, but seems unlikely to be relevant to TCos [24]. The likelihood, then, is for amalgamation of TCos constituting smaller control areas into larger units.

In this process, there may be more incentive to consider (or to be influenced by without overtly considering) whether there are optimal bounds (technical and/or economical) on control area size, and if so just what they might be. How will/should boundaries be determined? Will open access ultimately erode physical control area boundaries (i.e. relatively weak tie cut-sets)? Vojdani et al., in [25], have noted the factors that will lead to tighter interconnections, hence a more homogenous overall grid, with less discernible inter-area boundaries.

Control Structure The imperatives of deregulation drive toward decentralized control of generation by independent owners, but the physical conditions of effective system operation require very rapid continuous response to changing system conditions. Many new control structures will be possible, and it is not clear what structures actually will emerge. We assume that the following circumstances characterize a postulated restructured industry.

We assume that energy is obtained by DCos or RBs from various GCos using a mix of contracts, which we will denote, very simplistically as (i) base load, being energy that is provided and used at a constant, unvarying rate over a period of a month or more; (ii) diurnal load, being energy provided according to a schedule based on predicted load variations over a twenty-four hour period; (iii) fringe load, being energy provided on demand to cover non-monotonic fluctuations in actual demand.

We assume that the energy provided under these contracts will be increasingly costly, proceeding from (i) through (iii).

We assume further, on the basis of earlier conclusions, that the TCo will provide to users, under contract to appropriate GCos, default supply of energy to cover transmission system losses and default amounts of regulating energy (which may be fringe), and (iv) spinning reserve to cover contingent loss of generating units from the network.

Under the postulated restructured schema, the centralized (albeit hierarchical) traditional control would be replaced by a decentralized, uncoordinated control. AGC as it has been known will no longer be used. GCos (including cogenerators and other independent power producers) would provide base load and diurnal load generation to schedule. Logically other (whether or not physically other) GCos would provide fringe generation in response to difference signals between a customer group, whether a DCo or aggregator, and its contracted load. The result will constitute essentially autonomous closed control loops embedded within the network. Network control (i.e. what remains of the traditional AGC) will focus on control of the distribution of power (flows), of voltage, and of network security; provision of necessary reactive sources will be embedded in network capital costs. The automatic network control (ANC) function will observe system frequency, and control (whether directly or by contract) reserve generation (which physically might coincide with regulating generation) to compensate for system losses and provide default regulation (high gain control to frequency excursions outside a deadband).

Type of Generation	Control Responsibility	Control Objective	Control Horizon
Base load	GCo or Marketer	Contract	Contract term
Load following	GCo or Marketer	Contract	Contract term
Regulation	GCo or Marketer	Contract	>15 minutes
Losses & default regulation	Load serving entity	Feedback	1 - 15 minutes
Voltage support	TCo	Frequency feedback	10 - 60 seconds
Reserves	TCo	Security	On demand

The questions posed by such a structure include (i) stability (in the most general sense) of the system with a multiplicity of independent controllers who interact only through the data which they individually extract from the system, and (ii) to what extent overall system efficiency (which includes control effort and losses) is sacrificed by lack of overall central coordination. The former may require some innovative work in system (control structure) stability analysis, including definition of the necessary and sufficient control means that would enable the network controller to maintain stable operation. The latter is a question that may be posed in game theoretic terms.

It seems, then, that the really significant developments (from a systems engineering standpoint) are *dispersal of generation, increased loading of transmission* (perhaps to saturation, but ultimately perhaps to decreased loading as penetration of dispersed generation increases), and *diffusion of responsibility.* Each of these suggests a number of serious problems. In what follows, we consider in turn challenging problems in the control of generation and of the network.

Generation Control

The future always surprises, but it seems likely that ultimately most bulk power generaton will be owned by a number of large companies, each of whom may own many large units at widely dispersed sites around the country. A lesser amount of generation may consist of smaller units, perhaps locally owned, that will have built or acquired to meet local needs (including local fringe control). Most of the large units may either (i) be ramped to initiate or terminate contracts for blocks of load, and otherwise held at constant load, or (ii) will be controlled to follow contractual trajectories developed from short term load forecasts. In this way, an increasing proportion of total generation may be removed from centralized control, resulting in a largely decentralized multi-operator control structure. While unprecedented, and to those who have grown up with the traditional industry structure, unsettling, there is from the control engineering perspective no reason to rule out the practicability of such a structure. It has been shown, for instance [8], that dynamic dispatch of economic generation to follow short term predicted load trajectories actually decreases the need for regulating effort.

Wholesale Generators Owners of bulk power generators will require a variety of control capabilities. Consider two types of contracts: one, to provide constant base load, and two, to provide diurnal load following power following a contracted trajectory based on client load prediction. Control of both would be feedforward in nature. The former would be injection of a contracted amount of constant power, and would require only ramping control. The latter would be injection of power following a trajectory based on predicted diurnal variations in load. They will require data from their customers providing forecast load movement.

Owners of multiple plants will be faced with the traditional problem of minimizing production costs of their units. They will require traditional (or advanced, dynamic) economic dispatch capabilities, but under new circumstances. Their contractual commitments will provide assurance against most unforeseen drastic changes in load (except insofar as they are committed to spinning reserve contracts). On the other hand, they will much more constrained in shifting power between plants, since their contracts will, in one way or another, be tied to injection at delimited locations. Contractual definition of areas within which arbitrary dispatch will be permitted may be necessary, since switching of generation outside such areas may affect system security.

Suppliers of spinning reserve, probably under contract to the transmission entity, will be responsible to respond effectively to random requirements for considerable blocks of energy. It will almost certainly be necessary, in contracts establishing access by producers to the bulk power network, to require that generators be fitted with adequate, active, well maintained governors, otherwise this expense is unlikely to be accepted by them.

Stability of such a control structure would depend, inter alia, on bandwidth coordination of the several feedback controls (in terms of *control horizons*). Regulation would be the widest bandwidth control. Loss replacement, as an integral (reset) control would be a secondary control. Another possibility arises from the development of power electronic devices capable of providing rapid modulation of network parameters. Such technology may enable TCo to exert effective control of system transient stability without (or with minimal) resort to direct control of generation. Another development contributing to the same lessening of need for direct system stabilizing control of individual generators is represented by advances in excitation control [65].

Load Serving Entities There are a number of ways in which fringe control (of fluctuations of load about diurnal variations) could be provided. One way would be by provisions for direct load control, by distributors or aggregators, of their customers' loads, but in most cases this would be neither acceptable nor sufficient. Another way would be for the contracted diurnal energy supplier to provide regulating energy in addition. Effectiveness of this approach would depend on the willingness of the contracted supplier to provide responsive generation, and on the respective locations of the end user and the supplier: the more distant the supplier, the greater the intervening transmission losses and the less effective the control. Since contracted energy suppliers would be chosen on price, could be changed frequently, and in any case would not be likely to have energy that was responsive to rapid changes in loading, it would seem more effective to contract separately with local suppliers for regulating energy at the subtransmission level. Thus, regulation could become a local, distribution company or other load aggregator, responsibility, nor only for reasons of cost and efficiency, but also because of the impossibility of assigning responsibility and credit for necessary elements of regulation at the bulk system level. (A partial indication of the strength of considerations that will tend toward divesting production entities of responsibility for regulatation may be seen even historically in the operation of vertically integrated utilities; what we have termed *operator provincialism* often obstructed centralized control of plants that would have been very beneficial to overall system performance.) *Freeing the bulk power transmission system from flows of stochastic regulating power would certainly lead to its improved utilization.*

In the end, it would seem to make sense for regulation to be effected at the local level by distribution companies or other aggregators, using whatever mix of means seems attractive, whether under contract or under direct control by the latter of leased or owned generation. Regulating generation would be controlled in response to a measured difference between contracted and actual client load. Suppliers of fringe energy will require data adequate to follow variations of load about diurnal variations. Any residual difference would be provided by the network as as default provider, and billed to the designated distributor on a capacity basis.

Network Control

In view of the potentially radical restructuring that is now underway, still other practices that have become customary over the past century may now have to be questioned. Several of these are addressed in what follows - How, and to what extent, can power electronics contribute to effective control of power distribution on the network? - What are the fundamental criteria for specifying a voltage profile that may be considered optimal in a meaningful sense? - What is the proper level of regulation on an interconnected system? or, equivalently, What is the proper level of frequency control effort? - How can security of the network be maintained in a restructured, open access environment?

Network Flow Control If we accept the burden of the discussion thus far (distributed control of generation by individual owners), it appears likely that AGC as it has been known will no longer be used (although owners of multiple plants may want to dispatch their own units economically). What then remains of the traditional AGC will be an automatic network control (ANC), functioning within an overall network security control and focused on control of the distribution of power (flows), and of voltage. Provision of necessary reactive sources for voltage support will be embedded in network capital costs, with generator reactive capabilities being reserved for emergency response; this pattern already seems to be emerging in Great Britain. Freeing the bulk power transmission system from flows of stochastic regulating power would certainly lead to its improved utilization.

When attention is focused on the problem of controlling the distribution of the flow of power on the transmission network (to maximize the accommodation of scheduled delivery of blocks of energy while maintaining an adequate level of security), a new picture emerges. In this case, one is aware of the source and destination of the scheduled flows, and appropriate control actions may be taken to satisfy the objectives. The significant difference is that the choice of controls in this case may be somewhat arbitrary so long as the objectives are satisfied; economics are embedded in satisfaction of the contractual flows/deliveries, thus obviating any need (at the network level) for traditional economic dispatch. Transmission system operating objectives would then be confined to optimizing transmission loading, and maintaining network security. The related tasks are control of the distribution of power flows, and potential strategies include enhanced optimal power flow, redesigned voltage/VAr control, and newly conceived transmission system parameter control.

Reference to "optimization" of network loading may fudge a non-trivial consideration. What is "optimal" in network usage will depend on to whom the network operator is basically answerable - to generating entities who own the "wires," to "load serving entities" (distribution companies, retail brokers, load aggregators) whose customers have paid (and are continuing to pay) for the wires, or to himself as an independent transmission company owning the "wires." *In the latter, perhaps ultimately most probable, case, optimization will mean maximum usage; in the first case, optimization might, for instance, imply trading in "access rights," which would not necessarily maximize loading.*

In any event, given distributed control of generation, automatic network control (ANC, the successor to AGC) could be realized by a much cleaner control regime than is possible for providing AGC under the current transitional industry structure (partly regulated, partly market driven). The major factor will be the removal of responsibility for dealing with production costs; the ANC regime will be focused sharply on "optimal" use of the network, however defined, subject to security constraints.

It is likely that by the time a new industry structure becomes predominant, a variety of power electronics (FACTS) devices will be available and affordable. For as long as transmission companies are still regional, path control as well as network flow optimization will still be necessary. A possible complicating factor will be assignment of responsibility for stability problems involving interaction of machine and network parameters, e.g. subsynchronous resonance. Power system stabilizers, presumably, will still be installed on machines, but network control of line parameters will exacerbate the problem and may require coordination between the two control regimes - plant and system.

Network Voltage Control Currently prominent among power system control problems are those related to voltage stability. Until relatively recently, voltage control was not a major issue: systems were built robustly, to a degree that often resulted in lightly loaded transmission, and generally were either transient- stability limited or thermally limited. Moreover, reactive support is limited in range, thus conducive to local control especially on sparse, lightly loaded systems. For many years voltage control meant local control; manual control of voltage levels on the bulk power network was the rule, and often was treated almost casually. Thus, system voltage control did not receive the same attention, nor has it evolved to the same level of maturity, as has automatic generation control.

Increased loading of transmission has brought into prominence effects of nonlinear phenomena which, earlier having been largely dormant, were neither well understood nor studied. Subsequently, considerable study has been invested in understanding system voltage phenomena under heavy loading. In the wake has come an appreciation of the need for a more comprehensive approach to control of system voltage. This is a significant problem within the context of traditional system control structures. It becomes even more interesting in the context of an emerging restructured industry with its as yet unforeseen control structures. Irrespective, then, of system restructuring, there is need for more comprehensive, coordinated control of voltage as systems become tightly interconnected and more heavily loaded.

Under the same conditions, the restricted range of effectiveness of reactive injections for affecting voltage levels means that voltage control is almost ideally suited to hierarchical implementation, even more so than is active power control. The hierarchical nature of the voltage control problem has long been recognized by theorists [27,28,29], and to some extent has been implemented in practice by a few utilities [30,31,32].

The major purpose of hierarchical voltage control has been overall optimization of system performance. In such case, the effectiveness of the control depends on the criteria that are prescribed for the optimization. Unfortunately, while maintenance of a desirable system voltage profile, one that is secure and economically sound, is becoming increasingly important, discussion of what constitutes the optimality of a voltage profile has been confined mostly to *obiter dicta*, with little serious investigation and evaluation of possible criteria. The problem of determination of the desirable ("optimal") voltage profile of a transmission network has never been fully addressed, even in the context of the traditional industry structure. It is becoming more acute as new structures evolve.

Voltage profile criteria. For many years, control of voltage was discussed within the context of overall economic system control, as a generalization of economic dispatch [33,34], predating concerns with security control. Until recently, this objective has generally dominated discussion of the topic. A recent CIGRE 38.02 Task Force Report [11] notes that "The criterion most often used is that of minimizing active power loss." However, an earlier 39.02 report [35] concluded, inter alia, "that voltage and reactive power control is mainly related to service quality and electric system security, and that economic considerations are a secondary target."

The broader question has not been entirely neglected, of course, and surfaced in the literature some years ago [36,37]. Capasso et al. [37] stated the problem as "determination of the reactive control variables in such way that the system is in the best state to cope with perturbances (load increase, generator or line tripping) while still complying with specific constraints (minimum and maximum voltages, minimum and maximum reactive power injections). In other words, the system must have available the maximum reactive power margins, and these margins must be distributed in an appropriate way among the generators." To achieve these objectives, they evaluate the performance of four objective functions on two types of networks. The objective functions considered are: (i) minimization of the sum of the absolute values of the reactive injections from the generators; (ii) minimization of the weighted reactive injections; (iii) minimization of active transmission losses; (iv) minimization of the deviation of voltages from preselected values. They conclude, *inter alia*, that the second of these objectives is most effective in ensuring a desirable distribution of reserves. The very recent CIGRE report [11] concurs: "Security is improved by maximizing the global and local dynamic reactive power reserves. Such operation can also reduce active power losses. An objective function which aims for large global and local reactive reserves leads, as a natural consequence, to a high operating voltage (near maximum limit)."

Other commentators chose to maximize reactive power reserve margins [11], or to minimize total reactive output of generators, "with the generators sharing, in so far as possible, the reactive powers proportional to their reactive power capability limits" [38]. Obviously, a wider variety of objective functions could be considered. Possible objective functions in addition to those evaluated by Capasso et al. might include, e.g. minimum VAr flow, and concentration of VAr reserves in most vulnerable areas of system.

What constitutes an optimal voltage profile *for operation of a given existing system*? It seems clear that, from a *planning* perspective, the ideal profile is high and flat. A high profile minimizes the flow of active power, and a flat profile minimizes the flow of reactive power, thus minimizing losses and maximizing reserves, so that neither security nor operating costs are compromised. It is in the operating arena that the question becomes difficult. The ideal system is always too costly, the desirable system that is planned is never fully realized, and the system as built is never fully operational - there are nearly always major equipments out of service for one reason or another. It is in this context, then, of the system actually at the disposal of the operator at any point in time, that the question of optimality must be answered.

Thus, it appears that, (1) In general, the "ideal" voltage profile with limited reactive resources will be a function of the particular use that is being made of the network at any point in time; (2) Reactive margins and their distribution are not critical problems under light and medium load conditions; (3) The relative effectiveness of reactive resources *vis-a-vis* site specific requirements of system security must always be taken into account. (4) In the literature that has been reviewed, a restricted variety of objective functions have been considered. Possible additional objective functions might include, e.g. minimum VAr flow, and concentration of VAr reserves in most vulnerable areas of system.

There are several open questions: (1) *How should the tradeoff between loss minimization and security be resolved*? (2) *What distribution of reactive reserves provides the best overall security, and under what conditions*? (3) *How much loss is incurred by giving priority to security considerations*? i.e., if reactive injections are minimized, *then* how much additional savings in losses could be realized under a minimum loss criterion?

Network Frequency Control Much difficulty arises from problems inherent in regulating power (fringe control). Within the traditional, vertically integrated structure, it has been learned that it is inefficient to supply regulation without distinguishing it from economic generation [8,9], but just how much regulation is necessary to good system performance has never been studied. CEGB, for instance, never exercised load-frequency control. It may well be appropriate to ask to what extent might frequency regulation be relaxed in the future, in a restructured industry?

Automatic generation control (AGC) has been with us for nearly half a century, is widely accepted, and is well entrenched in practice in many countries. Nevertheless, its performance has never been viewed as fully satisfactory, and there is a considerable literature that discusses its deficiencies and suggests improvements [e.g. 38,39]. There are significant reasons why AGC performance might be less than ideal, reasons that are rooted in the basic characteristics of large scale power systems. What is perhaps remarkable is that traditional AGC has performed and continues to perform as well as it does. Reasons for less than ideal performance include, inter alia, the large scale of power systems, their micro-heterogeneity, their pervasive nonlinearity, their stochastic operating environment, and their widely distributed (albeit largely hierarchical) control structures. A possible reason contributing to the general acceptability of their performance may well be that last, parenthetically noted, hierarchical nature of current control structures. A significant reason for reviewing this problem at this time is that, as discussed above, restructuring may lead to disassembly of current hierarchical control structures.

The necessity and essence of frequency control. A fundamental requirement of alternating current power systems is that the generators supplying the system operate in synchronism - a necessary condition for the flow of power on the network. A corollary requirement is that generators and associated equipment must be operated at speeds very near their design frequency in order to avoid catastrophic damage. A necessary condition for maintaining constant speed operation is balance between load and generation; in other words total generator power input to the system at any point in time must equal total power withdrawn from the system at that same time. Excess power generation results in acceleration of the machines (increasing frequency); deficiency in generated power results in deceleration (decreasing frequency). Thus, the necessary and sufficient condition for maintaining constant frequency is to maintain generation output equal to load consumption.

In our discussion of ancillary services, it was maintained that active-power related ancillary services, which were identified as default regulation and transmission loss compensation, were system-specific requirements. The preceding paragraph is the inescapable reason for this conclusion. Stable, reliable system operation requires that the system operator maintain a stable (quasi-constant) frequency, and in order to do this, he must control sufficient generation to compensate for default deficiencies in regulation of load-frequency balance by the other (independent) control agents on the system, and constantly supply the fluctuating amounts of energy being lost in the transmission process. These deficiencies are conjointly apparent through frequency fluctuations.

A consequence of this view is that the TCo should simultaneously be providing both loss compensation and default fringe control. That losses and regulating energy thus be de facto bundled may seem questionable to an economist in view of the inherent differences. It also raises the question as to how should magnitude of losses be determined, and how should they be distinguished from regulating energy? How could each be metered? or observed?

One way of handling this might be to monitor aggregated users - distribution companies (DCo), retail brokers (RBr), large wholesale customers, etc. - netted with their respective fringe generation, in order to charge them for their usage. Perhaps residual fringe energy might be distinguished from loss-supply energy, to an adequate degree, by virtue of considering the former to be composed of a zero-mean component. Loss compensation would then become the steady state component. Assignment of a higher cost to transmission of regulating energy should follow naturally, and equitably, from consideration of the fact that the transmission capacity "band" required to transmit such energy is inescapably greater than twice the band required to transmit an equal amount of non-fluctuating, i.e. constant, energy. Thus, equitable assignment of costs should encourage local regulation.

Under the scenario developed here, default energy for frequency control and compensation of system losses could be procured by TCo under contract for such energy on a periodic (weekly or monthly) basis through a formal bid process by GCos. It would cover both system losses and net deficiencies in total system (control area) regulation, based on integral control of area control error (ACE) deviation (on an isolated system, this would be integral control of frequency). ACE might be defined for the entire TCo area, or, if the TCo area becomes too large, for individual transmission zones defined within the TCo area. *TCo generation would respond only to ACE deviations outside of defined bands.*

System frequency regulation is mandatory. At the same time, it is costly, and assignement of responsibility for causing frequency flucttuations is extremely difficult. A most important question, then, remains to be considered: *how tightly must frequency be controlled?*

Conventional power system frequency control. Under what we are now to view as the "bundled" paradigm of the past, power system control practice comprehended a number of objectives, conveniently categorized as security, economy, and more recently, environmental impact. All of these are addressed, at least partially, within AGC. Frequency control is effected, on different time scales, by machine inertia, governor response, and AGC action on governor motors. (AGC itself has comprehended frequency bias, tie-line flow control, and economic dispatch.)

Speed governors were developed as a necessary adjunct to rotating machines in order to ensure their stable operation. Viewed at its simplest, supplemental load/frequency control was subsequently required in the context of electric energy system operation, inter alia, in order to maintain machine governors within operating range. Automatic generation control was required in order to integrate such (otherwise uncoordinated) load-frequency control within the larger objectives of maintaining an economic (or other desired) distribution of generation among the diverse machines operating on a given system, to control power on tie-lines at desired levels, and at times to achieve other objectives.

On conventional power systems, operating with this hierarchical control structure, the continual matching of total generation output to continuously varying total system load requirements is implicit. Any given generator may be operated at constant output, and in practice, for a variety of reasons, many machines are thus operated. Nevertheless, at all times, a sufficient number of machines (i.e. machines having sufficient output capacity and response rate capability) must be under automatic control to maintain stable system operation. Insufficient capacity to respond to cumulative changes in system load would ultimately, if uncorrected, result in system disintegration, with machines being taken off-line by operation of under- or over-speed protective relays.

Within the lower and upper limits on system frequency that must be observed in order to avoid such system instability [40], stricter limits have come to be observed on most modern interconnected systems in order to satisfy additional requirements of interconnected system operation. Flows on limited-capacity ties connecting large systems have had to be controlled in order to avoid their being lost due to inadvertent overloading. As ties proliferated and became stronger, tie-line/frequency-bias control became the means whereby each system could meet its obligation for supplying its own load. Observance of general performance objectives was encouraged by development, by the North American Electric Reliability Council (NERC) Operating Committee, of the NERC-OC criteria for interconnected system operation.

In practice, the performance of AGC generally, and tie-line/frequency-bias control in particular, has been far enough from ideal to remain a source of chronic irritation to system operators [e.g. 39,41,42] and a challenge to theoreticians. Some of the deficiencies represent physical limitations on achievable control capabilities; others are manifested by persistent, poorly understood phenomena.

To what extent is it necessary, or advantageous, to control system frequency? Among the many beneficent practices of the traditional industry were a variety of implicit and explicit gentlemen's agreements among utilities, including cooperation in contributing to a high level of quality in delivered energy. Such attitudes and practices cannot be expected to flourish within the emergent restructured industry.

The bounds to which frequency should be controlled have never been fully justified on either engineering or economic grounds. (We should note that, in addressing bounds on frequency control effort, it is necessary to distinguish between, and discuss separately, bounds on both the magnitude of frequency departures from scheduled frequency, and the spectral bandwidth of such fluctuations (considering the frequency itself as a stochastic process in time). An early EPRI report [40] addressed some implications of the former; some authors have suggested heuristic bounds on the latter [e,g., 43]). The intent of industry operation over many years, as evidenced by the NERC-OC criteria, has been to do as good as might be practical. This is in contrast to the practice in those other countries (e.g. Great Britain) wherein AGC has never been implemented. In retrospect, it is arguable that an emphasis on controlling higher frequency fluctuations poses a control problem that is in large measure both unachievable and unnecessary.

Implications for frequency control within a restructured industry. Whatever the rationale of past practice, it is likely that, sooner or later, a cost-driven trend will emerge toward looser control of frequency. It is possible that this may proceed by a trial-and-error process, pushing the envelope until system performance starts to suffer. (An increasing scope of interconnections has resulted in tighter frequency bands due to greater inertia of systems [44]. Will a trend to smaller machines decrease overall system inertia?) It would be more prudent, however, to seek a better understanding of what frequency control effort is desirable, of what performance should, at least, be maintained, rather than focusing on what, at most, can be achieved. While it might have been debatable during one period in the past whether or not the industry was in the business of selling "time," that product is no longer on the table as an ancillary service. It is true that diverse end users will have diverse requirements on the quality of their delivered power. The fact is, however, that there is no practical, inexpensive way of providing diverse levels of frequency invariance on a common, synchronous system, and most end users have no need for rock-solid frequency in delivered power.

With the U.S. FERC's designation of fringe control (which they term "load following") as an ancillary service (which implies that it should be provided and paid for only to the extent required by individual end users), basic frequency control (which FERC also designates as an ancillary service, but which is necessary to ensure stable system operation) should rightly be considered on its own merits [45].

In view of this, it might be appropriate to phrase the question as "*what level of performance will be acceptable to most end users*?" In an open market, of course, this is a question to be determined by the market - How much is the market willing to pay for a given level of service? In responding to the market, however, more efficient, less costly ways of providing an attractive level of service become necessary. Are we able yet, after a century of studying power system dynamics, to specify how this might be achieved?

Network Security Control System security control may be viewed as the "covering" ancillary service (provision for operation of the network), the most basic, comprehensive, or even the ultimate one, to which the others are necessary but not sufficient contributors. Security is the resiliency of a system *vis-a-vis* potential (defined) significant disturbances. Formally it may be defined as a joint function of the contingent probability of violation of critical operating constraints and of the probability of disturbances [46,47]. Pragmatically, it may be seen as a matter of adequate reserve margins in critical system variables [48].

The mechanism of security control is that by which the findings of security assessment are brought to bear so as to ensure ongoing normal operation of the system. A direct consequence of any invoking of this function is an increase in system marginal cost, due to operating within constraints that are calculated to contain the system's exposure to major disruption of its operation.

Of the many approaches that might be taken to maintaining system security within a restructured industry, two of those being discussed may be suggested as compatible with the framework that has been discussed herein. One is that of direct curtailment by the network controller of any access to the network that would impair security margins viewed as hard constraints [49]. The other is that of indirect control of access by escalation of transmission tariffs as a function of diminishing margins viewed as soft constraints [50]. Each of these is subject to *a priori* objection. Depending on whatever specific criteria might be developed, the former could be viewed as arbitrary in curtailing access on fortuitous grounds, while the latter could be viewed as arbitrary in assigning prohibitive prices bearing no relation to actual transmission operating cost. (It should be noted that a certain degree of objective increase of transmission cost with increased congestion would follow from the use of FACTS devices to permit increased loading beyond that achievable in their absence. Distortion of natural flow patterns (which effectively minimize losses) would result in an increase of losses, with attendent increased costs.) The latter would, however, have the advantage of discouraging use of the network by less "valuable" transactions, and in order to avoid having the transmission company profit from congestion, and to encourage building of necessary transmission reinforcement, the escalated portion of "congestion" rates could be placed in escrow for new transmission capacity. The portion of construction paid for by escrow funds would <u>not</u> then be added to the transmission company rate base. In either case, justification would have to be provided by regulatory approval of judiciously developed criteria for curtailment of transmission access when necessary to maintain security of system operation.

Conclusions

Although things have always changed slowly in the electric energy industry, the restructuring of the industry that is now underway is so radical that it is unlikely that industry operations will be able to avoid equally dramatic changes in the long run. From a very pragmatic viewpoint, if the industry does indeed fragment into independent generation, transmission, and and load aggregation entities (very broadly conceived), it is quite likely that control responsibility and authority will follow ownership. This implies a decentralized control pattern very unlike what has been seen in the past. This seems startling to those of us who grew up in a vertically integrated and centralized, hierarchically controlled industry, but from a control theoretic standpoint, there is no reason the industry could not be successfully (and reliably) controlled in this very different way.

There will of course be technical challenges to be faced in implementing any changes in power system control. As a matter of fact, there have always been significant challenges in keeping control capabilities abreast of changing conditions even within the slowly evolving industry of the past. But as long as we're trying to anticipate technical challenges, we might as well consider those circumstances that would appear to be the most challenging. That is what this chapter has been all about.

Considering then that the industry may indeed fragment, and that control will follow ownership, we are led to the supposition that many incentives will lead to self-interested actions - i.e. that generators will be motivated to control as little as possible while fulfilling their contractual responsibilities and providing for safety of their machines during disturbances. Load aggregators will be motivated to minimize costs, and may well find local power factor correction, load management, and some degree of local regulation to be attractive. Network controllers will put great emphasis on security control, but will be pressed by generators to justify every requirement they might wish to impose for tight control of frequency. The challenge to control engineers is to develop control regimes that will accommodate all these obectives within a robust, reliable system.

References

1. L.J.Hill: Is Policy Leading Analysis in Electric Restructuring?; *Elec. Jour.*, v.10/6, July 1997, pp.50-61
2. L.H.Fink, M.D.Ilic, F.D.Galiana: Transmission Access Rights and Tariffs; *Elec. Power Syst. Res.*, (to appear)
3. K.Anderson: Transmission Services Under Restructuring: Getting the Incentives (Almost) Right; *Elect. Jour.*, v.10/5, June 1997, pp.14-21.
4. N.E.Nilsson, J.Mercurio: Synchronous Generator Capability Curve Testing and Evaluation; *IEEE Trans.* v.PD-9/1, Jan. 1994, pp.414-24.
5. FERC: Promoting Wholesale Competition Through Open Access Non-Discriminatory Transmission Services; U.S. Fed. Energy Reg. Comm., 70 FERC 61,357; p.175, note 269.
6. C.Nichols: Techniques in Handling Load-Regulating Problems on Interconnected Power Systems; *AIEE Trans.* v.72/6, 1953, pp.447-60.
7. N.Jaleeli et al.: Understanding Automatic Generation Control; *IEEE Trans.* v.PS-7/3, 1992, pp.1106-22.
8. L.H.Fink, I.Erkmen: Economic Dispatch to Match Actual Data to the Actual Problem; *Proc. EPRI Power Plant Perf. Monitoring and System Disp. Conf.*, Washington D.C., 12-14 November 1986, EPRI CS/EL-5251-SR, July 1987, pp.4/23-43.

446 POWER SYSTEMS RESTRUCTURING

9. H.G.Kwatny, T.A.Athay: Coordination of Economic Dispatch and Load-Frequency Control in Electric Power Systems; *Proc. IEEE/CSS Conf. on Dec. and Cont.*, 1979, pp.703-14.
10. F.F.Wu, P.Varaiya: Coordinated Multilateral Trades for Electric Power Networks - Theory and Implementation; Rpt. PWP-031, June 1995, Univ. of Cal. Energy Inst., Berkeley, Cal.
11. CIGRE Task Force 38.02.12 (ed. C.W.Taylor): Criteria and Countermeasures for Voltage Collapse; CIGRE, 1995.
12. P.Nedwick et al.: Reactive Management - A Key to Survival in the 1990s; *IEEE Trans.* v.PS-10, n.2, May 1995, pp.1036-43.
13. A.F.Mistr,Jr.: Reactive Planning and Operation on the Virginia Electric and Power Co. System; *Proc. EPRI/NERC Forum on Voltage Stability*, Breckenridge, Colorado, September 1992, (EPRI TR-102222), pp.2/15-24.
14. L.H.Fink: Security - Its Meaning and Objectives; *Proc. Int'l Symp. on Power System Security Assessment*, Ames, Iowa, 27-29 April, pp.35-41.
15. N.Balu et al.: On-Line Power System Security Analysis; *Proc. IEEE*, v.80/2, Feb. 1992, pp.262-80.
16. B.Avramovic, L.H.Fink: Real-Time Reactive Security Monitoring; *IEEE Trans.* v.PS-7/1, 1992, pp.432-37.
16. M.D.Ilic, F.Graves: Optimal Use of Ancillary Generation under Open Access and its Possible Implementation; TR 95-006, Mass. Inst. of Tech., Lab. for Electromagnetic and Electronic Systems, August 1995.
17. A.Haurie et al.: A Two-Player Game Model of Power Cogeneration in New England; *IEEE Trans.* v.AC-37/9, Sept. 1992, pp.1451-56.
18. Y.Tsukamoto, I.Iyoda: Allocation of Fixed Transmission Cost to Wheeling Transactions by Cooperative Game Theory; *Proc. IEEE/PES PICA*, May 1995, Denver, pp.3-10.
19. J.F.Nash: Noncooperative Games; *Ann. Math.*, v.54/2, 1951, pp.286-95.
20. G.P.Papavassilopoulos, J.B.Cruz: Nonclassical Control Problems and Stackelberg Games; *IEEE Trans.* v.AC-24/2, April 1979, pp.155-66.
21. Y.C.Ho, D.Teneketzis: On the Interactions of Incentive and Information Structures; *IEEE Trans.* v.AC-29/7, July 1984, pp.647-50.
22. U.S.FERC Docket No. RM95-9-000: "Real-Time Information Networks," Notice of Technical Conference and Request for Comments, (March 29, 1995).
23. "PacifiCorp leading talks on creating transmission company for entire West;" *Elect. Util. Week*, 7 August 1995, p.1
24. A.F.Vojdani, C.F.Imparato, N.K.Saini, B.F.Wollenberg, H.H.Happ: Transmission Access Issues; 95WM121-4PWRS

25. J.W.Chapman et al.: Stabilizing a Multimachine Power System via Decentralized Feedback Linearizing Excitation Control; *IEEE Trans.* v.PS-8/3, August 1993, pp.830-39.
26. K.Kumai, K.Ode: Power System Voltage Control by Using a Process Control Computer; *IEEE Trans.* v.PAS-87 n.12, 1968, pp.1985-90.
27. S.Narita, M.S.A.A.Hammam: Multicomputer Control of Voltage and Reactive Power in Multiarea Power Pools; IEEE/PES 1971 Summer Meeting, Portland, Oregon, paper 71 CP 597-PWR.
28. H.Kobayashi, Y.Tamura, S.Narita, M.S.A.A.Hammam: Hierarchical Control of System Voltage and Reactive Power; IEEE/PES 1976 Winter Meeting, New York, paper A 76 212-1.
29. V.Arcidiacono: Automatic Voltage and Reactive Power Control in Transmission Systems; *Proc. CIGRE-IFAC Symp. 39-83*, Florence 1983, survey paper E
30. S.Corsi et al.: Coordination Between the Reactive Power Scheduling Function and the Hierarchical Voltage Control of the EHV ENEL System; IEEE/PES Summer Meeting, San Francisco, 24-28 July 1994 (94SM 584-3 PWRS)
31. X.Liu, C.Vialas, M.Ilic, M.Athans, B.Heilbronn: A New Concept for Tertiary Coordination of Secondary Voltage Control on a Large Power Network; *Proc. 11^{the} PSCC*, Avignon, France, Aug./Sept. 1993, pp.995-1002.
32. H.M.Smith,Jr.,S-Y.Tong: Minimizing Power Transmission Losses by Reactive - Volt-Ampere Control; *AIEE Trans.* v.82, 1963, pp.542-44.
33. W.O.Stadlin: Criteria for the Control and Economic Dispatch of Reactive Volt Amperes; IEEE/PES Summer Meeting, New Orleans, July 1966, 31PP66-332.
34. CIGRE Task Force 39.02: Voltage and Reactive Power Control; *CIGRE* 1992 (39-203)
35. L.Franchi et al.: Evaluation of Economy and/or Security Oriented Objective Functions for Reactive Power Scheduling in Large Scale Systems; *IEEE Trans.* v.PAS-102/10, 1983, pp.3481-88
36. A.Capasso, E.Mariani, C.Sabelli: On the Objective Functions for Reactive Power Optimization; IEEE/PES Winter Meeting, 1980 (paper A80 090-1; abstract appeared in *IEEE Trans.* v.PAS-99, n.4, July/August 1980, p.1326)
37. CIGRE Task Force 38.02.03: Improvement of Voltage Control; *Electra*, n.135, April 1991, pp.117-27.
38. D.N.Ewart: Automatic Generation Control, Performance under normal conditions; *Systems Engineering for Power - Status and Prospects*, U.S.-E.R.D.A. Conf.-750867 (76-66), 1975, pp.1-14.
39. F.R.Schleif: Interconnected Power Systems Operation at Below Normal Frequency; EPRI Report EL-976, February 1979.

40. L.S.VanSlyck, N.Jaleeli, W.R.Kelley: A Comprehensive Shakedown of an Automatic Generation Control Process; *IEEE Trans.* v.PWRS-4 n.2, 1989, pp.771-81.
41. IEEE/PES Working Group: Current Operational Problems Associated with Automatic Generation Control; *IEEE Trans.* v.PAS-98 n.1, Jan. 1979, pp.88-96.
42. J.Zaborszky, J.Singh: A Reevaluation of the Normal Operating State Control of the Power System Using Computer Control and System Theory; *Proc. 11the IEEE/PES PICA*, 1979, pp.205-13.
43. C.Concordia: "Effect of Prime-Mover Speed Control Characteristics on Electric Power System Performance," IEEE Trans. v.PAS-88 n.5, May 1969, pp.752-54.
44. L.H.Fink: Ancillary Transmission Services; *Elec. Jour.*, v,9/5, June 1996, pp.18-25..
45. L.H.Fink: Security - Its Meaning and Objectives; *Proc. Int'l Symp. on Power System Security Assessment*, Ames, Iowa, 27-29 April, pp.35-41.
46. N.Balu et al.: On-Line Power System Security Analysis; *Proc. IEEE*, v.80/2, Feb. 1992, pp.262-80.
47. B.Avramovic, L.H.Fink: Real-Time Reactive Security Monitoring; *IEEE Trans.* v.PS-7/1, 1992, pp.432-37.
48. F.F.Wu, P.Varaiya: Coordinated Multilateral Trades for Electric Power Networks - Theory and Implementation; Rpt. PWP-031, June 1995, Univ. of Cal. Energy Inst., Berkeley, Cal.
49. M.D.Ilic, F.Graves: Optimal Use of Ancillary Generation under Open Access and its Possible Implementation; TR 95-006, Mass. Inst. of Tech., Lab. for Electromagnetic and Electronic Systems, August 1995.

---

[1] FERCs fundamental theory of Commission ratemaking states: costs should be recovered in the rates of those customers who utilize the facilities and thus cause the costs to be incurred. [5]

[2] In other words, the total system load is an amalgam of a <u>very</u> large number of fluctuating loads which are varying rapidly at rates ranging from <u>very</u> slow to <u>very</u> fast.

[3] It is not uncommon for a given load bus to be largely dependent on a single generating bus for effective support of voltage. The specification of a market mechanism under which buyers are effectively limited to choice from a single supplier seems somewhat incongruous.

[4] In general, the flow of power along a transmission line between two buses is proportional to the difference in the voltage <u>magnitudes</u> at those two buses. In addition, reactive power consumption (loss) in transmission is significantly greater than that of active power, since transmission lines have high reactance and low resistance. Accordingly, reacctive power can be drawn from the network only at the

NEW CONTROL PARADIGMS FOR DEREGULATION 449

expense of significant losses in transit and depressed voltage at the receiving bus: the greater the reactive requirement, the greater the voltage depression.

5 The use of FACTS devices to permit increased loading, beyond that achievable in their absence, to ease congestion, would result in a certain degree of objective increase of transmission cost. Distortion of natural flow patterns, which effective minimize losses, would result in an unnatural escalation of losses, with attendent increased costs.

6 In order to avoid having the transmission company profit from congestion, and to encourage the building of necessary transmission reinforcement, the escalated portion of congestion rates could be placed in escrow for new transmission capacity. The portion of construction paid for by escrowed funds would not then be added to the TCo rate base.

7 If it were possible, it is even conceivable that owners might prefer to interface with the grid through back-to-back DC conversion, to avoid exposure of their machines to the vicissitudes of system operation.

8 It is assumed that the basic provision for maintenance of steady state stability, i.e., effective governor action, will be prerequisite to connectin of any generator to the network.

9 In the first instance [18], the situation considered is that of the effect of PURPA rules on the relationship of a utility and qualifying facilities (QF). In the second [19], the focus is on cost allocation.

12 THE CONTROL AND OPERATION OF DISTRIBUTED GENERATION IN A COMPETITIVE ELECTRIC MARKET

Judith B. Cardell* and Marija Ilić**

*Office of Economic Policy
Federal Energy Regulatory Commission
Washington, DC 20426

judith.cardell@ferc.fed.us

**Department of Electrical Engineering and Computer Science
Massachusetts Institute of Technology
Cambridge, MA 02139

ilic@mit.edu

INTRODUCTION

Small scale power generating technologies, such as gas turbines, small hydro turbines, photovoltaics, wind turbines and fuel cells, are gradually replacing conventional generating technologies in various applications, in the electric power system. These distributed technologies have many benefits, such as high fuel efficiency, short construction lead time, modular installation, and low capital expense, which all contribute to their growing popularity. The prospect of independent ownership for distributed and other new generators, as encouraged by the current deregulation of the generation sector, further broadens their appeal. In addition, the industry restructuring process is moving the power sector in general away from the traditional vertical integration and cost-based regulation and toward increased exposure to market forces. Competitive structures for generation and alternative regulatory structures for transmission and distribution are emerging from this restructuring process.

These changes introduce a set of significant uncertainties. How will the siting of numerous small scale generators in distribution feeders impact the technical operations and control of the distribution system, a system designed to operate with a small number of large, central generating facilities? How will the power system architecture evolve as a result of both technological advances and competitive market forces? In response to the new and potentially conflicting

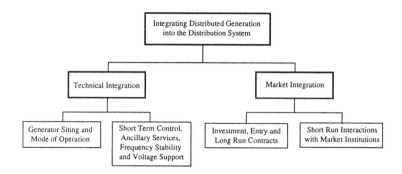

Figure 12.1 Integration of Distributed Generation into the Distribution System

economic and technical demands of a growing number of independent generators, what balance between market forces and real-time control will be found to coordinate distribution system operations? How will ancillary services be maintained in the new environment?

The development of a competitive market for generation creates the opportunity for distributed generators (as with all generators) to participate in energy and related markets as independent producers. Two factors influence the ability of small generators to participate in the emerging markets.

- The first factor is the legal and regulatory structure which may constrain the extent to which small generators are *allowed* to participate in these markets (entry). In particular, the growth of retail competition, and the designation of distributed generation (DG) as part of the distribution versus the generation sector will both shape the role of DG in the emerging competitive markets.

- A second issue is the day to day, and minute to minute operation of these technologies, and the creation of a competitive market to coordinate such short run operation within the distribution system. This short run coordination can be performed by a market based control signal. Whereas the regulatory structure discussed above is important to the extent that it defines entry criteria into the market, a market- or price-based control signal is important in that it facilitates operation in the markets.

For a system with distributed generation to operate reliably and efficiently, the system's operation and control strategy must accommodate both the engineering need to maintain collective system services and the economic push for independent and decentralized decision making. This chapter addresses both technical and economic issues associated with integrating numerous small scale generators into the distribution system, in a competitive electric market. The main issues addressed in this chapter are the short-term control and short-term market interactions as identified in Figure 12.1. To accommodate the

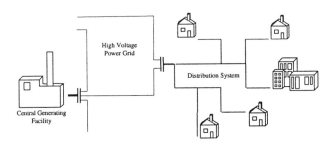

Figure 12.2 The Historical Structure of the Distribution System

expanded use of distributed generation in the near term, the first part of this chapter discusses integrating distributed generators into the general operations of the system, and maintaining system performance and stability as defined by engineering criteria. The second part of this chapter discusses issues for distributed generation that deal with the ability of small generators to participate in the competitive markets which will be established as generation is deregulated and the industry is restructured.

POWER SYSTEM EVOLUTION

To introduce the issues investigated in this chapter, this section traces the anticipated evolution of the distribution system, as driven by both an increasing penetration of distributed generation and the industry restructuring process.

Traditional Distribution System Structure

The traditional industry structure is that of a vertically integrated, regulated industry, operated without the price-based incentives of a competitive market to guide customer and producer decisions. In the traditional power system, the load in the distribution system is supplied exclusively by power delivered through the substation as shown in Figure 12.2.[1] In such a system power flow is unidirectional, frequency does not fluctuate significantly, and most of the control effort in the distribution system is focused on maintaining the desired local voltage profile.

Retail Competition

The industry is currently moving away from this traditional structure. Much of the initial effort in the restructuring process has focused on creating wholesale markets, with the result that the deregulation of generation and the creation of new institutions such as the Independent System Operator, ISO, and the Power Exchange, PX, at the transmission level are reasonably well defined and accepted by the industry. The potential need to extend the competitive institutions to the retail level, creating a competitive retail market and allowing direct access to customers—referred to as retail competition—is less well accepted. Such development of retail competition will be a critical factor in determining the role of distributed generating resources in the power industry.

There are two basic institutional models for the institutional structure of the distribution system. The first is simply to continue with the status quo, where the distribution system and all the services and operations within it remain together as a regulated distribution utility. In this structure therefore, distributed generation will be part of the regulated utility rather than the competitive marketplace.

The second basic model, expressed in current proposals for retail competition, point out that there is no economic or technological justification for services at the distribution level to remain bundled as the exclusive responsibility of the distribution utility [15]. The first step in developing retail competition is to open services in the revenue cycle, such as metering and billing, to competition. Proponents of retail competition seek access to the wealth of customer information that would become available through direct access to customers.

This step is only the beginning. From the perspective of distributed generation, retail competition represents much more extensive changes to the structure of the distribution system than simply opening the revenue cycle. More significanly changes will come as the commodity portion of the distribution system functions are separated from those functions that are a natural monopoly. The power lines that compose the distribution system will remain a regulated utility while all other products and services that are provided will be opened to competition. In such a system distributed generators will be free to participate in the emerging competitive markets both through direct access to the spot market and through contracting to provide services to the ISO.

Initial Phases of Restructuring

The extent to which retail competition encourages the increased use of distributed generation is likely to be limited at first, with only a few small scale generators installed in distribution systems. Instead, initial restructuring efforts in the distribution system are likely to focus on capturing the price elasticity or price responsiveness of customers, since efforts in this area will not be as complex or time consuming as building new power plants.

As more customers are charged based on the actual cost of supply, variations in local frequency and voltage may exceed the ranges common in the current system. These variations will be self-correcting to some extent due to the response of frequency sensitive load and local VAR support equipment. The point of interest is the possibility that the variations in local frequency and voltage may exceed those typical in the traditional distribution system, even before any small scale generation is installed in the system (simply as a result of customers responding to market incentives). If this does occur, it will be the first indication of a need to establish market driven methods for ensuring the desired level of power quality and reliability in the distribution system. In such

CONTROL AND OPERATION OF DISTRIBUTED GENERATION   455

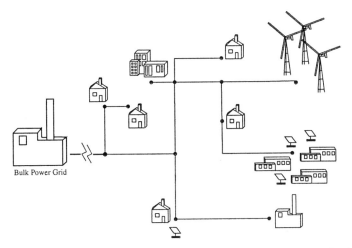

Figure 12.3   The Future Structure of the Distribution System with Multiple Distributed Generators

a system it will be customers, via market forces, who determine the allowable deviations of frequency and voltage from the nominal values.

The Future Distributed Utility

As the industry continues to evolve under the restructuring process it is likely that multiple distributed generators will be sited in each distribution system as shown in Figure 12.3, which shows customers and different types of generators distributed throughout the system. It is this anticipated direction of power system evolution—one which assumes a significant role for distributed generation—which raises the questions addressed in this chapter.

One set of questions focuses on the behavior of small scale generators within a single distribution system and the performance of the distribution system itself. The industry restructuring process raises engineering concerns of maintaining system performance levels (frequency and voltage in particular), as a growing number of active devices with diverse characteristics are sited within the distribution system. The potential affects on distribution system behavior introduced by distributed generation may demand that the issue of distribution system stability be revisited. Strengthening the technical capability for decentralized control and dispatch of generation, to parallel the growing potential for independent ownership will also be of interest in the restructured industry. Assuming that it does not compromise stability, decentralized control is desirable because it will facilitate non-utility ownership by allowing non-utility generators to be more fully independent from the local utility.

As more generators become independent and distributed technologies are better understood, the power system may evolve into what can be referred to as a distributed utility, as shown in Figure 12.4. Distributed generators are only one component of the distributed utility concept—a concept that anticipates

456 POWER SYSTEMS RESTRUCTURING

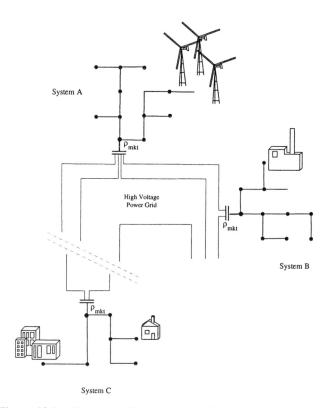

Figure 12.4 Distributed Generation in a Future Distributed Utility

an increased use of distributed resources in general to increase the efficiency of the power system.[2]

Assuming this path for system evolution, this chapter examines the performance of distributed generators operating in a competitive market. Legitimate concerns are raised over the extent to which market forces can replace the traditional centralized command and control structure. Will a system controlled in a decentralized manner consistently have access to the resources required to meet system demand and respond to system fluctuations? When should criteria other than market efficiency take precedence in operating decisions? Can a price signal be used to coordinate energy transactions? Can a price signal be used for system regulation[3] as well as for bulk energy exchange?

The first part of the chapter focuses on issues of technical integration. In the second part of this chapter a price signal is proposed that facilitates the desired industry transition toward increased independent and decentralized generator decision making. This signal conveys efficiency incentives to generators through market mechanisms, and is designed both to maintain desired system performance and to coordinate distributed generators as they participate in both the short run energy market and the ancillary services[4] market. The simulations with the price model demonstrate that this type of signal can be used to coordinate energy transactions and meet system regulation needs.

To model the operation of a future distributed utility, this chapter assumes that distributed resources are located throughout the system, and are free to contract to supply load anywhere in the system, as shown in Figure 12.4— they are not restricted to operate only within their local distribution system. In this future phase of the restructuring process a closed loop price signal as developed in this chapter is particularly important. The price signal is shown to be effective in coordinating generator operations in a setting where the generator operating decisions are based on private economic incentives, and are not controlled by a central authority.

SYSTEM FREQUENCY PERFORMANCE AND DISTRIBUTED GENERATION

The considerations involved in the smooth integration of distributed generation into the distribution system range from long term siting questions to concerns over maintaining frequency stability and the desired voltage profile. Once location and mode of operation are decided, and the necessary protection equipment is installed, the small generators will be able to supply power to customers, whether by contracting directly with customers, a power marketer or the system operator. The modeling for these bulk interactions involves using well established static models such as load flow and optimal power flow models, and does not raise new engineering questions.

Supplying bulk power is only one of the possible functions open to these small generators. The question of participation in the short run energy and ancillary services markets within the distribution system is also of interest, and is an issue for both market structure and engineering concerns. For example, in the case of a potential outage, local DG capacity could fill the role of spinning reserve by maintaining a continuous power supply to customers when they may otherwise have experienced a blackout. DG can also be used to maintain frequency within the local distribution system when an instability is caused locally by a fluctuation in connected load. (Frequency on the HV transmission grid will remain the sole responsibility of generation connected to the grid.) This section addresses the engineering aspects of the issue of system frequency regulation and analyzes what impacts, if any, distributed generation may have on the reliability and stability of the power supply in the distribution system.

Concern over frequency stability is a *new* issue related to the relatively recent increased interest in distributed generation. Previously there was only a single substation supplying power to the distribution system (or possibly one other small generator at a customer site), but now the distribution system faces the possibility of having multiple generators seeking to supply multiple customers each. The introduction of numerous active, generating sources in the distribution system could cause frequency to go unstable in some situations. For example, a recent study by Lee et. al. [28] found that the installation of a cogeneration plant in a distribution system could cause low frequency oscillations. In this and other new scenarios, the distribution system has the potential of having frequency drift (as on the transmission grid when there is no tertiary control), or even of losing synchronism.

In light of these concerns, in this section we model and present simulations of frequency behavior in a distribution system with multiple distributed generators. The generator and system dynamics which correspond to system regulation functions, such as those ensuring frequency stability, occur at the primary and secondary dynamics levels (see the chapter on operations). Small-signal models, developed as part of this chapter, are used for examining the system regulation questions and analyzing the frequency stability in the distribution system in particular.[5] Before presenting the models, the next section reveals the need for such models by discussing the physical phenomena that can cause frequency deviations and potentially lead to instability in the distribution system.

Causes and Impacts of Frequency Deviations

The frequency dynamics that are modeled and analyzed in this chapter are driven by deviations from the scheduled demand (and equivalent scheduled generation), equal to

$$P_{mismatch} = \frac{P_{gen} - P_{load}}{P_{gen}} \quad (12.1)$$

These deviations are shown in the sequence of figures, from Figure 12.5 to 12.9. Figure 12.5 shows the demand in a hypothetical distribution system for one day, from three sources—residential air conditioning, commercial air conditioning and street lighting. Figure 12.6 shows the output from three generators as scheduled to meet this demand, along with the total generation. Figure 12.7 next graphs the (exaggerated) result of deviations in both the supply and de-

CONTROL AND OPERATION OF DISTRIBUTED GENERATION   459

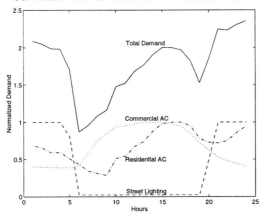

Figure 12.5 Scheduled Demand

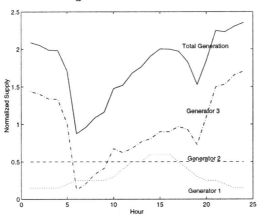

Figure 12.6 Scheduled Generation

mand from the schedule. The total hourly mismatch between the supply and demand for one day is shown as the bottom line in this figure.

The analysis in this chapter is based on power deviations which follow a similar pattern but are focused on a shorter time scale. Figure 12.8 mirrors Figure 12.7, for a period of one hour, plotting only the scheduled power flow and the mismatch. Note that this figure shows idealized behavior based on the assumption that disturbances will only occur at five minute intervals, and that between these disturbances, power flow is constant. Finally, this same data is graphed in Figure 12.9 which plots the "mismatch" explicitly as a deviation from the scheduled power flows. Each step change in the scheduled power flow is assumed to be the result of a disturbance, which is caused either by a change in demand or a change in output from a non-dispatchable generator (such as wind or photovoltaics). It is this series of disturbances and their affect on distribution system performance that are analyzed with the models presented in this section. These models and relevant assumptions are presented next.

460   POWER SYSTEMS RESTRUCTURING

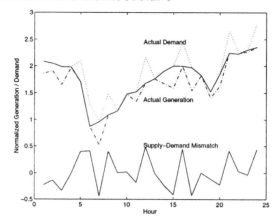

Figure 12.7 Mismatch Between Scheduled Generation and Scheduled Demand for One Day

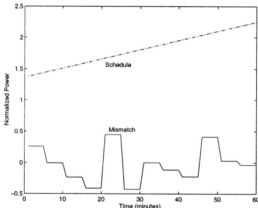

Figure 12.8 Mismatch Between Scheduled Generation and Scheduled Demand During One Hour

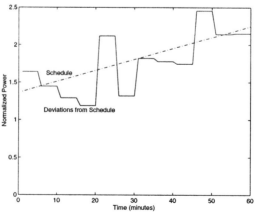

Figure 12.9 Mismatch Represented as Deviations from the Schedule

System Models with Distributed Generation

Model Specification and Assumptions. The frequency dynamics within the radial distribution system are the focus of this analysis—the bulk power grid behind the local substation is grouped together and modeled as a "very-large" bus, filling the role of the infinite bus for the system. Within the distribution system itself, load is distributed throughout the system, and generators are located at specified buses. To simulate the dynamic behavior of the system, disturbances are specified as either load or stochastic fluctuations, and are assumed to be small enough in magnitude to allow the use of small-signal, linear models.

A system model is defined by specifying the distribution system topology, the location and size of loads and the location, size and type of the generators. The inputs to the models are the system disturbances, represented as the input vector to the system of state equations. This vector is defined by specifying the location and the timing of the system disturbances. For the non-dispatchable technologies such as wind turbines a fluctuation in the wind resource is a system disturbance, otherwise the disturbance is a small increase or decrease in demand.

The model is used to simulate the dynamics due to both the disturbances and the specified control actions. The output from the simulation is the dynamic behavior of all the state variables, with frequency and real power output typically being of greater interest than the others. The frequency stability can be assessed by monitoring the frequency at each bus. Interactions of rotating machines with each other are analyzed by altering the system definition in terms of the generators and their location in the system. Different control strategies can also be specified and analyzed in order to determine their effectiveness in maintaining frequency stability. The next sections explain the derivation of the models, and their use in exploring frequency stability.

Generator and System Model Development. The goals in developing models for analyzing frequency behavior are to represent the dynamics of distributed generators in response to system disturbances, and to propose and analyze the effectiveness of different control strategies designed to ensure system stability.

State space models developed for this chapter include models for steam turbines, hydroelectric turbines, combustion turbines, combined cycle plants, and wind turbines. Numerous dynamic models exist for each of these technologies, however the majority are very complex, involving a large number of state variables. In developing the models for this chapter, the objective is to represent each generator with a small number of state variables (three to four) so that interconnected system models, which each include a number of the distributed generators, will not be overly complex. A second objective is to develop each set of local state equations such that they incorporate P_G as the system coupling variable, ensuring that they will be mutually compatible when modeled together in the extended state space. The traditional system coupling variable is rotor angle, δ.[6]

The emphasis of the modeling in this section is on decoupled real power/frequency dynamics. One major reason for this emphasis is that the frequency dynamics of a radial distribution system with distributed generators, and the possibility of these units participating in the supply of ancillary services such as frequency stability and spinning reserve, are relatively new issues. Voltage is more of a local issue, affecting power quality at load sites (i.e. in the distribution system), and is not a new concern for the distribution system.

The small-signal, dynamic models for each generator are shown in Table 12.1. Note that the state vector for each generator model includes ω_G, the generator frequency, and P_G, the system coupling variable. The models are all small-signal models, and so are useful for analyzing the system dynamics in a small range around an operating point, which is found by running a load flow program. All the variables in the linearized generator models represent *deviations* from the equilibrium or operating point.[7]

Each individual generator model from Table 12.1 can be written in matrix form as

$$\dot{x}_{LC} = \mathbf{A}_{LC} x_{LC} + \mathbf{C}_M P_G \qquad (12.2)$$

where x_{LC} is the local state vector, and $\dot{x}_{LC}$ is the time derivative of this vector representing the time evolution of the state variables. The bold variables represent matrices, where $\mathbf{A}_{LC}$ in particular is referred to as the local system matrix, whose elements consist of the linear coefficients of the generator parameters.

After developing the individual generator models, the next step is to specify the distribution system model. There are two aspects to the modeling of the radial distribution systems. The first is the actual topology—the number of buses and the structure of the systems. The second is the mathematical representation of the systems. The test systems used in this chapter are all taken from the literature on modeling and simulating radial distribution systems. A number of test systems were developed by an IEEE Working Group [19]. Others were developed for specific projects, based on actual systems, and have subsequently been used by a number of different authors [9, 11, 25, 36].

The distribution test system that is used for the majority of the simulations in this chapter is shown in Figure 12.10. The data for this system is presented in the Appendix, and can also be found in [11, 36].

The mathematical representation of the distribution system is simply the set of load flow equations

$$P_i = \sum_{j=1}^{n} |V_i||V_j|[g_{ij}cos(\delta_i - \delta_j) + b_{ij}sin(\delta_i - \delta_j)]$$

$$Q_i = \sum_{j=1}^{n} |V_i||V_j|[g_{ij}sin(\delta_i - \delta_j) - b_{ij}cos(\delta_i - \delta_j)] \qquad (12.3)$$

where P_i is the real power at each bus, Q_i is the reactive power, $|V_i|, |V_j|$ is bus voltage magnitude, g_{ij} and b_{ij} are the line admittance parameters, and $\delta_{i,j}$ is

Hydro-Turbine-Generator

$$M\dot{\omega}_G = -(e_H + D)\omega_G + k_q q - k_w a - P_G$$
$$\dot{q} = \omega_G/T_f - q/T_q + a/T_w$$
$$T_e \dot{v} = -v + r'a$$
$$T_s \dot{a} = -\omega_G + v - (r_h + r')a$$

Steam-Turbine-Generator

$$M\dot{\omega}_G = (e_t - D)\omega_G + P_t - P_G$$
$$T_u \dot{P}_t = -P_t + k_t a$$
$$T_g \dot{a} = -\omega_G - ra$$

Combustion-Turbine-Generator

$$M\dot{\omega}_G = -D\omega_G + cW_F - P_G$$
$$b\dot{V}_{CE} = -K_D \omega_G - V_{CE}$$
$$\dot{W}_F = W_F dot$$
$$\alpha \dot{W}_F dot = aV_{CE} - \delta W_F - \beta W_F dot$$

Combined Cycle Plant

$$M\dot{\omega}_G = -D\omega_G + (f_2 + P_{ST}) - P_G$$
$$b\dot{V}_{CE} = -K_D \omega_G - V_{CE}$$
$$\dot{W}_F = W_F dot$$
$$\alpha \dot{W}_F dot = aV_{CE} - \gamma W_F - \beta W_F dot$$
$$T_v \dot{W}_{air} = d\omega_G + V_{CE} - W_{air}$$
$$\dot{P}_{ST} = P_{ST dot}$$
$$(T_M T_B)\dot{P}_{ST dot} = -p\omega_G + nW_F + mW_{air} - P_{ST} - (T_M + T_B)P_{ST dot}$$

Wind Turbine – Induction Generator

$$M_G \dot{\omega}_G = -(D_G - D_T)\omega_G + (D_G - D_T)\omega_T + T_w - P_G$$
$$\dot{\delta} = -\omega_G + \omega_T$$
$$M_T \dot{\omega}_T = D_T \omega_G - K - D_T \omega_T + T_w$$

Table 12.1 Small-Signal Generator State Space Models

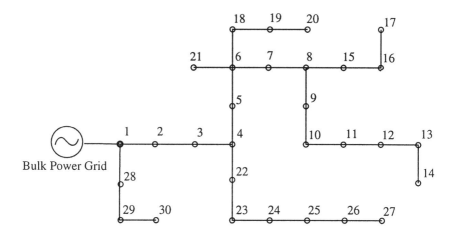

Figure 12.10 30 Bus Radial Distribution Test System

the rotor angle. The incidence and admittance matrices and line parameters in these equations will differ depending on the specific test system being modeled.

Finally, to build the complete system model, the individual generator models are coupled to each other via the distribution system. Mathematically, the local state space of each individual generator must be extended to include the system coupling variable, which allows the dynamics at one point on the system to be transmitted to all other points. This coupling variable is selected to be power output, or P_G, the state equation for which is

$$\dot{P}_G = \mathbf{K}_P \omega_G + \mathbf{D}_P \dot{P}_L \tag{12.4}$$

The two matrices in this equation, $\mathbf{K}_P$ and $\mathbf{D}_P$ are derived from the Jacobian matrix. $\dot{P}_L$ represents a load disturbance and is the input variable to the system. In this form, the variable P_G can be included in the local generator state spaces to form what is referred to as the extended state space. This equation for P_G was first developed in [21, 23].

With Equation (12.4) added to the dynamic models, the system model has the form

$$\dot{x}_{ext} = \mathbf{A} x_{ext} + \mathbf{D}_P \dot{P}_L \tag{12.5}$$

where x_{ext} is the vector of extended state space variables (local variables plus the system coupling variable, P_G), and $\mathbf{A}$ is the partitioned system matrix.

The control input is $u[k]$, and this signal controls the variable ω^{ref}, which is the reference frequency for the governor. Note that in the time scale of the primary dynamics ω^{ref} is constant and so is not included in the small-signal equations of Table 12.1. For the secondary dynamics ω^{ref} is variable, and so is included in Equation (12.2).

Distribution System Frequency Simulations

This model is now used to simulate the performance of a radial distribution system with distributed generators. In particular, the system performance is analyzed in terms of frequency behavior, with particular attention paid to the question of: Under what circumstances, if any, can the frequency in the system as modeled either drift noticeably from the desired nominal value, or even go unstable?

Maintaining a stable frequency has not traditionally been a concern in the distribution system, because the distribution system traditionally has had no, or very few power plants. This question is important for the distribution system in the restructured industry because the presence of distributed generators will change some basic properties and operating characteristics of the distribution system. For example, with multiple distributed generators it will be possible for the generators to lose synchronism, as can happen now at the transmission level. It may also be desirable to extend secondary controls to distributed generators, so that after a system disturbance the frequency is restored to its nominal value as quickly as possible.

Since secondary level controls are required for the transmission system, it is not surprising that they may also be required in a distribution system with small generators. Concerns of frequency *stability* are less expected since this issue is well understood and has been successfully addressed in the control strategies for the transmission system. The mere presence of distributed generators in the distribution system does not explain why the frequency behavior may differ from that on the distribution system today. Nonetheless, the simulations in this section demonstrate that the frequency in a distribution system with distributed generators may in fact exhibit instability. The details of this analysis follow the simulations, which are presented next. The first simulation presents the desired system performance, with respect to system stability. This first example also demonstrates the need for secondary level frequency controls in the distribution system when distributed generators are present.

Frequency Drift and Secondary Controls. The distribution system modeled in the first set of examples is the 30 bus system shown in Figure 12.10. Total load on the system is 15 p.u. and the total capacity from distributed generation varies from 0.7 p.u. to 2.5 p.u. in the examples presented. The first example discussed has a 0.7 p.u. combustion turbine (700 kW) at bus 24 (as well as a slack bus at the substation). The load disturbance is a 0.1 p.u. increase in demand at bus 21, at time equals 2 seconds.

Figure 12.11 shows the *frequency deviation* from the equilibrium point for this system.[8] The rotor frequency for the small combustion turbine is seen to oscillate around the nominal 60Hz frequency, and settle to a slightly slower value. The behavior demonstrated by the system in Figure 12.11 is the expected behavior, given that in this example the secondary level controls, which would return the system frequency to the nominal value, have been temporarily removed. The simulation is allowed to run without any secondary control

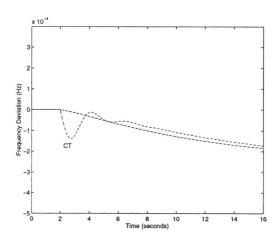

Figure 12.11 Frequency Deviation from Equilibrium for Single Combustion Turbine

action, well beyond the primary time scale of 2 to 5 seconds, in order to show the nature of the dynamic behavior.

This simulation shows that without the corrective action of secondary controls, the frequency may drift from the nominal, desired value after a system disturbance. Secondary controls if implemented will act to return the steady state frequency to the nominal 60Hz value, as shown in Figure 12.12. In this figure, simulated at the secondary time scale, the system disturbance causes system frequency to drop at time t = 8 seconds, and then the secondary control acts to return the system frequency at time t = 10 seconds. The models for the secondary dynamics are developed in [8, 23].

Frequency Stability. The next two examples explore the possibility of distributed generators causing frequency instability in a radial distribution system. As stated above, this is a relatively new concern, which has not been addressed previously for the simple reason that generators have not traditionally been sited in significant numbers in the distribution system. As the following examples show, different *types* of generators are more or less likely to cause instability, and a greater number of generators is more likely to cause instability than when only one or two DG units are operating.

For the first example, the system is modeled with four combustion turbines, ranging from 0.5 p.u. to 0.75 p.u., distributed throughout the system, at buses 10, 17, 24 and 29. The turbines have slightly different values for the controller gains, all within normal operating ranges. (See for example [14, 13, 34]). A small load disturbance of 0.1 pu at bus 21 occurs at time $t = 2$ seconds. The frequency deviation from equilibrium for two of these generators, along with the slack bus, is plotted in Figure 12.13. The frequency deviations of the remaining generators are not plotted to avoid confusion in the figure. This

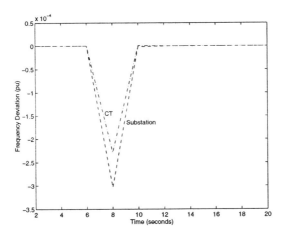

Figure 12.12 Frequency Deviation with Secondary Controls Active

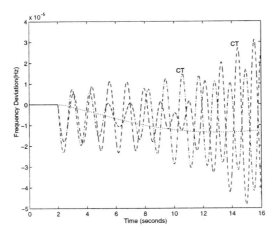

Figure 12.13 Frequency Deviation from Equilibrium with Four Combustion Turbines

figure clearly demonstrates that local system frequency becomes unstable, as a result of the same load disturbance as occurred in the system in the first example, Figure 12.11. It is significant to note that the system remains stable when only two combustion turbines are in the system. It is not until there are four generators that the instability is exhibited, suggesting that at least for frequency stability, technical problems are loosely a function of the number of distributed generators.

For the next example, if the distribution system is modeled with a single hydroelectric generator at bus 17 the frequency also becomes unstable. With a

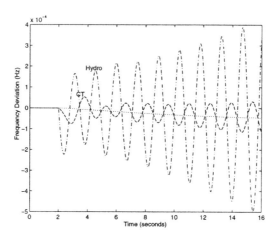

Figure 12.14 Frequency Deviation from Equilibrium for Hydroelectric and Combustion Turbines

combustion turbine added to the system at bus 24 (both generators of capacity 0.7 p.u.), the instability caused by the hydroelectric plant creates instability at the combustion turbine bus as well. See Figure 12.14. Note that the instability remains local to the distribution system in all examples; the slack bus frequency is unaffected as a result of the modeling assumption of a large inertia representing the bulk power system behind the substation.[9]

The instability found in the above example can be avoided by carefully tuning the generator to the specific system. Note though that the hydro plant as modeled, has all parameters set within the ranges as established for existing small hydro facilities. Therefore, the point of this example is not that hydro or any other small scale generating technology will automatically cause frequency instability, but rather that it is possible for them to do so unless close attention is paid to the new situation represented by siting numerous generators in a radial distribution system. The following section addresses the significant characteristics of this new situation in detail.

Analysis of Frequency Stability

Eigenvalue Analysis and Participation Factors. Eigenanalysis of the system matrices, $\mathbf{A}_{LCi}$ and $\mathbf{A}$, is used to begin identifying the cause of the instability. The eigenvalues for the individual generators and for the examples presented above are listed in Tables 12.2 and 12.3 respectively. (The eigenvalues are calculated for $\mathbf{A}_{LC}$ for each generator, and for $\mathbf{A}$ for two sample systems—one with four CTs and the other with one CT and one hydro generator. See Equations (12.2) and (12.5) for the definitions of the matrices.) For the examples presented here all $\mathbf{A}_{LC}$ system matrices are stable. In some scenarios though the interconnected system is found to be unstable. The ta-

Steam Turbine	Combustion Turbine	Hydro Turbine
$-0.50 + j1.63$	$-20.24 + j4.95$	$-0.03 + j1.48$
$-0.50 + j1.63$	$-20.24 - j4.95$	$-0.03 - j1.48$
-5.66	$-0.12 + j4.83$	-7.17
	$-0.12 + j4.83$	-0.36

Table 12.2 Eigenvalues of Individual Generator Models

4 CT System	4 CT con't	Hydro & CT
$-21.23 + j4.94$	$-0.46 + j2.95$	$-20.30 + j2.41$
$-21.23 - j4.94$	$-0.46 - j2.95$	$-20.30 - j2.41$
$-21.20 + j4.92$	-5.00	-6.62
$-21.20 - j4.92$	-0.67	$0.07 + j4.33$
$-20.31 + j2.41$	$-0.07 + j0.22$	$0.07 - j4.33$
$-20.31 - j2.41$	$-0.07 - j0.22$	$-0.47 + j2.68$
$-20.30 + j2.40$	-1.19	$-0.47 - j2.68$
$-20.30 - j2.40$	-0.19	-5.00
$0.18 + j5.72$	-1.61	-1.26
$0.18 - j5.72$	0.00	-0.90
$-0.06 + j4.97$		-0.17
$-0.06 - j4.97$		$-0.06 + j0.05$
$-0.25 + j3.77$		$-0.06 - j0.05$
$-0.25 - j3.77$		0.00

Table 12.3 Eigenvalues of 30 Bus System Examples

bles clearly show that each generator is individually stable, while the systems that include multiple CTs or a hydro plant can be unstable. (Note that the zero eigenvalue for each system is inherent to the structure of power systems, and does not represent a stability problem [23].) The unstable modes shown in Table 12.3 are the slow, electro-mechanical, or swing modes, which are related to the state variables ω_{Gi} (and δ_{Gi}, when δ is included in the state space as the system coupling variable). If the unstable eigenvalues can be uniquely associated with specific state variables, then the identified state variable could be directly controlled to regain system stability. However, the state variable(s) associated with the unstable modes vary with different system configurations, so other means must be investigated to stabilize the system.

Participation factors, developed fully in [31], can be used to associate individual state variables with specific modes. A participation factor, p_{ij}, is defined as

$$p_{ij} \equiv w_{ij} v_{ij} \tag{12.6}$$

where w_{ij} is the ith entry in the jth left eigenvector, and v_{ij} is analogous for the right eigenvector. The p_{ij} provide a measure of the contribution of the ith state variable to the jth eigenvalue and so can be used to correlate the contribution of each state variable to each mode.

Participation factors have been calculated for the unstable modes for the systems discussed in this section, as well as others with the generators or load disturbances located at different buses. This analysis leads to identifying different state variables, in particular ω_G from different generators (in addition to other state variables), as causing the instability for each different system configuration. These results show that the instability is not caused by a single state variable or a single generator, but is more appropriately identified as a system phenomenon.

Recognizing the instability as a characteristic of the system raises the question of what are the differences, as related to stability, between the high voltage network with large generators and a radial distribution system with smaller distributed generators? One distinction is that the generators on the high voltage grid are very large with correspondingly large inertias. In comparison, the distributed generators as modeled for this chapter have relatively small machine inertias, making the elements in matrix $\mathbf{C}_M$ (see Equation (12.2)) relatively large. This leads to stronger coupling between the local state space x_{LC} and the system coupling variable P_G than is common on the high voltage grid. This relationship can be seen by referring to the equation for the full interconnected system, Equation (12.2).

A second distinction is that the radial distribution system has relatively high line impedance, representing a basic change to the interconnecting network and its subsequent influence on local generator dynamics. When modeling the high voltage transmission system it is usually assumed that the local dynamics in x_{LC} are slow relative to the network dynamics, with the implication that any change in x_{LC} is instantaneously transmitted through the system, so that the network itself has no affect on the local generator dynamics. The impedance of the distribution lines affects elements of the Jacobian-based matrix $\mathbf{K}_P$, and so affects the coupling between the system and the local frequency dynamics.

The relatively large line impedances and relatively small machine inertias can be seen as acting together in the following manner:

1. The smaller inertias create stronger coupling between the system dynamics, P_G, and the local frequency, and also have relatively smaller damping and so are less effective in damping the oscillations rapidly.

2. The increased line resistance, though representing a larger dampening than the lines on the high voltage grid, is not enough to counter the affect of the small inertias.

These observations of the differences between the high voltage grid and the distribution system are not surprising. What is unexpected is that they may be significant enough to affect stability within the distribution system.

Stabilizing the System. The stability problem suggests that new attention may need to be paid to local control settings in order to ensure that stability will be maintained in a radial distribution system that has multiple distributed generators. Rather than suggesting new controls for a specific state variable, this more general approach aims at finding ranges for values of the parameters in the system matrix.

A general method for specifying ranges for the values of local control parameters, as defined in the local system matrix $\mathbf{A}_{LC}$, is to calculate eigenvalue sensitivity to the parameters, for the unstable system eigenvalues. This calculation is similar to that for the participation factors discussed earlier. The sensitivity matrix, S_i, for the ith eigenvalue is defined to be

$$S_i = [\partial \lambda_i / \partial a_{jk}] = w_i v_i' \tag{12.7}$$

where the λ_i are the eigenvalues of the system, the a_{jk} are the elements of the $\mathbf{A}_{LC}$ matrix, and w_i and v_i are the left and right eigenvectors respectively for the ith eigenvalue, where v_i' is a row vector. (Note that the diagonal elements of this matrix are identical to the participation factors.)

This matrix is calculated for the unstable eigenvalues for each system with instability, two of which are shown in Figures 12.13 and 12.14. The sensitivity matrix shows that for the systems with a hydroelectric plant, the unstable mode is most sensitive to the parameters in the equation for the gate position. The time constant T_s is a factor in each of these parameters, (see the hydroelectric model in Table (12.1)), suggesting that T_s would be a good value to adjust. Figure 12.15 shows the system of Figure 12.14, with the time constant for the gate opening of the hydro plant increased so that it can not react as quickly to a disturbance, preventing it from resonating with the oscillations. (The unlabeled, dotted line on this and the following two figures represents the substation.) Note that although this solution solves the stability problem, it also serves to challenge one of the anticipated benefits of distributed generation, specifically that the fast response capabilities of small generators would be beneficial in responding quickly to changes in demand and so help minimize any disturbance.

A second parameter found to significantly affect the stability is the inertia constant. Figure 12.16 shows the deviation from the equilibrium frequency for the same system as Figure 12.14, but with an increased inertia for the hydroelectric turbine-generator. Implementation of this change implies the need to specify a minimum inertia or size of plant installed.

For the system with only combustion turbines (Figure 12.13), the greatest sensitivity is found in the gain in the fuel system controller (see [34] for detailed explanation of these parameters). When this gain is decreased, the system is stabilized, as shown in Figure 12.17. Note that the system modeled for Figures 12.14 through 12.16 has both a hydro generator and a combustion turbine, and that the gain in the CT fuel system controller is not identified as the parameter to which the unstable mode is most sensitive—a finding consistent with the earlier assertion that the instability is a system phenomenon, and not caused by one generator or generator type.

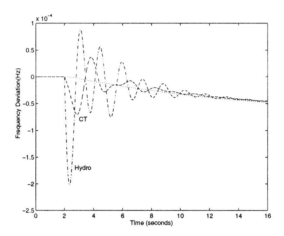

Figure 12.15 Hydro Gate Opening T_s Increased in System with Hydro and Combustion Turbine

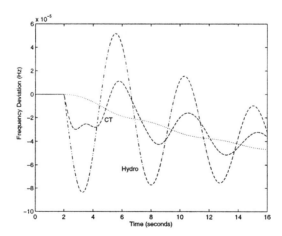

Figure 12.16 Hydro Inertia Increased in System with Hydro and Combustion Turbine

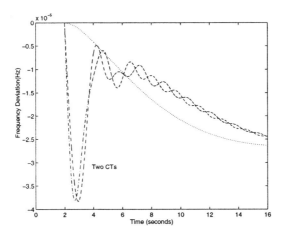

Figure 12.17 Fuel Controller Gain Decreased for System with Four Combustion Turbines

Summary. The first part of this section has described the modeling approach used to simulate the decoupled frequency dynamics for a distribution system with small, distributed generators. In the frequency analysis, instability at the primary dynamics level was found, and was shown to be a system level phenomenon rather than one caused by a single state variable. Examining the sensitivity matrix suggested various methods for stabilizing the system, requiring that close attention be paid to local control parameters—time constants and gains, or to generator selection—machine size or inertia. It was also demonstrated, that in some cases instability may only occur as the number of distributed generators in the distribution system increases.

The frequency issues raised in the previous sections are not new to power systems, but are new to the distribution system. One difference in the solutions suggested here from those currently implemented on the high voltage grid is the focus on using the generator governors to secure frequency stability. At the high voltage level, generator governors react more slowly and so are not relied upon for maintaining system stability. In contrast, the analysis in this section has shown that local generator governors *can* be used at the distribution level to ensure frequency stability. A drawback of this sensitivity to the governor settings is that at the distribution level generators may not be able to turn off their governors and drift with the system frequency as they can at the transmission level.[10]

A deregulated capacity market incorporating distributed generators is more consistent with decentralized than with centralized control. However, the methods for stabilizing the system introduced in this section do require some degree of centralized oversight in determining governor standards or in generator selection. It is important to point out that the frequency concerns for the distribution system raised here are easily addressed. It is vital that the extra stability analysis is performed though, as the penetration of distributed generators increases, so that the potential frequency problems are successfully avoided.

Voltage Support and Distributed Generation

Voltage support with distributed generation is presented in a brief discussion below. It is likely that the benefit from using distributed generators to provide voltage support will equal that of using them for frequency regulation—the lack of parallel treatment in this chapter is not intended to imply that voltage support is less important. Additional reasons for the focus on real power/frequency dynamics are presented in the Appendix.

Role of Distributed Generators in Providing Voltage Support. On the high voltage grid in quasi-steady state operation there is essentially a single system frequency which is observable throughout the system. Frequency control schemes such as automatic generation control (AGC) use this system-wide variable to regulate the system frequency to the desired value. In contrast to this system level approach for frequency regulation, voltage is regulated on a regional level, in countries such as France which do have automatic voltage control (AVC) (the United States currently has no equivalent system for voltage control). Since there is no single, system-wide voltage level, but rather a desired voltage profile across the system, multiple buses throughout the system are regulated to specified voltage levels to provide regional reference or pilot points. In the distribution system, which directly supplies customers, voltage regulation is much tighter than at the transmission level to avoid either adversely affecting customer load or causing annoyance from flickering lights.

Distributed generators, being sited within the distribution system, and therefore close to customers, are well suited to provide voltage support. There are two general ways in which these units can provide this support. First, these small generators improve the voltage profile simply by supplying power close to the load, which decreases the need to transmit power from more distant buses. Even in situations when most power is centrally generated, distributed generators could be operated as synchronous condensors to reduce the need for additional equipment in the distribution system that is dedicated to voltage support. This potential function of distributed generation is likely to become increasingly important as the power system is restructured and contracts are made between distant parties. Distributed generators will be in a position to compensate for the reactive power mismatch *near* the load centers, and so facilitate the long distance power sales.

The second means by which distributed generators can provide VAR support is by controlling the power electronics in the power conditioning equipment so that the voltage at the generator bus is held constant. The power conditioning equipment can be set to supply power with either a leading or a lagging power factor, and so can be used to improve the local reactive power mismatch. In this way distributed generators could be operated much as static VAR compensators, and provide a valuable service even when their real power output is zero. If distributed generator units are operated in this manner they could come to fill the same role as FACTS devices are expected to fill at the transmission level.

Industry Structure and Local VAR Support. The extent to which distributed generators are used for local voltage support will be significantly influenced by the restructuring process, and the mechanisms through which voltage support is provided. When operating in the competitive market, the small generators can be expected to supply voltage support only if they are compensated for it, suggesting a need to develop mechanisms to monitor and pay for voltage support that is provided locally. Developing both the policies to promote the unbundling of ancillary services, and the mechanisms for these services to be bought and sold in the competitive markets will be an important part of the restructuring and deregulation process. If this is not done, the ISO (Independent System Operator) will be required to purchase voltage support from other, contracted sources, or rely on the regulated distribution utility to purchase and provide it—breaking with the trend toward increased reliance on market forces.[11]

Distributed generation is likely to be introduced slowly, as the power system evolves toward a more distributed architecture. A competitive setting may come to facilitate an increased penetration of distributed generation, if the competitive environment is such that maintaining system reliability, voltage profile and frequency become the responsibility of the local provider rather than of the central authority. The development of market institutions and competitive markets at all levels are discussed in the following two sections.

INTEGRATING DISTRIBUTED GENERATORS INTO THE MARKET STRUCTURE

The previous section focused on the technical issues related to an increased penetration of small generators in the distribution system. This section shifts the focus of the chapter to the market integration of distributed generators, and the development of a price-based control signal.

In the existing distribution and transmission systems, only a skeleton of the institutional structure required for such market driven operation is in place. The growth of competition in the generation sector demands that this framework be expanded. To fully integrate small scale generators into the markets, this institutional framework must extend to all the markets in which distributed generators can participate—long run contract, wholesale, scheduling, and short run energy and services markets.

To coordinate distributed generator actions in the shorter term operations and control of the spot energy and ancillary services markets, this chapter proposes a new closed loop price signal. There is no mechanism for a closed loop price structure in the traditional, vertically integrated utility, since this need arises only with deregulation and the increased reliance on market forces. To the extent that price signals are included in existing power system operation they are exclusively open loop, feedforward signals. In the initial phases of the restructuring process any price signal will remain simply open loop, being determined by a central authority as a function of "real-time"[12] costs and then communicated to small generators and customers. Such a signal would influence generator and customer behavior as a function of their price elasticities, but would not be updated in response to feedback from these suppliers or customers. The effectiveness of an open loop signal is limited by the fact that it is not a function of local deviations in supply or demand. For a fully competitive market the signal needs to be a closed loop signal, where the market price itself is a function of the aggregate behavior of the generators and customers, and so will change in response to local deviations.

If the closed loop price signal can be developed to accurately capture the cost[13] associated with local deviations from scheduled power and energy, it will provide a stabilizing and least cost incentive to generators and customers. If however, the price is "wrong," a competitive retail market could prove very detrimental to stable distribution system operation.

This section first discusses the objectives in developing this closed loop signal and its anticipated role in the market. It next presents the mathematics for the closed loop price model, including the development and interpretation of a basic cost equation at the primary dynamics level, and the control law for the price model. Finally, this section then demonstrates the role of a closed loop price signal in coordinating both the engineering and the economic aspects of distributed generator operation in a restructured power system. Both dispatchable and non-dispatchable technologies are modeled. The simulations demonstrate the ability of the distributed generators to participate in the competitive energy and ancillary services markets, with the closed loop price signal coordinating system operation.

Distributed Generation in Spot Energy and Services Markets

Objectives of the Closed Loop Price Signal. The market structure envisioned in this chapter assumes that a competitive market will be developed at the distribution level and that distributed generators will be allowed not only to enter into contracts at the wholesale and retail levels, and participate in the Power Exchange, but also provide ancillary services to the ISO and local customers on a competitive basis.

One objective in introducing a closed loop price signal to the generation sector is to aid in the creation of the desired competitive market. Market based institutions must be purposefully created as regulatory oversight is decreased in the generation sector, or it is likely that the sector will simply become an unregulated monopoly rather than a competitive market. A price signal expresses to consumers and suppliers the efficient levels of demand and supply. A closed loop price signal will capture the market clearing dynamic of a competitive market in the dynamics of the feedback control, and so incorporate market prices into system *control* decisions as well as in siting and investment decisions.

A second goal of the price signal is to provide a decentralized control mechanism which allows each generator to operate independently while also providing an incentive for the generators in aggregate to produce at the efficient level. The price signal facilitates the creation of a decentralized system in which distributed generators are free to act independently, required neither to give control, nor any private information to a centralized authority.

Note that the price signal developed here is not designed to quantify an expected revenue stream for a distributed generator, which could then be used to promote investment. Instead, the objective of the price model is to *demonstrate* that a market-based price signal can be used in conjunction with the existing bulk flow market price to successfully control and coordinate a distribution system.

The Role of the Closed Loop Price Signal in the Market. The future power system is anticipated to have competitive markets for both energy and ancillary services. In the proposed price framework the basic piece of information in both of these markets, communicated to the distributed generators from the ISO and the PX, is the spot price of energy and/or services. The spot price corresponds to the price of the scheduled power flows as determined by the ISO and PX and may differ according to differences in the location and the product provided. In the price framework proposed in this chapter, the full price communicated to the distributed generators via the substation represents both the spot price (corresponding to the scheduled flows) and a component to account for deviations from the scheduled power flows.

The price variable introduced in this chapter represents this component for the deviation from equilibrium and *not* the full market or absolute value. The full price of energy in the market can thus be expressed as

$$\rho_{base} \pm \Delta\rho$$

where $\Delta\rho$ is the quantity determined by the price based control loop in this section and ρ_{base} is the spot price of the scheduled, bulk power flows. In the context of current power system operation, $\Delta\rho$ would likely be calculated *after* all flows and power output levels are known, or else forecasted using either expected or historical values. In contrast to this approach, the price control model derived in this section determines $\Delta\rho$ dynamically, via feedback, and without centralized control.

This price signal operates at the secondary price dynamics level. Every K minutes the market or system price, ρ_{mkt}, is updated to reflect the current price of power delivered to the distribution system. The time step K could be as long as 30 minutes or 1 hour, and so coincide with the spot market as typically defined in the ongoing industry restructuring debate. To capture system regulation needs, and provide market incentives for small generators to provide ancillary services though, K must be defined for a shorter time step, such as 5 minutes. A significant aspect of the proposed price control structure

478   POWER SYSTEMS RESTRUCTURING

is that the mathematical representation and corresponding system response are identical whether it is the real-time energy market or the services market that is being modeled. This mirrors events in the actual power system since *inside* the 30 minute or 1 hour window of the traditional spot market, a change in the demand for energy *is* the source of the system demand for ancillary services. At this time scale both the services and short term energy markets are driven by deviations from scheduled power flow, and are differentiated only in the length of the time step K, and also conceptually, in the cause of the system disturbance.

Price based controls are typically precluded from acting this quickly due to the longer time frame assumed necessary for market interactions. It is not a theoretical constraint however that prevents the price feedback from being implemented in the shorter time step—the price signal defined in this section is in fact capable of acting in this short time period. It is within this shorter time window that system regulation is an issue, and that controls act to regulate system frequency. The price control model in this section demonstrates that both the short run energy and the services markets can be operated competitively.

Anticipated Generator Response to Price Feedback. The closed loop price signal corresponds to the marginal revenue earned by a participating distributed generator, and as dictated by economic theory, the competitive suppliers will produce at the level where their marginal cost equals marginal revenue. The price model incorporates this economic objective ($MC = MR$) into the short run operating strategies of the individual distributed generators such that the generators respond automatically to changes in the system price by altering their output until their marginal costs of production equal the spot price.

Figures 12.18 through 12.20 demonstrate the anticipated generator response to the price signal. Figure 12.18 shows a system disturbance on the test system of Figure 12.10, occurring at time t = 8 minutes, and the resulting increase in generator output without the price signal implemented. To compare the system response with and without the price signal, Figure 12.19 first shows this system output and corresponding price deviations without the price feedback implemented. Figure 12.20 then shows the output and price deviations with the price signal implemented. The price signal, acting at time t = 10 minutes, causes the generators to adjust their output so that the final generation levels are all close to the system price (represented by the lower, dotted line on the graphs). Note that the objective of the price control signal is not to return the absolute price deviation to zero, but rather to adjust generator outputs so that the quantity ($\rho_i - \rho_{mkt}$) is equal to zero. The simulations and the control action will be analyzed more fully at the end of this section after the price model has been developed.

CONTROL AND OPERATION OF DISTRIBUTED GENERATION 479

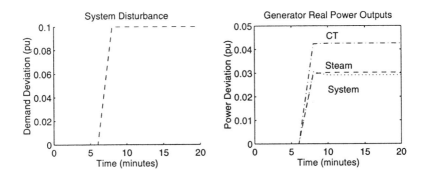

Figure 12.18 Load Disturbance with Corresponding Increase in Power Output (No Price Feedback)

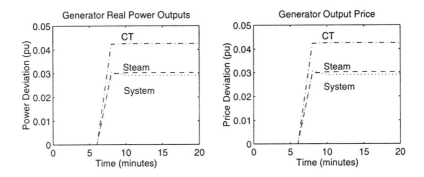

Figure 12.19 Power Deviation and Corresponding Price Deviation Without Price Feedback

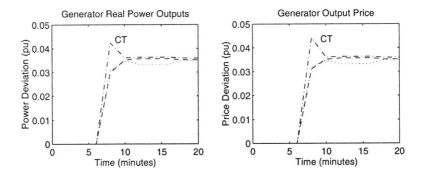

Figure 12.20 Power Deviation and Corresponding Price Deviation With Price Feedback

Closed Loop Price Signal Model

In the power system today, there is no closed loop market signal integrated into system operating decisions. Industry restructuring, and particularly the deregulation of generation, is opening the power sector to market forces. As part of this process, price-based market signals will be integrated into the operating decisions at all levels of the power system. An hourly spot market is currently being designed in the regulatory and policy arena, with extensive input from utility engineers and the academic community. There is at present however, little effort to make this hourly spot market a *closed loop* structure. Instead the spot market development is following the pattern established in other countries as well as in some areas of this country, by setting the hourly schedule a day in advance, and determining the price as an *open loop* signal. In addition to the lack of effort in designing a closed loop signal, there is not yet effort to integrate market forces into the operations and control decisions on a time scale shorter than one hour, such as every five or ten minutes, or even shorter as is consistent with the dynamics of system regulation.

This section develops, for the first time, a mathematical framework for a closed loop price signal, designed to coordinate distributed generators as they participate in both the short run energy market and the ancillary services market. A price signal of this form is of interest because it creates the means for competitive market forces to guide operating and control decisions in realtime. Assuming there are no market failures, the efficiency of the power system will improve as the reliance on market forces increases. Improving efficiency will be a long term process reflected through investment decisions as well as short run operating decisions—developing a closed loop price signal is one component of this evolution.

The closed loop price signal developed here is a contribution to the theory and process of integrating distributed generators into the power system because it demonstrates that a closed loop price signal can be effective in guiding distributed generator operating decisions, without compromising system stability. It is important to note that the price signal models presented in this section are idealized models developed for the specific application of distributed generation in a radial system, and are not intended to be the definitive answer for market driven operation decisions throughout the power system.

The price models presented below are for decoupled real power/frequency dynamics. The reason for this emphasis is two fold. First, with this emphasis, the modeling effort mirrors the pattern to date for developing a spot price or responsive price system, which usually focuses on pricing real power, since that is the major commodity of the industry. To be consistent with developments on the high voltage grid,[14] a price framework for distributed generators should also focus first on real power and the manner in which generators in the distribution system can be integrated into market structures being created for the power system.

A second reason for this emphasis of the modeling is that the use of distributed generators for voltage support is reasonably well accepted by the power industry, and many studies have been performed on this topic (see [10, 35, 40]).[15] In contrast, the frequency dynamics of distribution systems with distributed generation units, and the possibility of these units participating in the supply of ancillary services such as frequency stability and spinning reserve, are relatively new issues.

The scheduled, bulk power flows (large signal characteristics) are determined exogenously by the PXs as one of their prime functions. It is the deviations from this schedule that drive the short run energy and ancillary services markets, and which are the focus of this section. Small signal, linearized models are used for analyzing these markets.

Cost Output Equation. The development of the closed loop price model begins here by expressing the cost of power generation in terms of the state variables in the generator equations of Table 12.1. Cost can be incorporated into the state space generator models by writing an output equation to capture the variable costs associated with generating power from any given technology. Each state space model identifies the set of elements that together can reproduce the basic machine performance. The cost output equation is then based on the assumption that the sum of the marginal costs associated with each state variable will accurately represent the full marginal cost of generating with the technology. Referring to the dynamic generator models in Table 12.1, the cost equations for the different generator types would be written as

$$
\begin{aligned}
c_H &= c_{wH}\omega_G + c_q q + c_{vH} v + c_{aH} a + c_{gH} P_G \\
c_s &= c_{ws}\omega_G + c_p P_t + c_a a + c_{gs} P_G \\
c_{CT} &= c_{wCT}\omega_G + c_{vCT} V_{CE} + c_{fCT} W_F + c_{gCT} P_G \\
c_{CC} &= c_{wCC}\omega_G + c_{vCC} V_{CE} + c_{fCC} W_F + c_{aW} W_{air} + c_{pst} P_{ST} + c_{gCC} P_G
\end{aligned}
\qquad (12.8)
$$

The coefficients in these equations represent the marginal cost associated with each piece of equipment or process represented by the specified state variable. In particular, c_g is the marginal fuel cost. The *existence* and sign of c_w can be established, though it does not have as direct an interpretation as c_g.[16]

The significance of the values of the coefficients in the cost equation lies not in the absolute values chosen, but rather in the relative values of the coefficients between the different technologies and distributed generators. It is the relative cost values that capture the real-time differences in using one technology before another. This interpretation of the cost coefficients is valid for all generators modeled except the slack bus. The cost equation for the slack bus is interpreted as representing the cost to the bulk system (rather than to a single generator) of generating the power supplied to the distribution system (delivered to the substation).

With the addition of the output cost equation, the model for each generator can be expressed as

$$
\begin{aligned}
M\dot{\omega}_G &= -(e_H + D)\omega_G + k_q q - k_w a - P_G \\
\dot{q} &= \omega_G/T_f - q/T_q + a/T_w
\end{aligned}
$$

$$T_e \dot{v} = -v + r'a \qquad (12.9)$$
$$T_s \dot{a} = -w_G + v - (r_h + r')a + w^{ref}$$
$$c_H = c_{wH} w_G + c_q q + c_{vH} v + c_{aH} a + c_{gH} P_G$$

where this set of equations is for the hydro-turbine-generator. More generally, each set of equations now has the form

$$\dot{x}_{LC} = f(x_{LC}, P_G, w^{ref})$$
$$c = h(x_{LC}, P_G, w^{ref}) \qquad (12.10)$$

In the framework presented in this section cost and price are used as follows.[17] At the primary control level the output cost variable is introduced, and represents the actual cost of generating electricity. This was presented above. Marketplace interactions are based on price however, not cost, so the cost variable is translated to price before being integrated into the system model. In the simulations presented at the end of this section, the generation sector is modeled as a competitive industry, making this distinction less important ($P = MC$). When imperfect information and alternative market structures are modeled though, the distinction between cost and price can become non-trivial. With this relationship established, the price model can now be derived.

Discrete Time Price Model. The generators and the system will respond to the price signal at specific intervals, indicating that the closed loop price signal is best modeled in discrete time. The first step in developing this discrete time model is to assume the primary dynamics have settled, reducing all the generator models of the form in Equation (12.10) to a set of simultaneous, algebraic equations of the form

$$0 = f(x_{LC}, P_G, w^{ref})$$
$$c = h(x_{LC}, P_G, w^{ref}) \qquad (12.11)$$

Solving these equations for cost results in a discrete time cost equation of the form

$$c_H[K] = \gamma_1 w^{ref}[K] + \gamma_2 P_G[K] \qquad (12.12)$$

where K represents the discrete time index for the price control loop, and γ_1 and γ_2 are constant expressions of the generator parameters and cost coefficients. For the hydro generator, for example, these coefficients are of the form

$$\gamma_{1H} \equiv [(\frac{c_{gH} T_q}{T_w} \frac{1}{r_h} + (c_{vH} r' + c_{aH})\frac{1}{r_h}) - (c_{wH} + c_{gH} T_q(\frac{1}{T_f} - \frac{1}{T_w r_h})$$
$$-(c_{vH} r' + c_{aH})\frac{1}{r_h})\sigma_H c_2] \qquad (12.13)$$

$$\gamma_{2H} \equiv [c_{gH} - (c_{wH} + c_{gH} T_q(\frac{1}{T_f} - \frac{1}{T_w r_h}) - (c_{vH} r' + c_{aH})\frac{1}{r_h})\sigma_H]$$

The coefficients for the remaining technologies are developed in a parallel manner, and can be found in the Appendix.

The second step in developing the price model is to translate Equation (12.12) from the private cost equation to a market price equation. For a competitive market model this requires only a change of variable from cost to price, where ρ is the variable used to designate price. In a competitive market then the individual price equation is simply

$$\rho[K] = \gamma_1 \omega^{ref}[K] + \gamma_2 P_G[K] \tag{12.14}$$

The format of this equation is identical for all the other technologies, with the unique properties of each technology being expressed in the definitions of the coefficients, γ_1 and γ_2.

The third step in developing a price signal is to form the dynamic model, by writing Equation (12.14) for two sequential time steps and subtracting. The dynamic equation for the price of energy supplied at a generator is expressed as

$$\rho[K+1] = \rho[K] + \gamma_1(\omega^{ref}[K+1] - \omega^{ref}[K]) + \gamma_2(P_G[K+1] - P_G[K]) \tag{12.15}$$

As with the development of secondary frequency control models, the control variable for the price model is ω^{ref} which is seen to be implicit integral control, such that

$$u_\rho[K] \equiv \omega^{ref}[K+1] - \omega^{ref}[K] \tag{12.16}$$

or

$$\omega^{ref}[K+1] = \omega^{ref}[K] + u_\rho \tag{12.17}$$

where the time index is K. The significance of using ω^{ref} as the control variable is that the proposed price model integrates the existing local generator control (i.e. the governor for frequency control) into the closed loop price feedback structure.

As the fourth step, Equation (12.16) is substituted into Equation (12.15), leading to the following form for the dynamic price equation

$$\rho[K+1] = \rho[K] + \gamma_1 u_\rho[K] + \gamma_2(P_G[K+1] - P_G[K]) \tag{12.18}$$

This equation can now be used alone or with other state equations to form a complete closed loop price model. The only remaining step is the calculation of the gain in the feedback loop, which is addressed below.

To summarize, the dynamic price model can be written in a general format as

$$x_\rho[K+1] = x_\rho[K] + \gamma_1 u[K] + \gamma_2 z[K] \tag{12.19}$$

where x_ρ is the price-based state space, $u[K]$ is the control and $z[K]$ is the system input.

The dynamic form for the specific model as used in this chapter is

$$\rho[K+1] = \rho[K] + G_3 u_\rho[K] + G_4(\omega_G[K+1] - \omega_G[K]) \tag{12.20}$$

where ρ is the state vector consisting of ρ_i at each bus, and $\Delta \omega_G[K]$ is the input to the model. G_3 and G_4 are matrices whose elements are functions of

484   POWER SYSTEMS RESTRUCTURING

generator parameters and cost coefficients. The full derivation of this model, along with the equation for $\omega_G[K]$ and the coefficient matrices are contained in the Appendix.

The most important model characteristics, in terms of operating in a competitive market, are the ability to maintain controllability and ensure decoupling, mathematically, between the state variables for each generator. These are the most important characteristics for the following two reasons.

- First, a potential loss of controllability implies that in some situations the generators operating within price framework would lose the ability to control their price and output in response to price-based signals from the system. This loss of control renders the price framework useless since its basic purpose is to provide a means for generators to control output in response to a price signal.

- Second, coupling between state variables of different generators implies that generators would need access to information about other generators— information that may be considered standard now, but which may not be available in a competitive market. Thus a model with coupling between generators may not be implementable in a competitive market structure.

In the price model used in this chapter, Equation (12.20), the state variables for different generators are decoupled from each other and the system always remains controllable. This implies that the model of Equation (12.20) can be used by individual generators to make operating decisions in response to a closed loop price signal while operating in a competitive market setting.

Control Law for the Price-Based Model. The final component required for the price model is the closed loop control law, which is the mechanism that moves the system to the desired equilibrium while minimizing a specified performance index. The target equilibrium point for the price model is defined by the competitive market equilibrium. The market equilibrium is in turn defined by the actions of all the participating distributed generators which are assumed to be competitive price takers.

According to the competitive model the market price represents each generator's marginal revenue, MR. To maximize profit, each generator will produce to the level where marginal cost equals marginal revenue, $MC = MR$. This market dynamic is captured in the control law by defining the market price, ρ_{mkt}, as the equilibrium point to which each small generator matches its price ρ_i (and the corresponding output level, P_G). Note that stating $\rho_{mkt} = \rho_i$ is equivalent to stating $MR = MC$. Mathematically, the economic goal can be expressed either as

$$\rho_i \Rightarrow \rho_{mkt} \quad \forall i, \ i = 1, 2, \ldots n \qquad (12.21)$$

where n = the number of distributed generators, or equivalently as

$$(\rho_i - \rho_{mkt}) \Rightarrow 0 \qquad (12.22)$$

The second expression is the version used in the price control law, since mathematically power system models must include a reference bus.

To represent the model in terms of a reference bus, the price model of Equation (12.20) is written in matrix form as

$$\begin{bmatrix} \rho_1 \\ \rho_2 \\ \vdots \\ \rho_n \\ \rho_{mkt} \end{bmatrix}_{[K+1]} = \begin{bmatrix} \gamma_{3\_1} & & & & 0 \\ & \gamma_{3\_2} & & & \\ & & \ddots & & \\ 0 & & & \gamma_{3\_n} & \\ & & & & \gamma_{3\_mkt} \end{bmatrix} \begin{bmatrix} \rho_1 \\ \rho_2 \\ \vdots \\ \rho_n \\ \rho_{mkt} \end{bmatrix}_{[K]}$$

$$+ \begin{bmatrix} \gamma_{4\_1} & & & 0 \\ & \gamma_{4\_2} & & \\ & & \ddots & \\ 0 & & & \gamma_{4\_n} \\ & & & \gamma_{4\_mkt} \end{bmatrix} \Delta\omega_G \quad (12.23)$$

where ρ_{mkt} is the market price, expressed as the price at the reference bus. This set of equations is multiplied by the $n \times (n+1)$ transformation matrix

$$T = \begin{bmatrix} 1 & 0 & 0 & \cdots & -1 \\ 0 & 1 & 0 & \cdots & -1 \\ \vdots & & \ddots & & \vdots \\ 0 & \cdots & & & -1 \end{bmatrix} \quad (12.24)$$

in order to explicitly reference every bus to the slack bus. The state vector is thus transformed to

$$\begin{bmatrix} \rho_1 - \rho_{mkt} \\ \rho_2 - \rho_{mkt} \\ \vdots \\ \rho_n - \rho_{mkt} \end{bmatrix} \quad (12.25)$$

In this form the objective of the feedback control is clearly that of returning the state vector to the origin.

The control input for the price model is defined as

$$u_\rho[K] \equiv -K_\rho x_\rho \quad (12.26)$$

where x_ρ is the state space of the price model, and K_ρ is the gain in the feedback loop. The generator price, ρ_i, is then controlled to the market price, ρ_{mkt}, by updating ω^{ref} by means of

$$\omega^{ref}[K+1] = \omega^{ref}[K] + u_\rho[K] \quad (12.27)$$

The final step is the calculation of the gain, K_ρ. One common method is to use a linear quadratic regulator, LQR, which calculates the gain of the feedback loop, K_ρ, to optimize the performance function

$$J_\rho = \Sigma_0^\infty (x_\rho[K]'Qx_\rho[K] + u_\rho[K]'Ru_\rho[K]) \quad (12.28)$$

The relative magnitude of the weighting matrices, Q and R, are the design variables which change the definition of the performance function and so influence the calculation of the optimal gain matrix K_ρ.

This performance index, J_ρ, defines the square of the price deviations, $(\rho_i - \rho_{mkt})$ $\forall i$, as the quantity to be minimized. Since ρ_{mit} is the point of least cost operation for the system (as determined by the power exchange or other appropriate coordinating institution), the action of the control signal becomes that of maintaining the output of each generator at the point where $\rho_i = \rho_{mkt}$ or $MC = MR$, which thus maintains the least cost system operation.

In this price loop, u_ρ is assumed to act on a slower time scale (T_ρ) than that of the secondary frequency control (T_s), where $T_\rho > T_s$, and the corresponding discrete time indices $K > k$. The control u_ρ essentially acts as a correction to ω^{ref}, based on economic and market goals. This correction control signal acts on a longer time scale than the existing frequency control, which is based on strictly technical objectives.

Closed Loop Price Signal Analysis

The objective in developing a feedback price signal is to facilitate the operation and control of the power system by means of market forces and independent production decisions rather than by the control room of a vertically integrated utility. The existence of a market based signal is a prerequisite to the creation of a generation market with truly independent generators, since market based coordination removes the need for generators to divulge private information to a central authority. This section presents simulations demonstrating the use of the price signal in coordinating system operation and control.

Base Case – Competitive Market. The first example uses the sample distribution system shown in Figure 12.11, with one hydro turbine at bus 10 and one combustion turbine (CT) at bus 24. As before the model input is a small load disturbance at bus 9 occurring at time $t = 8$ minutes. Conceptually the model action is that the Power Exchange (or market coordinator) updates the system price in response to the disturbance, and then the distributed generators respond to this price change by altering their output such that the MC of generation equals the new MR (recall that the MR is defined as the market price since for now all the distributed generators are price takers).

Figure 12.21 plots the changes in power and relative price at each generator, without the price signal implemented, and Figure 12.22 with the price signal.

The first two graphs, without price feedback, show the generator outputs and purchases from the grid increasing in response to the increase in demand, and the resultant price increase at each generator. Note that the slack bus represents power flow at the substation and so is a proxy for purchases from the grid. The price offered at this bus is ρ_{mkt}, can be seen to change in response to the disturbance. Note once again that the objective of the price control signal is *not* to return the absolute price deviation to zero, but rather to adjust generator outputs so that the quantity $(\rho_i - \rho_{mkt})$ is equal to zero.

CONTROL AND OPERATION OF DISTRIBUTED GENERATION   487

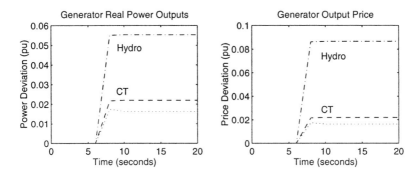

Figure 12.21 Power Deviation and Corresponding Price Deviation Without Price Feedback

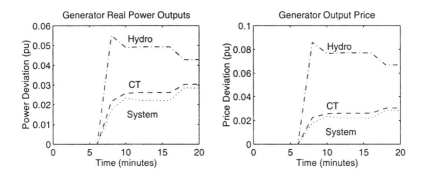

Figure 12.22 Power Deviation and Corresponding Price Deviation With Price Feedback

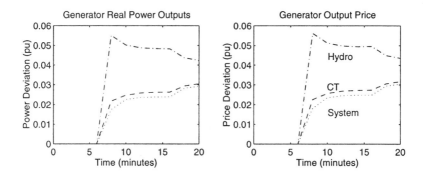

Figure 12.23 System with Price Signal but no AGC

The second two graphs show the same system operating in a competitive market setting with the price feedback implemented. The price signal is updated every ten minutes. The proportion of the increased demand met by each generator is now determined by each the individual economic objective of operating where $MC = MR$, as well as by system needs to maintain power balance and the nominal system frequency. The lower right graph demonstrates that the relative prices are now much closer than they were without price feedback (upper right graph). These values are not identical though as a result of the competing need to maintain system frequency as well as account for the small system losses.

The Price Signal in Conjunction with AGC. In the above graphs, the price signal is seen to act on a much slower time scale than the secondary frequency control, and can be interpreted as updating the secondary frequency control, with the objective of reaching an equilibrium point simultaneously for price and frequency. (The NERC control performance standard CPS_1 criterion is also a corrective control acting on a time scale slower than AGC.) Figure 12.23 shows the same system as in the base case above, when AGC is not implemented.

Clearly, the price signal alone can stabilize the system and move the generator outputs to an equilibrium point. This figure also demonstrates that when AGC is not implemented, the system converges to the price equilibrium point more rapidly than when both systems are implemented simultaneously—a situation where the two functions fight with each other to a small extent.

As the power system evolves to incorporate more extensive distributed automation, it is likely that AGC will be extended to the distribution system and distributed generators. It is also likely that a closed loop price signal as proposed in this chapter will be adopted more slowly than AGC, reflecting the fact that AGC is already well understood by industry, and the price signal is not. However, if no secondary control is implemented in the distribution system, with multiple distributed generators, it would be expected that the local system frequency would slowly drift, and eventually lose synchronism with the

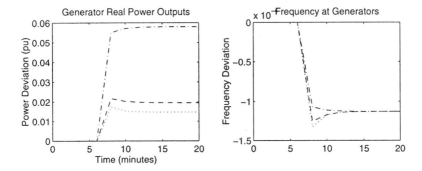

Figure 12.24 System with no Secondary Control

rest of the power system. This eventuality is shown in Figure 12.24, which graphs the power outputs and frequency and each generator in the test system after the system disturbance *when no secondary control is implemented*. The small negative frequency deviation shown in the graph is not necessarily a concern. With numerous system disturbances though, the frequency would be much more severe, and could result in damaging frequency sensitive loads, and disrupt systems dependent upon the 60Hz cycle.

The majority of simulations in this section are for a system with both AGC and the price signal implemented, to demonstrate that the systems can be operated together, especially during the restructuring process when not all generators would immediately participate in a price feedback structure, even if one were available. The figures in the remainder of this section demonstrate the system response and stability as controlled by the traditional, AGC framework, in conjunction with the market based price framework. An important point is that the generators will contribute to the energy and services markets differently depending upon whether they are pariticipating in either AGC, the price-based framework, or both.

Non-Participation in Price Feedback. The simplest market structure simulated with the price model is the competitive market example above where all the small generators are incorporated into the price control loop. It is likely however, that while the system is in the process of being restructured some generators will elect to not respond to the price signal, instead remaining under direct central control (and continuing to participate in AGC if they have historically done so). Figure 12.25 shows the output and corresponding prices in the test system when there are four combustion turbines installed, but only one has elected to participate in the price feedback framework. The solid line, lowest on the graph represents the system purchases and price, and the line just above the system (dot-dash line)represents the single combustion turbine (CT) that responds to the price signal.

The remaining three CTs have elected to not participate in the price feedback system, and as a result they do not reduce their output to match ρ_i to

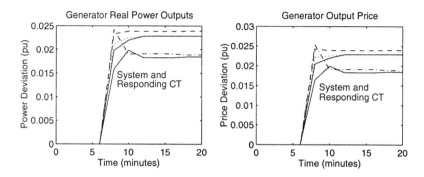

Figure 12.25 Generation and Price Deviations with Single CT Participating in Price Feedback

ρ^{mkt}. An important point to note though is that this *does not* imply that they are now receiving the higher price corresponding to the level on the righthand graph. The price they receive is determined exogenously by the central authority, and the righthand graph shows the price *at the generators* of producing at the given level, but not the price they receive. The generators not participating are seen to produce at a cost above the system marginal cost. This result can be interpreted as reflecting a suboptimal level of system efficiency and performance, due to the non-competitive decision making of three of the generators. This brings up a fundamental issue of how much the system deviates from the theoretical optimal performance if only some generators respond in a decentralized way. The system achieves the optimum if all participate.

Non-Dispatchable Technologies: Wind Turbines. In addition to the scenario introduced above, this mix of participating and non-participating generators can result when some of the generators are non-dispatchable technologies (NDTs) which do not have primary controllers, such as wind turbines. The stability of the system with such a mix of technologies is simulated next.

The first example with NDTs replaces the steam turbine from the previous example with a wind turbine. The wind turbine is a non-dispatchable technology (NDT), and is assumed not to participate in the price feedback framework. The small increase in wind turbine output after the load disturbance at $t = 8$ minutes is a consequence of the fact that system frequency is briefly disturbed from its nominal value, and so momentarily affects the output from the wind turbine. (The link between rotor frequency, system frequency and power output was mentioned above.) This example demonstrates the behavior of the system in general if one of the distributed generators is not participating in the price feedback. In such a situation, whether the generator is a non-dispatchable or a dispatchable technology, the output of the non-participating generator will not change in response to a change in the reference price except for a small deviation as the system finds its new equilibrium. The system does remain stable.

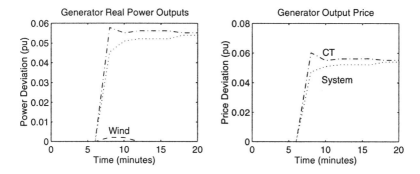

Figure 12.26 Generation and Price Deviations With Wind Turbine in System

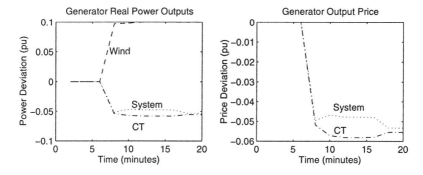

Figure 12.27 Generation and Price Deviations After Increase in Wind Turbine Output

For the second example, the system is the same as in Figure 12.26 but now the disturbance is an increase in output from the wind turbine at $t = 8$ minutes rather than a change in demand.[18] Figure 12.27 shows the changes in power and relative price after this disturbance. Both the output from the combustion turbine and the supply through the substation decrease to balance the increased output from the wind turbine (left hand graph). The system price and price of generation at the CT are both seen to decrease in the right hand graph. The interesting point from this example is the dynamic between the wind turbine and the neighboring combustion turbine. As output from the wind turbine increases, the CT is forced to decrease its output to maintain nominal system frequency, with a concurrent decrease in its revenue stream. System fluctuations driven by NDTs in small penetrations will most likely be indistinguishable from fluctuations caused by load changes. At larger penetrations NDTs may cause system fluctuations large enough to noticeably impact the revenue stream of other generators, which will create a tension between the system's need for dispatchable technologies to alter their output and those generators' financial objectives.

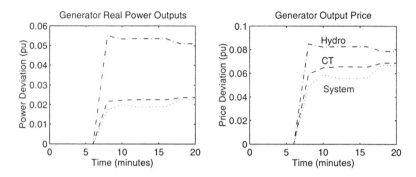

Figure 12.28 System with 1 Hydro and 1 CT: Uncertainty in Parameter Values

Imperfect Information: Uncertainty. The market organization itself is altered for the final category of market interactions. The first variation to the competitive market is a weakening of the assumption of perfect information. Imperfect information results both from uniform uncertainty in measurements and system values, and also from unequal access to system information. Unequal access can result from generators that were originally owned by a utility simply having greater operating experience than new, independent generators. It could also be the result of generators that contract to a power marketer, having access to more extensive, shared information than single units. In either case, one impact of such uncertainty in information will be that the independent generators will calculate their optimal control gain based on an estimated set of parameters, and will then operate in the actual distribution system. The estimated and actual values are likely to be different. Figure 12.28 shows the response of the system with one hydro and one CT when their estimated values (used to calculate the feedback gain) differ from the actual values (used to simulate system behavior) by 10% to 25%.

Figure 12.28 shows that the system remains stable even with this uncertainty. However, comparing this figure with Figure 12.22, when there is no uncertainty, reveals that the convergence of the output levels to the target equilibrium, as driven by the price signal, is much slower when there is uncertainty than when there is none. The right-hand graph in Figure 12.28 seems to imply that the price converges more quickly in face of uncertainty. This is incorrect since the prices plotted in this graph are functions of the incorrect (uncertain) values, and so do not reflect the actual prices associated with the generators. (Note that the values plotted for the generator output levels are the actual output levels, and are not directly functions of the uncertain parameters.)

Implementing a Closed Loop Price Signal

The first sections of this section have been devoted to developing the model for the price signal and demonstrating its use in the distribution system. This section addresses some issues relevant to implementing a closed loop price signal of the form proposed in this chapter. Information is an important factor in many aspects of the proposed framework. The information requirements are addressed first. Second, this section suggests one possible approach for adapting existing control hardware which will enable generators to sense and respond to the price signal. Finally, some general limitations of the proposed price framework are introduced.

Information Requirements. Identifying what data is needed, and the impacts on system performance when more or less information is available, is one dimension of the information question. A second dimension is obtaining this data, or establishing the means or technologies to measure and record the desired information.

Determining what data is needed and its impact on system performance is related to state space selection, which was addressed earlier (and again in the Appendix). In particular, criteria to guide state space selection are

1. The information required to calculate the feedback gain,

2. The convergence of the system to the equilibrium point, $x_\rho \equiv 0$ (i.e. $\rho_i = \rho_{mkt} \ \forall i$), and

3. The sensitivity of the control response to uncertainty in the gain calculation.

The first criterion above is simply the characteristic identified earlier—whether the state space is decoupled or not—rephrased from the perspective of an operator of a distributed generator. An operator or owner of a small generator will be interested in whether the individual feedback gain can be calculated strictly with private information and that publicly available on the system, or whether information on other generators or non-accessible system information is also required. Specifically, a price model in which the variables for any given generator are decoupled from those of the other generators, implies that generators need access only to local information to participate in the price framework.

With respect to the price model of Equation (12.20), the only information required by a generator to participate in the price system is the local cost data, expressed in matrices G_3 and G_4, and the local generator frequency. More complex price models could also require that generators have access to both network parameters and system configuration data, as well as the generator frequency at all generator buses. As demonstrated below, this increased information improves system performance, but at additional effort and cost.

The additional required information could pose significant problems. First, the system configuration in a distribution system is changed much more often than in a transmission system, as part of standard operating procedures. The costs associated with obtaining real-time system configuration data are much greater than those for using static data alone. Second, obtaining frequency data for all generator buses rather than only that for the local bus, increases the cost. The metering and coordination required to provide this additional data would increase the complexity and cost of implementing the price framework.[19]

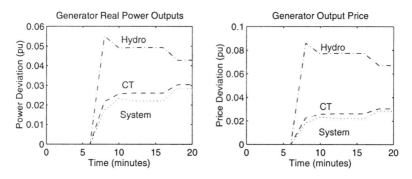

Figure 12.29 Power Deviation and Corresponding Price Deviation With Decoupled Price Model

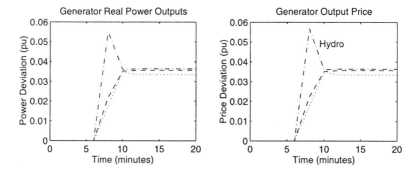

Figure 12.30 Power Deviation and Corresponding Price Deviation With Coupled Price Model

The speed of convergence of the system to the desired price equilibrium is related to the extent of available information and the cost of the control effort. If a more complex price model is implemented rather than that of Equation (12.20), more information would be required yet the system converges to the desired price equilibrium more rapidly. Figures 12.29 and 12.30 compare the time required for the output levels to converge to equilibrium, for the test system with one hydro generator and one CT. These graphs clearly show the tradeoff between cost (information) and performance.

The final criterion above, the sensitivity of the gain and subsequent system performance to uncertainty in parameter values, determines the accuracy required in measurements or estimates of the parameters, where a greater sensitivity implies increased effort and eventually cost in implementing a closed loop price system. The impact on system performance of this uncertainty was analyzed in the previous section (see Figure 12.28). The remainder of this section discusses the advances in metering technologies and methods used to obtain the data for the distribution system, which determine the amount and accuracy of the data available on the system.

As discussed above, the amount of data gathered in a typical distribution management system tends to be an order of magnitude greater than that for SCADA at the transmission level. Increases in the automation and control demands of the distribution system will only exacerbate this problem. In order to determine how much money is owed each day to the distributed generators which participate in the price feedback structure, extensive effort will be required to meter not only the output levels but also the *time of output* (in other words, not just energy but power as well).

Two recent articles present real-time metering schemes for the distribution system to help address this concern [2, 29]. The first study, by Baran et. al., suggests the use of state estimation, in conjunction with standard metering, to provide the required system data. Baran et. al. also point out that existing load forecasting techniques work well for estimating load in the aggregate, but can not be used to estimate individual loads accurately. The cost of metering prohibits extensive use of meters. To overcome this problem, they develop rules for meter placement, such that the real-time data gathered can be used along with forecasted data, to supply the information required by a real-time monitoring system, to be used for state estimation and system operation. The second study, by Lee et. al. [29], focuses on the potential of frequency oscillations in a distribution system after a cogeneration plant is installed. This study reports on the development of a real-time monitoring system, that can be used to help maintain system stability in a distribution system with distributed generation.

Adaptations to Control Hardware. The price feedback signal was developed assuming that ω^{ref} is an appropriate variable to use as the control variable. Referring to the governor equation for the hydro turbine generator model, the variable ω^{ref} is seen to control generator output by changing the governor valve position as

$$T_s \dot{a} = -\omega_G + v - (r_h + r')a + \omega^{ref} \tag{12.29}$$

Alternatively, a new control variable, ρ^{ref}, could be introduced via the above equation.

$$T_s \dot{a} = -\omega_G + v - (r_h + r')a + \omega_k^{ref} + \rho_K^{ref} \tag{12.30}$$

This new variable would not alter the derivations of the closed loop price signal presented above. In those equations, ρ^{ref} would simply replace ω^{ref}, as the frequency reference point set by the price loop, and updated on the slower time scale (K rather than k).

Equation (12.30) differentiates the reference variables updated by the two control loops, and explicitly identifies the variable with the control loop, based on time scale. One benefit from writing the equation in this form is that it emphasizes that it is the governor that should be designed to sense the price signal, in a manner parallel to its existing function of sensing the frequency reference value, ω^{ref}. Also, if a generator is participating in the price framework but for some reason decides to generate at constant output for a period of time, the ρ^{ref} parameter could be designed to be manually controlled. In this manner, a distributed generator operator could set ρ^{ref} to a constant value, and so control its output level to a constant value, without needing to alter other parameters in the system.

A second set of changes that will be required at the individual generator level, for implementing the price feedback framework, is the measurement, or metering, required to determine the values of the coefficients and state variables themselves in the cost output equation

$$c_H = c_{wH}\omega_G + c_q q + c_{vH} v + c_{aH} a + c_{gH} P_G \qquad (12.31)$$

This problem is greatly simplified if c_g alone is assumed to be non-zero. With the model simulated in this chapter, Equation (12.20), system performance is not greatly enhanced when all the cost coefficients are assumed to be non-zero. However, this may change with actual systems, or further optimization of the closed loop signal.

CONCLUSIONS

The siting of numerous small scale generators in distribution feeders raises questions about how these generators may impact the technical operations and control of the distribution system, a system designed to operate with a small number of large, central generating facilities. In response to the restructuring and deregulation process, the growing number of independent generators will also face new and potentially conflicting technical and economic incentives. Policies and institutions must be developed to balance the incentives provided by market forces with the continuing need for some centralized system coordination.

The integration of distributed generation into the distribution system, in terms of both engineering and market coordination issues, has been explored in this chapter. The modeling and simulation effort has emphasized the short-term primary and secondary dynamics, which represent phenomena evolving in a time period from seconds to five or ten minutes, as driven by deviations from the scheduled power flows. This section presents conclusions for the primary dynamics and engineering concerns first, followed by those for the secondary dynamics and market integration. Policy impacts are addressed at the end of this section.

The main conclusions from the analysis at the primary dynamics level, focusing first on frequency behavior, are:

- Local frequency in the distribution system can become unstable depending on the type and number of distributed generators in the the system. Altering the *location* of the generators does not affect the stability of the system. These results were demonstrated with models and system simulations.

- One method that was developed for ensuring frequency stability involves developing specific standards, or ranges, for governor setting on distributed generators. This proposed method to ensure local frequency

CONTROL AND OPERATION OF DISTRIBUTED GENERATION 497

stability is easily implemented, though it does represent a change from existing practice. Currently on the transmission system, generator governors on the large, central generating facilities react more slowly than those on small distributed generators, and so are not relied upon for maintaining system stability. The analysis in this chapter demonstrated that local generator governors *can* be used at the distribution level to ensure frequency stability.[20]

Continuing to focus on the engineering integration of distributed generators, general conclusions from the discussion with respect to voltage support in the distribution system are:

- Generators operated as synchronous condensors can provide local voltage support,

- Power conditioning and power electronic equipment increase the flexibility of distributed generators, allowing them to be used in the same manner as static VAR compensators and FACTS devices are, to provide local voltage support,

- Both of these possibilities have the potential to increase the value of DG units to the system, but unless this capability is recognized and compensated (financially), it is unlikely that distributed generators will provide this service, forcing the ISO to purchase local voltage support from other suppliers.

The second part of the chapter addressed the market integration of distributed generation. It focused primarily on secondary level dynamics and the development and analysis of a model for a closed loop price signal, used to coordinate the actions of distributed generators in a future competitive market. It was assumed that the owners of the new, small generators would to operate in the emerging competitive markets, independent of a central authority. These distributed generators would further require an incentive to supply ancillary services, or they would be likely to concentrate on the supply and demand market for real power.

The main conclusions from this part of the chapter are:

- A closed loop price signal, as designed in this chapter, will allow distributed generators to operate in a competitive market without depending upon the extensive information and centralized control structure of the traditional power system,

- The closed loop price signal was shown to function successfully in an ideal competitive market as well as other scenarios, such as:

    1. Not every distributed generator may choose to participate in the price framework. In this situation those generators not participating are likely to produce at a cost above the system marginal cost, reflecting a suboptimal level of efficiency and performance at the system level.

498   POWER SYSTEMS RESTRUCTURING

   2. If there is a mix of non-dispatchable (NDT) and dispatchable generation technologies in the distribution system, the dispatchable technologies may be forced to provide system balancing and ancillary services to compensate for the stochastic changes in NDT output, much as they compensate for the stochastic deviations in demand.

   3. When the generators operate with imperfect information—information critical to determining private operating decisions—the system converges to the price equilibrium more slowly than when perfect information is available,

- To maintain the desired power system performance, the future system may need to develop a coordinated transmission-distribution performance standards.

This chapter has demonstrated that if generators are to be sited in the distribution system in significant numbers, then operations and control issues that have historically been of concern only at the transmission level may become concerns for the distribution level as well. The final item above simply points out that if this does happen, then standards and operating procedures may need to be developed in a coordinated fashion for the transmission and distribution systems. In summary, this chapter demonstrated that a closed loop price signal can be used to coordinate generator actions in a competitive market while also maintaining the desired level of system reliability and stability.

Power System Evolution

Returning to the discussion early in this chapter, it is anticipated that the electric power system will evolve to a state where distributed generation is well established and freely operating in the competitive market structure. In this future system architecture distributed resources will be located throughout the system, and will be free to contract to supply load anywhere in the system—they will not be restricted to operate within their local distribution system. In this phase of the restructuring process the closed loop price signal as developed in this chapter is particularly important. The price signal will provide incentives for generator operations in a setting where the generator operating decisions are based on private economic incentives, and are not controlled by a central authority.

To understand the significance of the price framework proposed in this chapter, it is useful to consider the following example. Suppose that the small generator in System B of Figure 12.4 decides to supply a customer in the lower left of the figure, because the market price in System C is greater than that in System B. If the total generation in system B exceeds that system's demand, then there will be a power mismatch within that distribution system, which may have negative impacts on the system performance—in this case the local frequency will increase. In the power system today, there are sophisticated control strategies to counteract such supply-demand mismatches at the transmission level. The example presented for Figure 12.4 differs from the situation common in today's system for two reasons:

1. The existing hierarchical control strategies have been developed predominantly for the transmission system. Since there is little or no generation in the traditional distribution system, no parallel control strategies have been developed for that system, and

2. The example is set in the restructured, competitive market driven power system, not the traditional centrally controlled and regulated system.

The question then is what form of control structure will be developed for the future power system, such that bilateral transactions between *any* two distributed resources will not negatively impact power system performance (frequency behavior in this example)?

The closed loop price signal developed in this chapter presents one method for maintaining the desired system performance. In response to the excess generation in System B, the market price for ancillary services in that system would increase, or more specifically the price offered for the service of frequency stabilization, would increase. This price increase would trigger a decrease in power generation, until the price in System B rose to match that in System C. In this manner, the system frequency equilibrium would be restored via the price signal—i.e. the market mechanism. Note that the closed loop price signal is certainly not the only means to control the system in response to the hypothesized power mismatch. It does however, have the advantage of being consistent with the emerging competitive market structure, in contrast to traditional control strategies which are centralized and tend to rely on access to what will become private information.

Before the power system can evolve to this distributed structure, some more immediate issues must be addressed, dealing with entry and operations.

- Entry into Competitive Markets: The future penetration of distributed generators will be influenced by the extent to which the government policies open the emerging markets to distributed generators and extend the competitive institutions to the distribution system. The new industry rules will establish who will be allowed to participate in the competitive markets. In particular, state polices which constrain distributed generators to be owned by local distribution utilities will prevent these generators from becoming equal players in the competitive markets. Alternatively, if retail competition is encouraged and opened to distributed generators, the generators will gain an increased presence in the future electric power system.

- Operating in a Competitive Market: Competitive institutions at the distribution and retail levels will provide incentives to improve the system operating efficiency, which in turn will increase the demand for system automation and control equipment. At present the distribution system has very little dynamic control and automation. The development of distribution management systems (DMS) offers a general method for increasing the automation and control of generators and other equipment in the distribution system. This type of centralized automation system alone may not encourage the operation of distributed generators *in a competitive market*. For competitive operation generators must be free to make independent operating decisions. The development and implementation of a closed loop price signal as developed in this chapter—in conjunction with the scheduling for bulk power flows and the spot market institutions—will allow such independent decision making.

Once the means to operate in competitive markets is established, the next step is to identify the goods and services that the distributed generators will produce and sell. The institutions to facilitate distributed generator participation in scheduling and wholesale power markets will likely be the same institutions as those currently being created at the transmission level. When operating in the short run markets, the small generators can be expected to supply the local ancillary services only if they are compensated for them. The extent to which distributed generators are called upon to supply frequency regulation and local voltage support will thus be significantly influenced by the restructuring process and the mechanisms created for providing system services in the restructured industry. The short term coordination of distributed generators operating in competitive markets can be achieved with the new closed loop price signal proposed in this chapter.

Developing both the policies to promote the unbundling of ancillary services, and the mechanisms for these services to be bought and sold in the competitive markets is consistent with the restructuring and deregulation process. If such mechanisms are not developed, the ISO will be required to purchase ancillary services from other, contracted sources, or rely on the regulated distribution utility to purchase and provide them.[21]

In summary, the market driven control signal developed in this chapter is an important element of the industry restructuring process because it will promote:

- A method for determining the value, to customers and to the system, of frequency regulation,

- The development of competitive markets,

- The full integration of distributed generators into these markets, and

- An opportunity for distributed generators to provide and be compensated for both real power and ancillary services.

Acknowledgments

We would like to thank the Office of Utility Technologies in the US Department of Energy for support of this research. We would also like to thank Dr. Richard Tabors and Professor George Verghese at MIT for their invaluable advice and comments throughout this project. The authors claim full responsibility for any errors or omissions.

Appendix: Data and Model Development

Individual Generator Models

Modeling Goals and Assumptions. The modeling effort is based on building decoupled, linearized state space models[22] for each type of distributed generator, and coupling[23] them through a distribution system model. State space models have been developed for steam turbines, hydroelectric turbines, combustion turbines, combined cycle plants, wind turbines and inverters (to be used with fuel cells and photovoltaics). Numerous dynamic models exist for each of these technologies, however the majority are very complex, involving a large number of state variables. In developing the models for this thesis, the objective is to represent each generator with a small number of state variables (three to four) so that interconnected system models, which each include a number of the distributed generators, will not be overly complex. A second objective is to develop each set of local state equations such that they incorporate P_G as the system coupling variable. The traditional system coupling variable is rotor angle, δ. Regardless of which variable is used, all the models must include the same variable so that they will be mutually compatible when modeled together in the extended state space.

The models which include a synchronous generator all use a form of the swing equation as the generator state equation:

$$J\ddot{\delta} + D\dot{\delta} = P_m - P_e \qquad (12.\text{A}.1)$$

where $P_e \equiv P_G$, the electrical power output. Use of this equation facilitates the inclusion of the system coupling variable, P_G in each set of local state equations. The models for steam- hydro- and combustion-turbine generator plants all use this equation as the basis for the synchronous generator state equation, and are presented next. This generator equation differs for different technologies, since the mechanical power from the turbine, P_m, has a different representation for each turbine type.

Steam-Turbine-Generator. The simplest model of this form is for the steam turbine where P_m is equivalent to P_t, the local state variable for the turbine. The other state variables are ω_G for the generator (where $\omega_G \equiv \dot{\delta}$) and a for the governor. The full set of steam turbine-generator equations is:

$$\begin{aligned} M\dot{\omega}_G &= (e_t - D)\omega_G + P_t - P_G \\ T_u \dot{P}_t &= -P_t + k_t a \\ T_g \dot{a} &= -\omega_G - ra + \omega^{ref} \end{aligned} \qquad (12.\text{A}.2)$$

In these equations M is the inertia constant, e_T is a coefficient representing the turbine self-regulation, defined as $\partial P_t/\partial \omega_G$, D is the damping coefficient, T_u is the time constant representing the delay between the control valves and the turbine nozzles, k_t is a proportionality factor representing the control valve position variation relative to the turbine output variation, T_g is the time constant of the valve-servomotor-turbine gate system, and r is the permanent speed droop of the turbine. These parameters are defined in references [3, 16, 32]. ω^{ref} is the reference frequency set by the secondary controls, and so is assumed constant in the primary dynamics time scale. P_G is defined as an input to this system of equations.

Hydro-Turbine-Generator. A slightly more complex set of equations than that for the steam turbine is that for a hydro turbine-generator. This model follows the model for a low-head hydro facility developed in [3], with additional information for parameter values from [12, 38]. The state variables for this technology are ω_G for the generator equation, q for penstock flow, v for governor droop and a for gate position.

$$\begin{aligned}
M\dot{\omega}_G &= -(e_H + D)\omega_G + k_q q - k_w a - P_G \\
\dot{q} &= \omega_G/T_f - q/T_q + a/T_w \\
T_e \dot{v} &= -v + r'a \\
T_s \dot{a} &= -\omega_G + v - (r_h + r')a + \omega^{ref}
\end{aligned} \quad (12.A.3)$$

M and D are the inertia and damping constants as above. e_H, k_q and k_w are all ratios of constants from a standard hydro-turbine diagram referred to as the universal water turbine stead-state performance diagram (see for example Figure 8 in [3]), T_f, T_q, and T_w are also all ratios of constants from the same diagram, multiplied by T_c, the time constant of the penstock, T_e is the time constant of the valve-turbine gate system, T_s is the time constant of the servomotor gates, r_h is the permanent speed droop, and r' is the transient speed droop. These coefficients are contained in references [3, 17, 32].

Combustion-Turbine-Generator. The set of equations used for a combustion turbine are presented below. The equations represent the generator (ω_G), fuel controller (V_{CE}), and fuel flow (both W_F and $W_F dot$)

$$\begin{aligned}
M\dot{\omega}_G &= -D\omega_G + cW_F - P_G \\
b\dot{V}_{CE} &= -K_D\omega_G - V_{CE} + K_D\omega^{ref} \\
\dot{W}_F &= W_F dot \\
a\dot{W}_F dot &= aV_{CE} - \delta W_F - \beta W_F dot
\end{aligned} \quad (12.A.4)$$

These equations are derived from the equations and models found in [14, 13, 34]. M and D are the inertia and damping coefficients respectively. a, b and c are transfer function coefficients for the fuel system, and K_D is the governor gain. β and δ are algebraic functions of the parameters in the references, defined as $\beta \equiv b + c\tau_F$ and $\delta \equiv c + aK_F$, where τ_F is the fuel system time constant, and K_F is the fuel system feedback gain.

Combined Cycle Plant. A combined cycle combustion turbine, CCCT, plant has both a combustion turbine and steam turbine driving the synchronous generator. The hot exhaust gases from the combustion turbine, the first stage, are used to create steam in the boiler for the steam turbine. The model develop for the CCCT uses the equations for the fuel controller (V_{CE}), and the fuel flow (both W_F and $W_F dot$) from the CT model. The fourth equation represents the thermodynamic coupling between the turbines, using the air flow, W_{air} as the coupling variable. The fifth and sixth equations are for the steam turbine, where P_{ST} represents the mechanical power output from the steam turbine. The final equation is the generator output (swing equation), with the mechanical power from both the steam and combustion turbines as input.

$$M\dot{\omega}_G = -D\omega_G + (f_2 + P_{ST}) - P_G$$
$$b\dot{V}_{CE} = -K_D\omega_G - V_{CE} + K_D\omega^{ref}$$
$$\dot{W}_F = W_F dot$$
$$a\dot{W}_F dot = aV_{CE} - \gamma W_F - \beta W_F dot$$
$$T_v \dot{W}_{air} = d\omega_G + V_{CE} - W_{air} \quad (12.A.5)$$
$$\dot{P}_{ST} = P_{ST} dot$$
$$(T_M T_B)\dot{P}_{ST dot} = -p\omega_G + nW_F + mW_{air} - P_{ST} - (T_M + T_B)P_{ST dot}$$
$$(12.A.6)$$

The new parameters in this set of equations are T_v, the vane control time constant, d, the ratio of the fuel flow to rotor speed, T_M and T_B are time constants for a simplified steam turbine modeled in Figure 8 of [18], m and n represent the enthalpy in the mass flow of the air and fuel respectively, p is a function of the turbine exhaust temperature (see function f_1 in [34]), and the function f_2, also defined in [34], represents the turbine torque. This model was derived from the models in [6, 16, 18, 20, 34].

Wind Turbine – Induction Generator. The model for the wind turbine system is based substantially on the work in [5], which specifically developed a model to be used for dynamic studies of dispersed wind turbine applications. The model below differs from that model in that it has a single torque input, T_w (defined as the wind torque), rather than both T_w and $T_{turbine}$. Turbine torque is expressed in terms of the turbine inertia and wind torque.

The wind turbine system is modeled as two rotating masses—the turbine and generator rotors—coupled by a tortional spring. The three equations represent the induction generator, ω_G, the tortional spring, δ, and the wind turbine, ω_T. Note that the wind turbine system has no generator control, as in the other models, which is appropriate for a non-dispatchable technology.

$$\dot{\omega}_G = \frac{-(D_G - D_T)}{M_G}\omega_G + \frac{(D_G - D_T)}{M_G}\omega_T + \frac{1}{M_G}T_w - \frac{1}{M_G}P_G$$
$$\dot{\delta} = -\omega_G + \omega_T \quad (12.A.7)$$
$$\dot{\omega}_T = \frac{D_T}{M_T}\omega_G - \frac{K}{M_T} - \frac{D_T}{M_T}\omega_T + \frac{1}{M_T}T_w$$
$$(12.A.8)$$

Steam Turbine Parameters			
M	1.26	k_t	0.95
D	2.0	T_g	0.25
e_t	0.15	r	0.05
T_u	0.2		

Hydro Turbine Parameters			
M	1.5	T_q	0.72
D	2.0	T_w	0.76
e_h	-0.217	T_e	2.0
k_q	2.78	r'	0.40
k_w	1.52	T_s	0.10
T_f	-3.60	r_h	0.05

Combustion Turbine Parameters			
M	11.5	α	0.45
D	2.0	a	1.0
c	1.0	τ_F	0.40
K_D	25.0	K_F	0.0
b	0.05		

Wind Turbine Parameters			
M_G	5	D_T	1.0
M_T	11	K	400
D_G	0.8	s	-0.05

Table 12.A.1 Generator Model Parameters

M_G, M_T, D_G and D_T are the generator and turbine inertias and damping coefficients. T_w is the wind torque, and is an input to the system of equations, as is P_G, and K is the spring constant of the tortional spring used to model the drive train coupling between the two rotors. References [24, 41] were also used for developing this model.

Generator Model Parameter Values

The specific values for the parameters in the generator models, which are used in the system simulations in this thesis are presented in Table 12.A.1. In addition to the references cited in each individual section above, a number of people from industry assisted by providing parameter values, particularly for the inertias of the turbines and generators. These references are [1, 27, 33, 37, 39].

Bus i	Bus j	Branch Impedance $r_{ij}(\Omega)$	$x_{ij}(\Omega)$	Max. Load at Bus j $P(kW)$	$Q(kW)$
0	1	0.5096	1.7030	-	-
1	2	0.2191	0.0118	522	174
2	3	0.3485	0.3446	-	-
3	4	1.1750	1.0214	936	312
4	5	0.5530	0.4806	-	-
5	6	1.6625	0.9365	-	-
6	7	1.3506	0.7608	-	-
7	8	1.3506	0.7608	-	-
8	9	1.3259	0.7469	189	63
9	10	1.3259	0.7469	-	-
10	11	3.9709	2.2369	336	112
11	12	1.8549	1.0449	657	219
12	13	0.7557	0.4257	783	261
13	14	1.5389	0.8669	729	243
8	15	0.4752	0.4131	477	159
15	16	0.7282	0.4102	549	183
16	17	1.3053	0.7353	477	159
6	18	0.4838	0.4206	432	144
18	19	1.5898	1.3818	672	224
19	20	1.5389	0.8669	495	165
6	21	0.6048	0.5257	207	69
3	22	0.5639	0.5575	522	174
22	23	0.3432	0.3393	1917	639
23	24	0.5728	0.4979	-	-
24	25	1.4602	1.2692	1116	372
25	26	1.0627	0.9237	549	183
26	27	1.5114	0.8514	792	264
1	28	0.4659	0.051	82	294
28	29	1.6351	0.9211	882	294
29	30	1.1143	0.6277	882	294
$V_{rated} = 23kV$					

Table 12.A.2 Data for the 30 Bus Radial Distribution Test System

Distribution System Data

The data for the distribution test system that is used for the majority of the simulations in this thesis is presented in Table 12.A.2 and can be found in [11, 36]. The topology of the system is shown in Figure 12.10.

Dynamic Frequency Model

For the secondary dynamics, the state space for each generator is selected to consist of w_G and P_G. Models for the secondary dynamics are developed by assuming that the primary dynamics have settled, so that the time derivatives in the continuous time state equations are set equal to zero. Starting with the frequency model, the system of equations in (12.1) is solved with the left hand side set identically zero, resulting in

$$w_G = \sigma \frac{k_t}{r} w^{ref} - \sigma P_G \qquad (12.A.9)$$

Equation (12.A.9) is the droop equation for the steam turbine-generator. The droop coefficient, σ is defined as

$$\sigma \equiv \frac{r}{k_t + rD - re_t} \qquad (12.A.10)$$

Following the same procedure as for the steam turbine, the droop equation for the hydro turbine-generator plant has a similar format to that in Equation 12.A.9:

$$w_G = -\sigma_H c_2 w^{ref} - \sigma_H P_G \qquad (12.A.11)$$

with σ_H defined as

$$\sigma_H \equiv \frac{-1}{(D+e_H)} - \left(\frac{-k_q T_q}{T_f} + \frac{k_q T_q}{T_w r_H} + \frac{k_w}{r_h} \right) \qquad (12.A.12)$$

and c_2 defined as

$$c_2 \equiv \frac{-1}{r_h} \left(\frac{k_q T_q}{T_w} + k_W \right) \qquad (12.A.13)$$

The droop equation developed for the combustion turbine is

$$w_G = \left(\sigma_{CT} \frac{aK_D}{\gamma + aK_F} \right) w^{ref} - \sigma_{CT} P_G \qquad (12.A.14)$$

where

$$\sigma_{CT} \equiv \frac{\gamma + aK_F}{(\gamma + aK_F)D + aK_D} \qquad (12.A.15)$$

The droop equation for the combined cycle plant is

$$w_G = -\sigma_{cc} c_3 w^{ref} - \sigma_{cc} P_G \qquad (12.A.16)$$

where σ_{cc} is defined as

$$\sigma_{cc} \equiv \frac{1}{D - \frac{(1+n)aK_D}{\gamma + aK_F} + m(d - K_D) - p} \qquad (12.A.17)$$

and c_3 is

$$c_3 \equiv \frac{aK_D}{\gamma + aK_F} + mK_D + \frac{anK_D}{\gamma + aK_F} \qquad (12.A.18)$$

The droop equation for the wind model is

$$\omega_G = \sigma_W(T_w - P_G) \qquad (12.A.19)$$

with

$$\sigma_W \equiv \frac{-1}{s(D_G + D_T)} \qquad (12.A.20)$$

where s in Equation 12.A.20 is the slip of the induction generator, defined as

$$s \equiv \frac{\omega_G - \omega_T}{\omega_G} \qquad (12.A.21)$$

The next step is to develop the discrete time frequency model. Secondary dynamics evolve over a longer time frame than primary dynamics, with the controls acting only at specific time steps. As a result of this slower time evolution and response, secondary dynamics and control actions are modeled in discrete time, as opposed to the continuous time representation for primary dynamics. With the droop equations established, the next step is to develop the discrete time frequency equations. ω^{ref}, a constant at the primary dynamics time scale, is now a variable representing the secondary control. With $k = 0, 1, 2, \ldots$ representing the series of time steps at the secondary time scale, the droop equation (12.A.9) can be written in discrete time as

$$\omega_G[k] = \sigma \frac{k_t}{r}\omega^{ref}[k] - \sigma P_G[k] \qquad (12.A.22)$$

To bring the network interactions into the frequency model, the static form of the network coupling equation, Equation (12.4), is substituted into Equation (12.A.22), so that

$$\omega_G[k] = \Sigma \mathbf{r}^{-1}\mathbf{k_t}\omega^{ref}[k] - \Sigma \mathbf{K}_p \delta_G[k] - \sigma \mathbf{D}_p P_L[k] \qquad (12.A.23)$$

where Σ and the bolded coefficients $\mathbf{k_t}$ and $\mathbf{r}$ represent diagonal matrices of these coefficients for the individual generators.

Next, Equation (12.A.23) is subtracted at two sequential time steps to form the dynamic representation of the frequency model.

$$\begin{aligned}\omega_G[k+1] = {} & \omega_G[k] + \Sigma \mathbf{r}^{-1}\mathbf{k_t}(\omega^{ref}[k+1]) - \omega^{ref}[k]) \\ & - \Sigma \mathbf{K}_p(\delta_G[k+1] - \delta_G[k]) \\ & - \Sigma \mathbf{D}_p(P_L[k+1] - P_L[k])\end{aligned} \qquad (12.A.24)$$

Referring to Equation (12.A.24), the control signal is defined as

$$u[k] \equiv \omega^{ref}[k+1] - \omega^{ref}[k] \qquad (12.A.25)$$

where $u[k]$ is an implicit integral control. The change in load is defined as the system disturbance

$$d[k] \equiv P_L[k+1] - P_L[k] \qquad (12.A.26)$$

and finally, using the Euler approximation and defining T_s as the secondary time scale sampling interval

$$\omega_G[k+1] \approx \frac{\delta[k+1] - \delta[k]}{T_s} \qquad (12.\text{A}.27)$$

Using these definitions, Equation (12.A.24) can be rewritten as

$$\omega_G[k+1] + \Sigma \mathbf{K}_p T_s \omega_G[k+1] = \omega_G[k] + \Sigma \mathbf{r}^{-1} \mathbf{k_t}(u[k]) - \Sigma \mathbf{D}_p(d[k]) \quad (12.\text{A}.28)$$

The final set of definitions is

$$\mathbf{A}_s \equiv (1 + \Sigma \mathbf{K}_p T_s)^{-1} \qquad (12.\text{A}.29)$$
$$\mathbf{B}_s \equiv \mathbf{A}_s \Sigma \mathbf{r}^{-1} \mathbf{k_t} \qquad (12.\text{A}.30)$$

allowing the secondary dynamics frequency model to be expressed as

$$\omega_G[k+1] = \mathbf{A}_s \omega_G[k] + \mathbf{B}_s u[k] - \mathbf{A}_s \Sigma \mathbf{D}_p d[k] \qquad (12.\text{A}.31)$$

The subscript, s, designates the secondary time scale.

Identical derivations are performed for the other technologies, with the only differences being the form of the droop coefficient σ, and the matrix $\mathbf{B}_s$. These matrices for hydroelectric plants, are defined as

$$\mathbf{A}_{sH} \equiv (1 + \Sigma_H \mathbf{K}_p T_s)^{-1}$$
$$\mathbf{B}_{sH} \equiv \mathbf{A}_{sH}[-\Sigma_H(-r_h^{-1}(T_W^{-1} \mathbf{k_q} T_q + \mathbf{k}_W))] \qquad (12.\text{A}.32)$$

for combustion turbines as

$$\mathbf{A}_{sCT} \equiv (1 + \Sigma_{CT} \mathbf{K}_p T_s)^{-1}$$
$$\mathbf{B}_{sCT} \equiv \mathbf{A}_{sCT} \Sigma_{CT} \gamma + \mathbf{a} \mathbf{K}_F^{-1} \mathbf{a} \mathbf{K}_D \qquad (12.\text{A}.33)$$

for combined cycle facilities as

$$\mathbf{A}_{scc} \equiv (1 + \Sigma_{cc} \mathbf{K}_p T_s)^{-1}$$
$$\mathbf{B}_{scc} \equiv \mathbf{A}_{scc}(-\Sigma_{cc} c_3) \qquad (12.\text{A}.34)$$

and for wind turbines as

$$\mathbf{A}_{sW} \equiv (1 + \Sigma_W \mathbf{K}_p T_s)^{-1}$$
$$\mathbf{B}_{sW} \equiv 0 \qquad (12.\text{A}.35)$$

Note that since wind turbine systems do not have generator governor controls, the matrix $\mathbf{B}_{sW}$ is identically 0.

Real Power Model

The second state variable in this secondary level model is real power, P_G. To develop this model, Equation (12.A.22) is rearranged as

$$P_G[k] = \mathbf{r}^{-1} \mathbf{k_t} \omega^{ref}[k] - \Sigma^{-1} \omega_G[k] \qquad (12.\text{A}.36)$$

which can also be expressed as

$$P_G[k] = \Sigma^{-1}(\mathbf{A}_s^{-1}\mathbf{B}_s\omega^{ref}[k] - \omega_G[k]) \qquad (12.A.37)$$

Equation (12.A.37) is identical for each technology, using the appropriate matrices, $\mathbf{A}_s$ and $\mathbf{B}_s$, as defined above, and can be used to track changes in P_G that result from system disturbances and updates to ω^{ref}.

A dynamic model for real power is obtained by first writing the network coupling equation[24] at two consecutive time steps and subtracting. By using the approximation $\omega_G[k+1] \approx \frac{\delta[k+1]-\delta[k]}{T_s}$ and substituting the droop equation, (12.A.23) for ω_G, the model becomes

$$P_G[k+1] = (I - \mathbf{K}_p T_s \Sigma)P_G[k] + \mathbf{K}_p T_s \mathbf{A}_s^{-1}\mathbf{B}_s\omega^{ref}[k] + \mathbf{D}_p d[k] \qquad (12.A.38)$$

Discrete Time Price Models

The discrete time cost equation for the hydro electric generator was developed in the text. The model for the steam turbine is of the form

$$c_s[K] = (c_w - \frac{c_p k_t}{r} - \frac{c_a}{r})\omega_g[K] + (\frac{c_p k_t}{r} - \frac{c_a}{r})\omega^{ref}[K] + c_g P_G[K] \qquad (12.A.39)$$

where K represents the discrete time index for the price model. Substituting the generator droop equation, Equation (12.A.9), into (12.A.39) results in

$$\begin{aligned}c_s[K] &= (\sigma\frac{k_t}{r}c_w - \sigma\frac{k_t}{r}\frac{c_p k_t}{r} - \sigma\frac{k_t}{r}\frac{c_a}{r} + \frac{c_a}{r})\omega^{ref}[K] \\ &+ (c_g - \sigma c_w + \sigma\frac{c_p k_t}{r} + \sigma\frac{c_a}{r})P_G[K]\end{aligned} \qquad (12.A.40)$$

Defining

$$\begin{aligned}\gamma_{1s} &\equiv (\sigma\frac{k_t}{r}c_w - \sigma\frac{k_t}{r}\frac{c_p k_t}{r} - \sigma\frac{k_t}{r}\frac{c_a}{r} + \frac{c_a}{r}) \\ \gamma_{2s} &\equiv (c_g - \sigma c_w + \sigma\frac{c_p k_t}{r} + \sigma\frac{c_a}{r})\end{aligned} \qquad (12.A.41)$$

leads to Equation (12.A.40) being expressed as

$$c_s[K] = \gamma_{1s}\omega^{ref}[K] + \gamma_{2s}P_G[K] \qquad (12.A.42)$$

Equation (12.A.42) is almost of the form to be used in the closed loop price model. For the price model though, this equation is translated from the private cost equation to a market price equation, which for a competitive model requires only a change of variable from cost to price, where ρ is the variable used to designate price. For a competitive market the individual price equation is simply

$$\rho_s[K] = \gamma_{1s}\omega^{ref}[K] + \gamma_{2s}P_G[K] \qquad (12.A.43)$$

The format of this equation is identical for the other technologies, and differs only in the definitions of the coefficients, γ_1 and γ_2. For the other technologies these coefficients are defined as

$$\gamma_{1H} \equiv [(\frac{c_{gH}T_q}{T_w}\frac{1}{r_h} + (c_{vH}r' + c_{aH})\frac{1}{r_h}) - (c_{wH} + c_{gH}T_q(\frac{1}{T_f} - \frac{1}{T_w r_h})$$
$$-(c_{vH}r' + c_{aH})\frac{1}{r_h})\sigma_H c_2]$$

$$\gamma_{2H} \equiv [c_{gH} - (c_{wH} + c_{gH}T_q(\frac{1}{T_f} - \frac{1}{T_w r_h}) - (c_{vH}r' + c_{aH})\frac{1}{r_h})\sigma_H]$$

$$\gamma_{1CT} \equiv [(c_{wCT} - c_{vCT}K_D - \frac{c_{fCT}aK_D}{\gamma})(\sigma_{CT}\frac{aK_D}{\gamma})$$
$$+(c_{vCT}K_D + \frac{c_{fCT}aK_D}{\gamma})]$$

$$\gamma_{2CT} \equiv [c_{gCT} - \sigma_{CT}(c_{wCT} - c_{vCT}K_D - \frac{c_{fCT}aK_D}{\gamma})] \quad (12.\text{A}.44)$$

$$\gamma_{1CC} \equiv [c_{wcc} - c_{vcc}K_D - c_{fcc}\eta + c_{acc}(d - K_D)$$
$$+ c_p(md - mK_D - p - n\eta)]\sigma_{cc}c_3$$
$$+ [c_{vcc}K_D + c_{fcc}\eta + c_{acc}K_D + c_p m K_D + n\eta]$$

$$\gamma_{2CC} \equiv c_{gcc} - [c_{wcc} - c_{vcc}K_D - c_{fcc}\eta + c_{acc}(d - K_D)$$
$$+ c_p(md - mK_D - p - n\eta)]\sigma_{cc}$$

where η for the combined cycle plant is defined as $\frac{aK_D}{\gamma}$, and the other parameters are defined above. Note that no price control equation is developed for wind turbines since they do not have primary controls which could respond to the price feedback signal. They could be incorporated directly into the price framework using this same method, if they were to be equipped with primary control technology.

To form the dynamic model, Equation (12.A.43) is written for two sequential time steps and subtracted. The dynamic equation for the price of energy supplied at a generator, thus is

$$\rho[K+1] = \rho[K] + \gamma_1(\omega^{ref}[K+1] - \omega^{ref}[K]) + \gamma_2(P_G[K+1] - P_G[K]) \quad (12.\text{A}.45)$$

As with the secondary frequency control, the control for price through ω^{ref} is again seen to be implicit integral control, such that

$$u_\rho[K] \equiv \omega^{ref}[K+1] - \omega^{ref}[K] \quad (12.\text{A}.46)$$

or

$$\omega^{ref}[K+1] = \omega^{ref}[K] + u_\rho \quad (12.\text{A}.47)$$

The significance of using ω^{ref} as the control variable is that the proposed price model integrates the existing local generator control (i.e. the governor for frequency control) into the closed loop price feedback structure. Three variations of the dynamic price model can be developed from this point, differing in the selection of the state variables and the input variables.

In the first version of the price model, the state space is ρ, the vector of price variables from each generator, and the system input is the vector of the changes in real power, ΔP_G, at each generator. The deviations from the scheduled P_G at each bus result from exogenous system disturbances, such as a fluctuations in demand or stochastic resource inputs. This model is

$$\rho[K+1] = \rho[K] + G_1 u_\rho[K] + G_2(P_G[K+1] - P_G[K]) \qquad (12.A.48)$$

where the matrices G_1 and G_2 are diagonal matrices of the coefficients γ_1 and γ_2 for each generator.

The second possible dynamic price model is developed with both ρ and ω_G as state variables, and the actual system disturbance, e.g. ΔP_L or ΔT_w, as the input. To obtain this model, the network coupling equation is substituted into Equation (12.A.43). The dynamic form of the price equation for this model then becomes

$$\begin{aligned}\rho[K+1] &= \rho[K] + G_1 u_\rho[K] + G_2 \mathbf{K}_p(\delta_G[K+1] - \delta_G[K]) \\ &\quad + G_2 \mathbf{D}_p(P_L[K+1] - P_L[K])\end{aligned} \qquad (12.A.49)$$

Defining T_ρ as the sampling rate for the price signal, and using the approximation that $\omega_G[K+1] \approx \frac{\delta_G[K+1] - \delta_G[K]}{T_\rho}$, Equation (12.A.49) becomes

$$\begin{aligned}\rho[K+1] &= \rho[K] + G_2 \mathbf{K}_p T_\rho \mathbf{A}_s \omega_G[K] + (G_1 + G_2 \mathbf{K}_p T_\rho \mathbf{B}_s) u_\rho[K] \\ &\quad + G_2(I + \mathbf{K}_p T_\rho \mathbf{A}_s \Sigma) \mathbf{D}_p d[K]\end{aligned} \qquad (12.A.50)$$

The second state equation for this model is the frequency state equation, (12.A.31). The full model is then expressed as

$$\begin{bmatrix} \rho \\ \omega_G \end{bmatrix}_{[K+1]} = \begin{bmatrix} I & G_2 \mathbf{K}_p T_\rho \mathbf{A}_s \\ 0 & \mathbf{A}_s \end{bmatrix} \begin{bmatrix} \rho \\ \omega_G \end{bmatrix}_{[K]} + \begin{bmatrix} G_1 + G_2 \mathbf{K}_p T_\rho \mathbf{B}_s \\ \mathbf{B}_s \end{bmatrix} u_\rho[K]$$
$$+ \begin{bmatrix} G_2(I + \mathbf{K}_p T_\rho \mathbf{A}_s \Sigma) \mathbf{D}_p \\ \mathbf{A}_s \Sigma \mathbf{D}_p \end{bmatrix} d[K] \qquad (12.A.51)$$

or

$$\begin{bmatrix} \rho \\ \omega_G \end{bmatrix}_{[K+1]} = \mathbf{A}_r \begin{bmatrix} \rho \\ \omega_G \end{bmatrix}_{[K]} + \mathbf{B}_r u_\rho[K] + \mathbf{D}_r d[K] \qquad (12.A.52)$$

The third variation of the price model returns to ρ as the state variable, while changing the input variable to $\Delta \omega_G$. To obtain this model, for the steam turbine equations, Equation (12.A.37) is substituted into Equation (12.A.39) forming

$$\rho[K] = (\frac{c_p k_t}{r} + \frac{c_a}{r} + \frac{c_{gs} k_t}{r})\omega^{ref}[K] + (c_{ws} - \frac{c_p k_t}{r} + \frac{c_a}{r} + \frac{c_{gs}}{\sigma})\omega_G[K] \qquad (12.A.53)$$

By defining

$$\gamma_{3s} \equiv \frac{c_p k_t}{r} + \frac{c_a}{r} + \frac{c_{gs} k_t}{r}$$

$$\gamma_{4s} \equiv c_{ws} - \frac{c_p k_t}{r} + \frac{c_a}{r} + \frac{c_{gs}}{\sigma} \qquad (12.A.54)$$

and subtracting the equation at two consecutive time steps, Equation (12.A.53) can be written in dynamic form as

$$\rho[K+1] = \rho[K] + \gamma_{3s} u_\rho[K] + \gamma_{4s}(\omega_G[K+1] - \omega_G[K]) \qquad (12.A.55)$$

For the other technology types, the coefficients γ_3 and γ_4 are defined as

$$\gamma_{3H} \equiv \left(\frac{c_{gH} T_q}{T_w}\frac{1}{r_h} + (c_{vH} r' + c_{aH})\frac{1}{r_h}\right)c_{gH}c_2$$

$$\gamma_{4H} \equiv (c_{wH} + c_{gH} T_q(\frac{1}{T_f} - \frac{1}{T_w r_h}) - (c_{vH} r' + c_{aH}) - \frac{c_{gH}}{\sigma_H}$$

$$\gamma_{3CT} \equiv c_{vCT} K_D + \frac{c_{fCT} a K_D}{\gamma} + \frac{c_{gCT} a K_D}{\delta}$$

$$\gamma_{4CT} \equiv c_{wCT} - c_{vCT} K_D - \frac{c_{fCT} a K_D}{\gamma} - \frac{c_{gCT}}{\delta} \qquad (12.A.56)$$

$$\gamma_{3CC} \equiv c_{vcc} K_D - c_{fcc}\eta + c_{acc} K_D + c_p(md + n\eta) + c_{gcc} c_3$$

$$\gamma_{4CC} \equiv c_{wcc} - c_{vcc} K_D - c_{fcc}\eta + c_{acc}(d - K_D)$$

$$+ c_p(md - mK_D - p - n\eta) - \frac{c_{gcc}}{\sigma_{cc}}$$

For a system with more than one generator, the coefficients γ_3 and γ_4 are written as diagonal matrices G_3 and G_4.

Notes

1. Occasionally a larger customer may self-generate or cogenerate a portion of their own load.
2. A fully distributed utility architecture is one where no incentives can be found, whether to add a technology or implement an additional operating strategy, which will improve the system performance [22]. (Efficiency refers to both technological and economic efficiency.) New generating and control technologies will continually improve system efficiency, but at any given point in time there is a maximum achievable efficiency. The distributed utility architecture is not unique, but instead is defined for a specific system at a given point in time, as a function of the current system architecture and the existing technologies.
3. Regulation refers to the engineering control functions, and not the economic idea of industry regulation.
4. Ancillary services are compensation for losses, load frequency control, automatic generation control, voltage support, spinning reserves, scheduling and unit commitment, and monitoring and control.
5. These models, along with simulations and analysis of system behavior, are developed more fully in [30, 7, 8, 23, 4].
6. The reasons for selecting P_G are discussed in [23].

7. The state variables in these models and the parameter values are defined in detail in the Appendix.

8. All the graphs in this chapter are of *deviations* from equilibrium, and *not* absolute values. This is follows directly from the use of linearized state space models which represent small-signal dynamics around an equilibrium point. The deviations from equilibrium are the quantities of interest in this chapter, since the bulk power flows (the equilibrium) are assumed to be established by the system operator. The regulation and system stability questions addressed in this chapter are inherently small-signal issues, and so readily analyzed with linearized state space models.

9. In actual system operation with hydro-electric facilities, there are conditions which are known to cause unstable behavior. Two such situations are identified in [26], pages 396 and 752.

10. Observation with respect to central facility governors was made by Professor George Verghese, MIT.

11. If a DG unit is owned by a distribution utility, and so is not part of the competitive market, then it is possible that it will be used for voltage support regardless of whether there are industry mechanisms to pay it directly for this service.

12. The real time costs referred to here could be those calculated based on estimations of what the costs will be, or on actual real time data.

13. Including both the fuel cost and the costs to the system for compensating for the deviations, for example those costs associated with call contracts.

14. As part of FERC Order 888, the Federal Government is overseeing the development of an on-line information system called OASIS, Open Access Same-Time Information System, which may eventually serve as a type of bulletin board for spot prices for electric energy.

15. A methodolgy for pricing voltage support remains a complex, unresolved issue however.

16. At the time scale of primary dynamics, an increase in generator speed ω_G is correlated to a decrease in power output, P_G, and visa versa. Thus if the generator speed changes there is a non-zero affect on cost, linked through P_G. This inverse relationship between ω_G and cost, c, is represented as c_w.

17. Before proceeding with the development of the price model it is important to establish the relationship between cost and price. The total cost of producing a product is the sum of the actual cost to the firm of all the inputs, labor, equipment, maintenance, etc. In contrast, the price of a product is the amount charged by the firm, and which is seen by other participants in the industry. Price is related to cost to a greater or lesser degree depending on the nature of the industry. A competitive industry is identified by the fact that price is identical to the industry's marginal cost. For other industry structures though, price is determined based on other variables, with cost acting as a lower boundary.

18. Note that the plot for the price deviation of wind is not on the figure since wind does not participate in the closed loop price framework.

19. Some of this technical data could eventually be made available through a system such as OASIS, which would improve access to this required information.

20. Developing standards for primary control designs in general will remain a difficult research question for some time to come.

21. If a DG unit is owned by a distribution utility, and so is not part of the competitive market, then it is possible that it will be used to supply ancillary services regardless of whether there are industry mechanisms to pay it directly or not.

22. Decoupled here refers to the assumption that for small disturbances frequency and voltage dynamics are essentially independent, and are related to real power and reactive power respectively.

23. 'Coupling' here refers to the physical connection of the generators with each other by means of the distribution system.

24. This equation is the static form of Equation 12.4, $P_G = \mathbf{K}_P \delta_G + \mathbf{D}_P P_L$.

References

[1] Agee, J. (1996). United States Bureau of Reclamation, personal correspondence.

[2] Baran, M. E., Zhu, J., and Kelley, A. W. (1996). Meter placement for real-time monitoring of distributiong feeders. *IEEE Transactions on Power Systems*, 11(1).

[3] Calović, M. (1971). Dynamic state-space models of electric power systems. Technical report, University of Illinois, Urbana, Illinois.

[4] Cardell, J. (1995). Integrating small scale distributed generation into a deregulated market: Control strategies and price feedback. Technical Report LEES TR95-009, Massachusetts Institute of Technology. MIT PhD Thesis.

[5] Chedid, R., LeWhite, N., and Ilić, M. (1993). A comparative analysis of dynamic models for performance calculation of grid-connected wind turbine generators. *Wind Engineering*, 17(4).

[6] de Mello, F. P. (1991). Boiler models for system dynamic performance studies. *IEEE Transactions on Power Systems*, 6(1).

[7] Eidson, B. (1995). Estimation and hierarchical control of market- driven electric power systems. Technical Report LEES TR95-009, Massachusetts Institute of Technology. MIT PhD Thesis.

[8] Eidson, B. and Ilić, M. (1995). Advanced generation control: Technical enhancements, costs, and responses to market-driven demand. In *Proceedings of the American Power Conference*, Chicago, IL.

[9] El-Hawary, M. E. and Mbamalu, F. (1989). Stochastic optimal load flow using a combined quasi-Newton and conjugate gradient technique. *Electric Power and Energy Systems*, 11(2).

[10] Electric Power Research Institute (1993). Dispersed system impacts: Survey and requirements study, final report. Technical Report TR-103337, Electric Power Research Institute.

[11] Grainger, J. J. and Civanlar, S. (1985). Volt/var control on distribution systems with lateral branches using shunt capacitors and voltage regulators, Part III: The numerical results. *IEEE Transactions in Power Apparatus and Systems*, 104(11).

[12] Hannett, L. N., Feltes, J., and Fardanesh, B. (1994). Field tests to validate hydro-turbine governor model structure and parameters. *IEEE Transactions on Power Systems*, 9(4).

[13] Hannett, L. N., Jee, G., and Fardanesh, B. (1995). A governor/turbine model for a twin-shaft combustion turbine. *IEEE Transactions on Power Systems*, 10(1).

[14] Hannett, L. N. and Khan, A. (1993). Combustion turbine dynamic model validation from tests. *IEEE Transactions on Power Systems*, 8(1).

[15] Harvard (1997). Transcript from the Harvard Electricity Policy Group's Twelfth Plenary Session. Harvard Electricity Policy Group.

[16] IEEE (1991). Dynamic models for fossil fueled steam units in power system studies. *IEEE Transactions on Power Systems*, 6(2). IEEE Working

Group on Prime Mover and Energy Supply Models for System Dynamic Performance Studies.

[17] IEEE (1992). Hydraulic turbine and turbine control models for system dynamic studies. *IEEE Transactions on Power Systems*, 7(1). IEEE Working Group on Prime Mover and Energy Supply Models for System Dynamic Performance Studies.

[18] IEEE (1994). Dynamic models for combined cycle plants in power system studies. *IEEE Transactions on Power Systems*, 9(3). Working Group on Prime Mover and Energy Supply Models for System Dynamic Performance Studies.

[19] IEEE Distribution Planning (1991). Radial distribution test feeders. *IEEE Transactions on Power Systems*, 6(3). IEEE Distribution Planning Working Group Report.

[20] IEEE Working Group (1773). MW response or fossil fueled steam units. *IEEE Transactions in Power Apparatus and Systems*, PAS-92:455–463. IEEE Working Group on Powre Plant Response to Load Changes.

[21] Ilić, M. and Liu, X. (1993). A simple structural approach to modeling and analysis of the inter-area dynamics of the large electric power systems: Part I - linearized models of frequency dynamics. In *North American Power Symposium*, pages 560–569.

[22] Ilić, M., Tabors, R., and Chapman, J. (1994). Conceptual design of distributed utility system architecture: Final report. Technical report, Massachusetts Institute of Technology.

[23] Ilić, M. D. and Liu, S. X. (1996). *Hierarchical Power Systems Control: Its Value in a Changing Industry*. Springer-Verlag, London.

[24] Ju, P., Handschin, E., Wei, Z., and Schlucking, U. (1996). Sequential parameter estimation of a simplified induction motor load model. *IEEE Transactions on Power Systems*, 11(1).

[25] Kersting, W. and Phillips, W. H. (1992). Modeling and analysis of rurla electric distribution feeders. *IEEE Transactions on Industry Applications*, 28(4).

[26] Kundur, P. (1994). *Power system stability and control*. McGraw-Hill, The EPRI power system engineering series, New York.

[27] Ledger, D. (1996). General Electric Corporation, personal correspondence.

[28] Lee, W. J. et al. (1995). Dynamic stability analysis of an industrial power system. *IEEE Transactions on Industry Applications*, 31(4).

[29] Lee, W.-J., Gim, J.-H., Chen, M.-S., Wang, S.-P., and Li, R.-J. (1997). Development of a real-time power system dynamic performance monitoring system. *IEEE Transactions on Industry Applications*, 33(4).

[30] Liu, X. (1994). *Structural Modeling And Hierarchical Control of Large-Scale Electric Power Systems*. Doctor of Philosophy, Massachusetts Institute of Technology.

[31] Perez-Arriaga, I., Verghese, G., Pagola, L., Sancha, J. L., and Schweppe, F. (1990). Developments in selective modal analysis of small-signal stability in electric power systems. *Automatica*, 26(2).

[32] Report, I. C. (1973). Dynamic models for steam and hydro turbines in power system studies. *IEEE Transactions in Power Apparatus and Systems*, 92(6).

[33] Rini, M. (1996). ABB Power Corporation, personal correspondence.

[34] Rowen, W. I. (1983). Simplified mathematical representations of heavy-duty gas turbines. *Journal of Engineering for Power*, 105.

[35] Salman, S., Jiang, F., and Rogers, W. (1994). Effects of wind power generators on the voltage control of utility distribution networks. *Wind Engineering*, 18(4).

[36] Santoso, N. I. and Tan, O. T. (1989). Neural-net based real-time control of capacitors installed on distribution systems. *IEEE Transactions on Power Delivery*, 5(1).

[37] Simpson, J. (1996). Stone & Webster, personal correspondence.

[38] United States (1976). Selecting hydraulic reaction turbines. Water Resources Tecnical Publication Engineering Monograph No. 20, United States Department of the Interior Bureau of Reclamation.

[39] Widdinger, R. (1996). United States Army Corp of Engineers, personal correspondence.

[40] Willis, H. L. and Rackliffe, G. B. (1994). *Introduction to Integrated Resource T&D Planning*. ABB Power T&D Company Inc., Raleigh, NC.

[41] Xu, L. (1992). Dynamic model of an integral cycle controlled single-phase induction machine. *IEEE Transactions on Energy Conversion*, 7(4).

13 APPLICATION OF DYNAMIC GENERATION CONTROL FOR PREDATORY COMPETITIVE ADVANTAGE IN ELECTRIC POWER MARKETS

Thomas Gorski and Christopher DeMarco

Department of Electrical and Computer Engineering
University of Wisconsin-Madison
Madison, WI USA

ABSTRACT

This work extends investigations into the potential for coordinated control of a group of generators to act in a manner that destabilizes other machines, while maintaining nearly satisfactory performance within this control group. This control scheme raises the specter that dynamic control of generators might be used to gain predatory competitive advantage. To assess the risk of this behavior, it is useful to more carefully examine its technical feasibility. The work to be presented here will examine two questions in this context. First, to what extent is this control robust with respect to inclusion of realistic dynamic models, including the effect imprecisely known data describing the network and competitive generator units? Second, to what extent can this control be implemented using only local measurement information, such as that available from generator terminal measurements? This work demonstrates that the addition of governor and turbine dynamics into the model does not impede the ability of the feedback control to act in the predatory manner. It also demonstrates the possibility that the system state can be adequately recreated from a localized measurement set, such that this estimated state can act as a feedback source. Moreover, for the case study examined, the combined state observer and feedback design are shown to be quite robust, even when the design is based upon a model having parameter errors on the order of 20%.

INTRODUCTION

Evolving regulatory and business changes for electric utilities in the US and world-wide seek to force generating units to operate in a competitive manner. With this move towards competition among generators, it is generally assumed that the effects related to dynamic control operate on a time scale for which economic effects are less significant, and that relevant aspects of control performance can be monitored in a straightforward fashion by an independent system operator or equivalent entity. However, the role of a system operator in monitoring the performance of generators remains an on-going topic of discussion in the US. This debate focuses in part on market mechanisms to encourage certain types of performance, and in part on enforceable, system related control performance standards. At the risk of stereotyping these two approaches, the former may be largely associated with those individuals possessing a market economist's viewpoint, the latter with individuals having the perspective of the traditional utility engineer.

Against this backdrop, this work is intended as a cautionary note to indicate that both perspectives may be overlooking the danger that exists for a subtle, high risk form of anti-competitive behavior. In many ways, a synchronous electric power grid presents dynamic features unlike those of any other market. In a power network, two rival sets of generating units interact not only through the market itself, but their electro-mechanical performance can be tightly coupled through the electric grid. Therefore, the dynamic governor and excitation control exercised at one machine can have large impact on the dynamic response of other generating units, and on the network as a whole. Previous work [1] has demonstrated that, in simplified system models at least, it is plausible that a predatory market participant could design controls that target a competitors to experience unstable oscillations, while maintaining acceptable performance within the machines under this participant's control. In particular, that work demonstrated an algorithm for constructing a linear feedback control that destabilizes one mode of the system in such a way that (almost) all machines in the group exercising the control have zero participation in this mode. This behavior would be high risk, but if cleverly deployed, it could be extremely difficult to police against. Hence, it is important to more accurately assess its technical feasibility, and more generally, the plausibility that such control could be implemented.

The work to be presented here will examine two questions in this context. First, to what extent is this control robust with respect to more realistic dynamic models, including the effect imprecisely known data describing the network and competitive generator units? Second, to what extent can this control be implemented using only local measurement information, such as that available from generator terminal measurements? More specifically, we will examine the impact of incorporating governor and turbine dynamics into the classical linearized model, and what effect if any this has on the eigenvalue/eigenvector placement algorithm of [1],[2].

As its other focus, this work will examine the implementation of the control scheme based on a limited measurement subset. Whereas the prior work established that use of a dynamic state estimator in state feedback should preserve the components of the eigenvectors, this work explores the definition of a minimum practical measurement set and the impact of parameter errors in the estimator model. It also explores the extent to which the introduction of such errors causes a "smearing" of the eigenvector for the unstable mode onto previously non-participating machines.

TEST NETWORK AND MODEL

The test network employed in this work is the same as that used in [1], a (slightly modified) IEEE 14-bus network, as shown in Figure 1. Generators appear at buses 1, 2, 3, 6, and 8. No infinite bus is employed. Generator parameters are hypothetical, and are shown in Table 1. The network is specified on a 100 MVA base, with a total active load of 1.85 pu plus losses, allocated as shown in Table 2. The widest total spread of machine angles across the network is approximately 17 degrees, between buses 1 and 9, indicating a very light overall network loading. It is from this steady state operating point that the linearized model is produced. The reader is encouraged to note this light loading, as its implies that the destabilization to be achieved is not dependent upon a highly stressed operating condition.

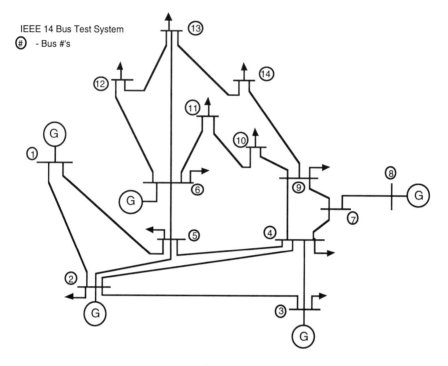

Figure 1: One-Line Diagram of 14 Bus System

Bus	H (MJ/S$_{base}$)	T$_G$ (sec)	T$_T$ (sec)
1	34	0.1	0.3
2	25	0.1	0.3
3	25	0.1	0.18
6	10	0.09	0.15
8	10	0.08	0.2

Table 1: Simulated Machine Parameters in 14-bus Network, on 100 MVA Base

Bus	Voltage (p.u.)	P (p.u.)	Q (p.u.)
1	1.02	-0.56	0.14
2	1.02	-0.56	0.13
3	1.01	-0.39	0.29
4	1.02	0.30	-0.55
5	1.01	0.30	0.48
6	1.02	-0.22	-0.25
7	1.01	0.30	-0.12
8	1.02	-0.22	-0.062
9	1.00	0.25	0.23
10	1.00	0.25	-0.15
11	0.99	0.25	0.071
12	1.02	0.05	-0.017
13	1.02	0.05	-0.040
14	1.02	0.15	-0.15

Table 2: Steady State Operating Point for 14-bus Network, Positive Load Convention, 100 MVA Base

LINEAR DYNAMIC MODEL

Figure 2 shows a block diagram for the linearized state model. First order linear blocks are used to model governor and turbine dynamics, as in sec. 9.5 of [3]. Relevant time constants T_G and T_T used here are shown in Table 1. The state dimension is twenty, consisting of five governor deviations $\Delta\phi$, five turbine deviations $\Delta\theta$, five frequency deviations $\Delta\omega$, and five machine angle deviations $\Delta\delta$, all ordered consistent with bus numbering. Note that absolute bus angle deviations are maintained as states. For this study, the control group consists of the machines at buses 1, 2, and 3.

In the block diagram, the matrix **N** represents the linearized relation between machine angle deviations and power drawn by the network from that machine (as obtained via suitable block manipulations on the power flow Jacobian). This matrix is a reduced-state representation of the full network, in which the dependent variables (load bus δ's and V's) are eliminated by using the algebraic active and reactive power balance constraints at those buses. The **R** matrix represents the droop coefficients, indicating the use of a "distributed slack" control. In addition to the basic dynamic model, the block diagram shows the locations of the four state feedback blocks, F_ω, F_δ, F_θ, and F_ϕ.

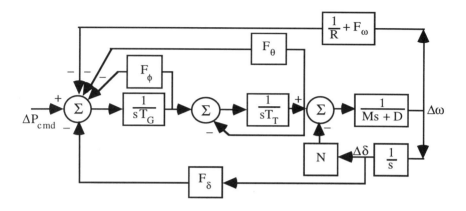

Figure 2: Linearized Model of Generation, Including State Feedback (F) Paths

STATE EQUATIONS

Figure 3 shows the dynamic equations governing plant behavior for the full state space, prior to adding state feedback. Diagonal matrices representing machine parameters, such as droop (**R**), normalized inertia (**M**), normalized damping (**D**), and time constants (T_T^{-1} and T_G^{-1}), are of size 5 x 5, corresponding to the five synchronous machines of the sample system. Only **N**, which represents the network behavior, is non-diagonal. Matrix $\hat{T}_G^{-1}$ multiplies the input function. It consists of the first three rows and columns of T_G^{-1} corresponding to the three machines within the control group.

$$\begin{bmatrix} \Delta\dot{\phi} \\ \Delta\dot{\theta} \\ \Delta\dot{\omega} \\ \Delta\dot{\delta} \end{bmatrix} = \begin{bmatrix} -T_G^{-1} & 0 & -T_G^{-1}R^{-1} & 0 \\ T_T^{-1} & -T_T^{-1} & 0 & 0 \\ 0 & M^{-1} & -M^{-1}D & -M^{-1}N \\ 0 & 0 & I & 0 \end{bmatrix} \begin{bmatrix} \Delta\phi \\ \Delta\theta \\ \Delta\omega \\ \Delta\delta \end{bmatrix} + \begin{bmatrix} \hat{T}_G^{-1} \\ 0 \\ 0 \\ 0 \end{bmatrix} \Delta P_{cmd}$$

Figure 3: Dynamic State Equations for Augmented Machine Model

Table 3 shows the eigenvalues for the linearized system without feedback. Note the presence of a zero eigenvalue. It appears as a result of the choice to maintain absolute versus relative machine angles, and is caused by the inherent rank deficiency in the network (**N**) matrix.

CONTROLLABILITY

A necessary step prior to the construction of the state feedback matrix is to establish that the modes of interest are controllable. Once controllability has been established, it should then be straightforward to apply the algorithm to generate the state feedback vector.

Perhaps the most common "textbook" approach for establishing controllability [4], [5] is the rank test:

$$\text{Rank } \mathbf{M}_c = \text{Rank } [\mathbf{B}|\mathbf{AB}| \ldots |\mathbf{A}^{n-1}\mathbf{B}] = n$$

where **A** is the state matrix and **B** the input weighting matrix, defined for our example in Figure 3. However, this approach proves inconvenient here due to the choice of coordinate system, which introduces an inherent rank deficiency in **A**.

Eigenvalues
-0.0531 +/- 6.7086j
-0.0425 +/- 5.8794j
-0.0730 +/- 5.3507j
-0.1017 +/- 4.6457j
0.0000
-1.2950
-2.3026
-5.6884
-3-9384
-4.4121
-4.5256
-10.2948
-10.7425
-10.5825
-11.9072
-12.9369

Table 3: Eigenvalues for Linearized System Prior to Addition of State Feedback

The alternative, appropriate to diagonalizable ("semi-simple") **A**, transforms **A** into block diagonal canonical form through the transformation

$$\mathbf{A'} = \mathbf{T}^{-1}\mathbf{A}\mathbf{T} \tag{1}$$

where the columns of **T** are obtained from the eigenvectors of A as follows. For the ith eigenvalue λ_i real, $\mathbf{t}_i$, the ith column of **T**, is simply the corresponding eigenvector $\mathbf{v}_i$. For λ_i complex, with positive imaginary part, $\mathbf{t}_i := \text{Re}[\mathbf{v}_i]$. For λ_i complex with negative imaginary part, $\mathbf{t}_i := -\text{Im}[\mathbf{v}_i]$. The resulting transformed matrix is of the form

$$\mathbf{A'} = \begin{bmatrix} \sigma_1 & \omega_1 & & & & 0 \\ -\omega_1 & \sigma_1 & & & & \\ & & \sigma_2 & \omega_2 & & \\ & & -\omega_2 & \sigma_2 & & \\ & & & & \ddots & \\ 0 & & & & & \lambda_n \end{bmatrix} \quad (2)$$

Each transformed state or pair of states represents a real or complex mode and can be directly affected by the input ΔP_{cmd} only if the corresponding row or rows in the transformed matrix $\mathbf{B'} = \mathbf{T}^{-1}\mathbf{B}$ are non-zero. Thus, this transformation has the useful property of directly identifying the uncontrollable modes. Applying this transformation to our example system, one finds that the largest 2-norm of any row of $\mathbf{B'}$ is approximately 40, while the smallest norm is approximately 0.05. These results indicate that all modes of the augmented linearized model are indeed controllable.

STATE FEEDBACK FORMULATION FOR EIGENVECTOR PLACEMENT

Having obtained a state matrix $\mathbf{A}$ for the linearized model and established that it is fully controllable, the algorithm for generating the state feedback may now be deployed. In the eigenvalue/eigenvector placement algorithm introduced in [1], the state space contained only machine frequency and angle deviations, with the assumption that the "sacrificial machine" frequency variable was assigned index 1. Furthermore, it was noted there that if the state feedback causes a frequency deviation to have zero participation in the unstable mode, then the angle deviation must have zero participation as well.

In work reported here, the control objective remains that of maintaining zero participation of the frequency and angle deviations in unstable mode for all control machines other than the sacrificial machine. As a matter of bookkeeping, we note that the frequency deviation variables are not ordered as in the previous work. The change in the nature of the algorithm is minor. Rather than choosing the submatrix $\mathbf{N}_{\hat{\lambda}_{1c}}$, which consists of the first m rows of $\mathbf{N}_{\hat{\lambda}_1}$, we simply select those m rows in $\mathbf{N}_{\hat{\lambda}_1}$ that correspond to the indices of the control group frequency deviations.

MODE SELECTION

Given the comparatively small size of the network and control machines in this group, it is possible to locate a good candidate mode and sacrificial machine combination through sampling the small set of possibilities (12 total). In larger networks, it is readily foreseeable that such "trial-and-error" techniques could impose an impractically large computational burden.

With that in mind, the following enhancement to the trial-and-error approach was examined, with generally good results. Let

$$\mathbf{A}(\gamma) = \mathbf{A} - \gamma \mathbf{BF} \qquad (3)$$

where $\mathbf{F}$, the state feedback matrix, is computed for a chosen candidate sacrificial machine and mode. $\mathbf{A}(0)$ therefore represents the state matrix without feedback, and $\mathbf{A}(I)$ the state matrix with feedback applied. An estimate of the eigenvectors for the unstable mode of $\mathbf{A}(I)$ is calculated as a first-order expansion about $\mathbf{A}(0)$, using the known eigenvectors of $\mathbf{A}(0)$, $(\partial \mathbf{A}/\partial \gamma) = -\mathbf{BF}$, and standard results (using the implicit function theorem) allow the computation of the eigenvector derivatives with respect to γ. This approximation can act as a pre-filter to predict those mode/machine combinations that might produce the desired degree of participation.

Bus	Before Feedback	After Feedback	$\partial \mathbf{v}/\partial \gamma$
1	0.4651∠-137.1°	0.2891∠-166.0°	0.7974∠-123.7°
2	0.1335∠-147.6°	0	0.2225∠+97.33°
3	0.8369∠+41.24°	0	0.7455∠-153.3°
6	0.0701∠-162.1°	0.5001∠+160.3°	2.345∠-153.9°
8	0.06877∠-170.9°	0.6677∠+157.2°	3.163∠-153.8°

Table 4: Eigenvectors for Frequency Deviations before and after Destabilization of Selected Mode $\lambda = -0.1017$ +/- $4.6457j$

Table 4 shows the eigenvectors before and after the addition of state feedback as well as the computed $\partial \mathbf{v}/\partial \gamma$ values for the frequency deviation states, when the mode λ = -0.1017 +/- 4.6457j is destabilized to λ = +0.1 +/- 4.6457j, with bus 1 selected as the sacrificial machine. Prior to state feedback application, the eigenvector magnitudes for the machines at buses 6 and 8 are comparatively small, indicating a low degree of participation in this mode as compared to the control group machines. By itself this would suggest that this is not a good candidate mode. However, with state feedback applied, we observe that the eigenvector magnitudes for these same machines increase dramatically, to over twice that of the sacrificial machine. In examining the directions and magnitudes of movement projected by the eigenvalue derivatives, we observe that for the sacrificial machine, movement is in the same general direction, with moderate velocity. For the other machines in the control group, movement is generally towards the origin, with an amplitude roughly proportional to the component size. For the machines outside the control group, movement is about 30° clockwise to that for the sacrificial machine, but at dramatically larger velocity. From the original eigenvector and derivatives. we would therefore predict an eigenvector with state feedback that displays components for machines 6 and 8 whose complex values show angles of approximately 30° clockwise relative to that for machine 1, and indeed this is the case.

As an additional check, on this prediction technique, we examine the eigenvalue derivatives. Linearizing about $\mathbf{A}(0)$, we can approximate the eigenvalue displacement as

$$\Delta \Lambda \cong \Delta \gamma \frac{\partial \Lambda}{\partial \gamma} = -\mathbf{V}^{-1}\mathbf{BFV}$$

where $\Delta \gamma$ = 1. Here $\mathbf{V}$ is the known matrix whose columns are composed of right eigenvectors of $\mathbf{A}$. The resulting $\Delta \Lambda$ is zero everywhere except in the location corresponding to the destabilized modes, where it has the value $\Delta \lambda$ = 0.2017 +/- 0.0358j. Noting that $\mathbf{F}$ was designed to move λ exactly 0.2017 to the right, with no change in frequency, we conclude that a first-order expansion based on eq. 3 has considerable utility in predicting how modes will change as a result of the application of state feedback both in frequency and in participation.

State Variable	Before Feedback	After Feedback
$\Delta\phi_1$	0.0355+/-0.0077j	0.0424+/-1802j
$\Delta\phi_2$	0.0107+/-0.0021j	0.0748+/-0.2131j
$\Delta\phi_3$	-0.0717+/-0.0010j	0.0347+/-0.1193j
$\Delta\phi_4$	0.0025+/-0.0008j	0.0105+/-0.0146j
$\Delta\phi_5$	0.0019+/-0.0013j	0.0090+/-0,0194j
$\Delta\theta_1$	0.0264+/-0.0202j	-0.0331+/-0.1632j
$\Delta\theta_2$	0.0081+/-0.0059j	-0.0190+/-0.2023j
$\Delta\theta_3$	-0.0598+/-0.0270j	-0.0165+/-0.1105j
$\Delta\theta_4$	0.0018+/-0.0015j	0.0038+/-0.0160j
$\Delta\theta_5$	0.0012+/-0.001j	0.0016+/-0.0198j
$\Delta\omega_1$	-0.3407+/-0.3166j	-0.2367+/-0.0696j
$\Delta\omega_2$	-0.1127+/-0.0715j	0
$\Delta\omega_3$	0.6293+/-0.5516j	0
$\Delta\omega_4$	-0.0667+/-0.0216j	-0.4715+/-0.1686j
$\Delta\omega_5$	-0-679+/-0.0109j	-0.6158+/-0.2582j
$\Delta\delta_1$	-0.0665+/-0.0748j	-0.0163+/-0.0601j
$\Delta\delta_2$	-0.0149+/-0.0246j	0
$\Delta\delta_3$	0.1157+/-1380j	0
$\Delta\delta_4$	-0.0043+/-0.0145j	0.0341+/- 0.1022j
$\Delta\delta_5$	-0.0020+/-0.0147j	0.0527+/-0.1337j

Table 5: Eigenvectors for all States Before and After State Feedback Application

To complete the discussion of mode selection, Table 5 shows the eigenvector for the full state space before and after the application of state feedback. The mode selected is -0.1017 +/- 4.6457j, moved to λ = +0.1 +/- 4.6457j, with machine 1 designated as the sacrificial machine. Note that while "external" states of frequency and angle deviations for machines 2 and 3 do not participate in the unstable mode, the internal states of governor and turbine deviations for those machines do, at a magnitude roughly equivalent to that of the sacrificial machine, which is to say less than seen by machines 6 and 8. An intuitive explanation for this effect can be obtained by examining the block model in Figure 2, particularly the node where turbine output power and network load power deviations sum as input to the synchronous machine block. In order for a given machine (at bus 2 or 3) to not participate in an unstable mode, it is necessary for any presence of that mode which is coupled to the node via the network to be exactly canceled by a corresponding output from the turbine.

State Variable	Bus 1	Bus 2	Bus 3
$\Delta\phi_1$	-0.2534	-0.2780	-0.1856
$\Delta\phi_2$	-0-0579	-0.0636	-0.0431
$\Delta\phi_3$	0.1225	0.1416	0.1367
$\Delta\phi_4$	-0.0317	-0.0341	-0.0224
$\Delta\phi_5$	-0-0623	-0.0615	-0.0399
$\Delta\theta_1$	0.2480	0.2616	0.1139
$\Delta\theta_2$	0.0519	0.0543	0.0213
$\Delta\theta_3$	-0.4285	-0.4516	-0.1944
$\Delta\theta_4$	0.0129	0.0132	0.0033
$\Delta\theta_5$	0.0078	0.0078	0.0006
$\Delta\omega_1$	0.7338	0.7801	0.3771
$\Delta\omega_2$	0.1368	0.1451	0.0679
$\Delta\omega_3$	-0.9559	-1-0159	-0.4888
$\Delta\omega_4$	0.0218	0.0230	0.0100
$\Delta\omega_5$	0.0192	0.0202	0.0083
$\Delta\delta_1$	1.7280	1.8784	1.1549
$\Delta\delta_2$	0.4856	0.5246	0.3031
$\Delta\delta_3$	-2.4944	-2.7046	-1.6227
$\Delta\delta_4$	0.1309	0.1408	0.0774
$\Delta\delta_5$	0.1499	0.1609	0.0873

Table 6: State Feedback Matrix **F** for Selected Mode

FEEDBACK MATRIX

Table 6 shows the state feedback coefficients for the selected mode. Examining the size of elements of **F**, we notice that while the largest participants in state feedback are those states associated with control machine frequency and angle deviations. with select turbine and governor deviations also show large coefficients. Additionally, even while machine 1 is designated the sacrificial machine, feedback is applied to all control group machines in roughly equal proportion. In short, an examination of this **F** offers no obvious opportunities for omission of variables in the state feedback. Thus, for operational use, those states which cannot be directly measured must be obtained by another means (such as state estimation).

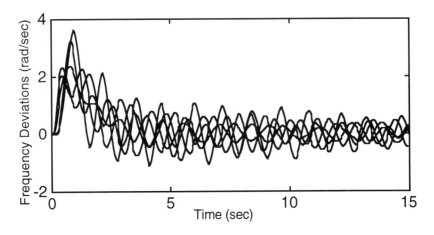

Figure 4: Response of Machine Frequency Deviations without Destabilizing State Feedback

TIME-DOMAIN RESULTS

Figure 4 shows the time domain response of the five machine frequency deviations to a pulse perturbation occurring at t = 0 s. The resulting waveforms show the presence of multiple oscillatory modes for each frequency as would be expected. Figure 5 shows the machine frequency deviations for the same excitation when state feedback is in effect, with the assumption made for the time being that all states are available. As suggested by the eigenanalysis, machines 1, 6, and 8 show significant participation in the unstable mode, and are moving with machine 1 approximately 30° ahead of machines 6 and 8, at about one-half the amplitude.

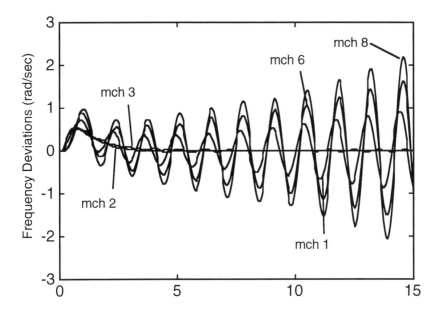

Figure 5: Perturbation Response of Frequency Deviations with State Feedback Applied

STATE FEEDBACK USING DYNAMIC STATE ESTIMATION

In a practical application of this control, the assumption that all states are available for state feedback will not be valid. To recreate those states, use of a dynamic state estimator (state "observer") is examined. The estimator is driven by both the inputs to the physical system plus a measurement set of physical system outputs. As long as the eigenvalues of the estimator feedback matrix **A - LC** are stable, the estimator error will asymptotically approach zero, and estimate $\hat{x}$ will approach physical state **x**.

EIGENANALYSIS OF ESTIMATOR WITH STATE FEEDBACK

In considering the use of a state estimator to provide the state feedback, questions naturally arise as to whether or not the desired eigenvalues and eigenvectors are preserved for the physical system states. The separation property, as discussed in [5], assures us that eigenvalues are preserved. In [1], it is argued that a consequence of this property is that the relevant eigenvector components are preserved as well. To illustrate these results, we perform the following analysis on the physical system augmented by an asymptotic state estimator which acts as the source of state feedback. We impose the constraint that all eigenvalues of the estimator feedback matrix **A–LC** are non-repeating, and distinct from those of the original matrix with state feedback, **A–BF**.

Consider the following state equation for the combined system

$$\begin{bmatrix} \dot{x} \\ \dot{\hat{x}} \end{bmatrix} = \begin{bmatrix} A & -BF \\ LC & (A - LC - BF) \end{bmatrix} \begin{bmatrix} x \\ \hat{x} \end{bmatrix} + \begin{bmatrix} B \\ B \end{bmatrix} u \tag{4}$$

Next consider the linear transformation $e = \hat{x} - x$. as shown in eq. 5.

$$\begin{bmatrix} x \\ e \end{bmatrix} = Q \begin{bmatrix} x \\ \hat{x} \end{bmatrix} = \begin{bmatrix} I & 0 \\ -I & I \end{bmatrix} \begin{bmatrix} x \\ \hat{x} \end{bmatrix} \tag{5}$$

Applying this transform to eq. 4, we obtain eqs. 6 and 7:

$$\begin{bmatrix} \dot{x} \\ \dot{e} \end{bmatrix} = Q \begin{bmatrix} A & -BF \\ LC & (A - LC - BF) \end{bmatrix} Q^{-1} \begin{bmatrix} x \\ e \end{bmatrix} + Q \begin{bmatrix} B \\ B \end{bmatrix} u \tag{6}$$

$$\begin{bmatrix} \dot{x} \\ \dot{e} \end{bmatrix} = \begin{bmatrix} (A-BF) & -BF \\ 0 & (A-LC) \end{bmatrix} \begin{bmatrix} x \\ e \end{bmatrix} + \begin{bmatrix} B \\ 0 \end{bmatrix} u = \hat{A} \begin{bmatrix} x \\ e \end{bmatrix} + \begin{bmatrix} B \\ 0 \end{bmatrix} u \tag{7}$$

From eq. 7, we observe that the eigenvalues for the combined system will be the union of those for the original system with state feedback plus those of the estimator, as predicted by the separation property. We also note that the error state vector **e** is unaffected by the input **u**, being governed only by the eigenvalues of **A - LC**, thus clearly showing the tracking ability of this estimator, and that the estimator modes are uncontrollable.

To address the question of eigenvector preservation, consider λ_p and v_p to be an eigenvalue and associated eigenvector of **A - BF**. λ_p must therefore also be an eigenvalue of $\hat{A}$. The question is, does v_p appear as a component to the associated eigenvector for $\hat{A}$?

Let $\hat{v}_p$ be a candidate eigenvector of $\hat{A}$ fr λ_p, with its first partition known, as v_p, and its second partition unknown, denoted as **q**, as shown in eq. 8. Writing the eigenvalue/eigenvector relations, it is apparent that because λ_p is not an eigenvalue of the estimator subspace governed by **A** - **LC** (since we have defined estimator eigenvalues to be distinct from those of the physical system under state feedback), **q** must equal 0. Using this value of **q** in the candidate eigenvector $\hat{v}_p$, one obtains the eigenvalue/eigenvector relation of eq. 9.

$$\hat{v}_p = \begin{bmatrix} v_p \\ q \end{bmatrix} \quad (8)$$

$$\hat{A}\hat{v}_p = \lambda_p \hat{v}_p = \begin{bmatrix} (A-BF) & -BF \\ 0 & (A-LC) \end{bmatrix} \begin{bmatrix} v_p \\ 0 \end{bmatrix} = \lambda_p \begin{bmatrix} v_p \\ 0 \end{bmatrix} \quad (9)$$

This example illustrates the point made in [1], that the eigenvector of the original system is preserved under conditions of state feedback from a state estimator. Another result suggested the above analysis is that as long as eigenvalues of the estimator matrix **A** - **LC** are distinct from those of the physical system, it should be possible to create an estimated of the state vector **x** from even only one measurement.

SYSTEM OBSERVABILITY

In section II.C we established that the physical system without state feedback was completely controllable. In the previous section we transformed the combined original system with a state estimator and showed that the estimator modes were uncontrollable. At this point we desire to know the observability of the system. As with controllability, the standard approach of testing the rank of an observability matrix is somewhat inconvenient to use, given the inherent rank deficiency of our state matrix. An alternate method is the dual of that used earlier in this work to establish controllability. The same transformation on **A** as is described in eqs. 1 and 2 is applied, providing states or pairs of states that correspond to natural modes for the system. To determine observability, the output matrix

534 POWER SYSTEMS RESTRUCTURING

C is transformed to **C'** = **CT**. Any zero columns or pairs of columns (for complex modes) correspond to unobservable modes in the system.

Using this approach the system was tested for observability when exactly one state is selected for output. It was found that if that state is any frequency deviation, then all modes are observable except for the $\lambda = 0$ mode which shows participation only from the angle deviations. If, however, the single measured state is any angle deviation, then all modes become observable.

As state feedback can affect observability, the logical next step is to test the observability of the zero eigenvalue (which remains) when state feedback is applied, and only frequency deviations are measured. As expected, the zero eigenvalue remains unobservable. This can be confirmed by examining the five coefficients for $\Delta\delta$ for any row of **F** in Table 6. Note that for any of the three rows, the five coefficients sum to zero.

Two conclusions may be drawn:

1. If it is desired to exactly reproduce the machine angle deviations in the estimator, then it is necessary to include at least one machine angle deviation in the measurement set.
2. From the viewpoint of state feedback, the presence of a DC error across set of the estimator angle deviations does not affect the resulting feedback term.

Another impact of the presence or absence of a angle deviation term in the measurement set involves the placement of estimator eigenvalues, and is discussed in the next section.

ESTIMATOR EIGENVALUE PLACEMENT

In choosing eigenvalues for the estimator, it is desirable that they be stable, and distinct from those for **A** - **BF**, as shown in section IV.A. For this work, the set of eigenvalues Λ_L chosen as follows. Noting that many of the eigenvalues, as shown in Table 3, have sufficiently high rates of decay, the estimator eigenvalues are chosen as

$$\Lambda_L = -\alpha_L \mathbf{I} + \Lambda.$$

APPLICATION OF DYNAMIC GENERATION CONTROL... 535

where Λ is the set of eigenvalues for **A**, and α_L is a positive, real-valued shifting factor. This method has the advantage of moving the poorly-damped eigenvalues by a proportionally large amount in the estimator, as well establishing a "baseline" rate of convergence for the estimator.

With Λ_L set, standard eigenvalue placement techniques (and in particular, the MATLAB [6] **PLACE** function) are used to compute a matrix **L** that returns the desired eigenvalues for **A** - **LC**. In conducting a number of numerical experiments, the following observations were obtained, consistent with the expected properties of estimator pole placement algorithms:

1. The algorithm implemented in the MATLAB **PLACE** routine was successfully able to compute **L** for larger values of α_L as more measurements were added to the measurement set. Table 7 summarizes the norms (induced 2-norm) of **L**'s computed for various measurement set sizes, denoted by m, and values of α_L. The results suggest that for decay rates of $\alpha_L=5$ or less, three measurements (including one shaft angle deviation) are adequate.

2. If no angle deviation is included in the measurement set, then the spectrum Λ_L must include a zero eigenvalue. While this creates a repeated eigenvalue in the composite state matrix, the associated eigenvectors remain linearly independent of one other.

	Set Size (m)		
	m = 1	m = 3	m = 5
$\alpha_L = 1$	10^4	10^1	10^1
$\alpha_L = 5$	10^8	10^4	10^4
$\alpha_L = 10$	–	10^7	10^5
$\alpha_L = 15$	-	10^{10}	10^6

Table 7: Approximate IILII Values for Different Measurement Set Sizes and Estimator Damping Factors

TIME DOMAIN EXAMPLES

Figure 6 shows the performance of the estimator as a source of destabilizing state feedback when only the sacrificial machine angle deviation is supplied as a measurement for $\alpha_L = 2$. The system has a zero initial condition, and gaussian noise is supplied on the inputs, with $\sigma = 0.001$. Note that the machine and estimator frequency deviations are identical, verifying the analytical result that suggests that one measurement is sufficient to operate the estimator. Also note that while multiple modes are stimulated, the participation in the unstable mode is limited to machines 1, 6, and 8, as suggested in figure 5.

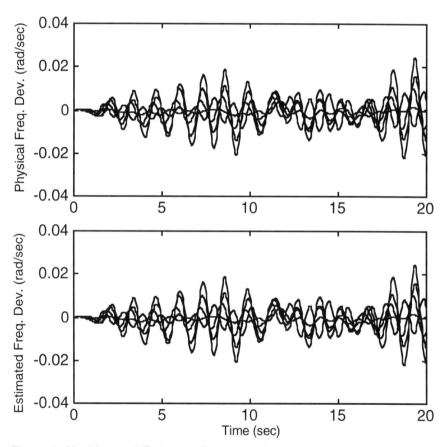

Figure 6: Machine and Estimator Frequency Deviations for Random Noise Input with $\sigma=0.001$, Measuring Sacrificial Machine $\Delta\delta$ Only.

Figure 7 shows the response of the machine and estimator angle deviations to an input pulse at $t = 0$ sec. The measurement set consists of the three control group frequency deviations, with $\alpha_L = 4$. The machine angle deviations are set to an initial condition of 2, while the estimator angle deviations are set to -1. The observed behavior confirms the observations that:

1. In the absence of a machine angle in the measurement set, the zero eigenvalue for the physical system is unobservable, and;
2. under the same circumstances, the estimator must have a zero eigenvalue associated with its own angle deviations.

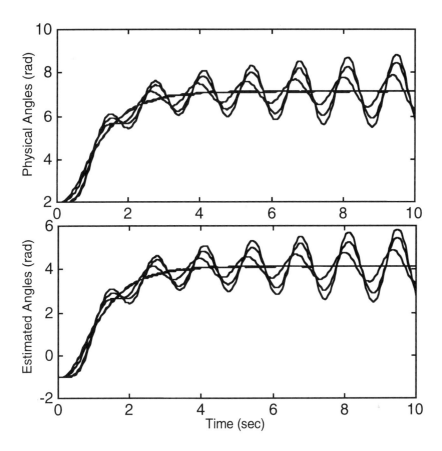

Figure 7: Pulse Response with Control Group Frequency Deviations (Only) Measured, Different Initial Conditions on Angle Deviations

At time t = 0 s, the input pulse perturbs the system, and excites the machine modes. As concluded, the DC offset of the estimator machine angles as a group does not affect the proper operation of the state feedback. This result suggests that under conditions of an ideal estimator. frequency deviations are sufficient to achieve the desired state feedback; i.e., the unobservable subspace (associated with uniform motion of angles) is irrelevant to the desired eigenvalue/eigenvector placement.

The above time domain results suggest that for a sufficiently well-modeled system, it should be possible to build an identical asymptotic estimator at each machine in the control group, providing state feedback to that machine based on measurement of its frequency and (possibly) angle deviations. Theoretically, each estimator will reproduce the same estimate for the full system, and therefore no interactions between estimators should occur.

IMPACT OF PARAMETER ERRORS

To this point, we have assumed that the linearized model state matrix, A, was exactly known, and therefore available for incorporation into a state estimator. In this section we conduct numerical experiments to examine the impact of using an inexact representation of A.

Let A be a function of a scalar parameter γ. Let the estimator state matrix, $A_e = A(\gamma_0)$, where γ_0 is the estimated value of the parameter. Let the physical state matrix, $A_p = A(\gamma_0 + \varepsilon)$, where ε is the (unknown) deviation in γ from γ_0. With these definitions we rewrite eq. 4 as follows:

$$\begin{bmatrix} \dot{x} \\ \dot{\hat{x}} \end{bmatrix} = \begin{bmatrix} A_p & -BF \\ LC & (A_p - LC - BF) \end{bmatrix} \begin{bmatrix} x \\ \hat{x} \end{bmatrix} + \begin{bmatrix} B \\ B \end{bmatrix} u \quad (10)$$

Performing the same transformation as in eq. 7, we obtain

$$\begin{bmatrix} \dot{x} \\ \dot{e} \end{bmatrix} = \begin{bmatrix} (A_p - BF) & -BF \\ A_e - A_p & (A_e - LC) \end{bmatrix} \begin{bmatrix} x \\ e \end{bmatrix} + \begin{bmatrix} B \\ 0 \end{bmatrix} u \quad (11)$$

For $\varepsilon \neq 0$, the presence of the non-zero coupling term in the lower left region indicates that the separation principle between physical and estimated eigenvalues no longer holds. The two subsystems can now interact in such a way as to affect the eigenvalues and eigenvectors of both subspaces. When we consider eq. 11 without state feedback, but with $\varepsilon \neq 0$, the upper right coupling term is eliminated. The eigenvalues for the system are the union of those for

A_p and A_e–LC. The error **e** becomes coupled to the physical state **x** by the mismatch term A_e–LC, which means it no longer asymptotically approaches zero while **x** ≠ **0**. As long as A_p remains stable, however, the error remains stable.

In selecting the group of parameter errors to study, a number of assumptions are made. First, that the entity or entities operating the control group have detailed knowledge of the network, and second, that they have accurate models of machines within the control group. For this study, this means that the inertia and time constants for machines at buses 6 and 8 are the quantities most likely to be in error. Our efforts will therefore be focused there.

SENSITIVITY ANALYSIS

In eq. 11, the matrix A_p appears in the upper and lower left locations to alter the eigenvalues and eigenvectors from the design targets. In the upper left location, it interacts directly with the state feedback - **BF**. In the lower left location, it interacts with A_e to provide a non-zero cross-coupling term between the estimator and physical system.

Ideally, one might hope to use first order sensitivity methods to predict changes in the combined system due to A_p. In practice, however, while the changes predicted for physical modes are consistent with experimental results, the changes for estimator modes are not. Specifically, the sensitivity calculation predicts very large variations in estimator λ's where very small ones are observed experimentally.

The explanation offered for these phenomena are twofold. First, that the coupling A_e-A_p is sufficiently weak as to make the calculation of estimator eigenvalue and eigenvector derivatives numerically undependable, and, second, that changes in the physical system eigenvalues and eigenvectors are dominated by the presence of A_p in the upper left block. This suggests a partitioning of the problem as follows. Sensitivity analysis may be performed on the upper left block only., ignoring the state estimator, on the assumption that the estimator error remains sufficiently small for an adequate length of time. Estimator performance can then in turn be tested with time domain experiments.

Sensitivity calculations were performed on $\mathbf{A_p}$-**BF** using the inertia and time constants for machines 6 and 8 as the control parameters. Tables 8 and 9 show the oscillator eigenvalues and their $\partial \lambda/\partial \gamma$ values for each parameter. For each mode, the size of ε, as a percentage of γ_0, needed to move (by linear projection) to the $j\omega$ axis is given. For the real-valued modes, the changes required in ε to do this were sufficiently large in comparison to the oscillatory modes, so may be ignored for practical purposes. In examining the table, we see that when the control parameter is the machine 8 inertia, linearized projection indicates that an error where m_8 is 37% larger than estimated is enough to push the previously stable mode 2 to the $j\omega$ axis. A 50% error is sufficient to make the unstable mode 4 borderline stable. We also notice that as a group, the effect of inertia errors on the eigenvalues is smaller than that for time constants. Interesting to note is the fact that the unstable mode seems quite resistant to errors in all four machine time constants. This is consistent with the earlier observation that machines outside the control group saw lower participation in the unstable mode in their governors and turbines. This experience suggests, significantly, that the components in the machines which are most difficult to model might also be the most immune to quantitative errors.

Index	λ
1	-0.05313 + 6.7086j
2	0.04249 + 5.8794j
3	0.07303 + 5.3507j
4	0.1000 + 4.6457j

Table 8: Oscillatory Modes for **A - BF**

APPLICATION OF DYNAMIC GENERATION CONTROL... 541

γ_0	Index	$\partial\lambda/\partial\gamma$	ε_{lim} (as % of γ)
m_6	1	-0.2555 - 48.58j	-392.1
	2	0.3350 - 0.2434j	239.1
	3	0.6940 - 6.641j	198.3
	4	-2.319 - 1.934j	81.28
m_8	1	-0.9674 - 13.18j	-103.5
	2	2.1476 - 10.840j	37.29
	3	1.000 - 24.86j	137.7
	4	-3.715 - 2.183j	50.75
T_{G6}	1	0.7801 - 0.4931j	45.41
	2	0.007484 + 0.005137j	3785
	3	0.1625 - 0.02298j	299.6
	4	0.04728 - 0.05893j	-1410
T_{G8}	1	0.1195 - 0.1432j	222.2
	2	0.1720 - 0.07154j	123.5
	3	0.4337 - 0.2065j	84.19
	4	0.02015 - 0.08397j	-2482
T_{T6}	1	1.065 - 0.3515j	55.44
	2	0.007371 + 0.007179j	6405
	3	0.1883 + 0.01610j	430.8
	4	0.07065 - 0.05518j	-1573
T_{T8}	1	0.2496 - 0.1130j	266.0
	2	0.2610 + 0.001710j	203.5
	3	0.6438 - 0.01995j	141.8
	4	0.07366 - 0.09146j	-1697

Table 9: Oscillatory Mode Eigenvalue Derivatives and Predicted ε Values Necessary to Move Eigenvalue to the $j\omega$-Axis (linear estimate)

Table 10 shows the eigenvector derivatives for the frequency deviations for the unstable mode for the same set of control parameters, assuming that the physical parameter is 10% larger than estimated. It shows that for errors of this magnitude, the linearized estimate of "mode smearing" across the machines (i.e., appearance of undesired non-zero eigenvector components among the control machines) is very small. In fact, for both inertias, errors where the actual value is underestimated actually cause the participation within the control group overall to decrease. We also note that the amount of smearing observed for time constant errors is very small, on the order of less than 1% for parameter errors of 10%.

542 POWER SYSTEMS RESTRUCTURING

γ_0	Index	$\partial v/\partial g$	$\lvert v_0 \rvert - \%$	$\lvert v_{est} \rvert - \%$
m_6	ω_1	6.053 - 1.323j	32.73	29.68
	ω_2	0.8279 + 0.11807j	0	0.5064
	ω_3	2.414 - 0.9199j	0	1.564
	ω_6	-4.205 + 2.694j	56.69	60.11
	ω_8	3.000 - 1.517j	75.59	74.18
m_8	ω_1	13.51 - 4j		26.32
	ω_2	1.765 - 0.2003j		1.088
	ω_3	4.311 - 2.206j		2.968
	ω_6	7.717 - 3.807j		52.62
	ω_8	-5.055 + 3.392j		80.8
T_{G6}	ω_1	0.0395 + 0.1606j		32.61
	ω_2	-0.001498 + 0.02283j		0.03886
	ω_3	0.02768 + 0.06453j		0.1193
	ω_6	-0.07539 + -0.1089j		56.76
	ω_8	0.04088 + 0.08119j		75.6
T_{G8}	ω_1	0.1585 + 0.261j		32.29
	ω_2	0.01523 + 0.03692j		0.09056
	ω_3	0.07362 + 0.07859j		0
	ω_6	0.1227 + 0.1429j		0.2442
	ω_8	-0.09959 - 0.08313j		56.62
				75.84
T_{T6}	ω_1	0.006244 + 0.186j		32.68
	ω_2	-0.007184 + 0.0237j		0.02524
	ω_3	0.01379 + 0.07608j		0.07878
	ω_6	-0.05687 - 0.1387j		56.7
	ω_8	0.02777 + 0.0983j		75.61
T_{T8}	ω_1	0.06346 + 0.3858j		32.61
	ω_2	0.0003432 + 0.04914j		0.04453
	ω_3	0.04579 + 0.126j		0.1215
	ω_6	0.07795 + 0.2269j		56.72
	ω_8	-0.07632 - 0.1491j		75.63

Table 10: Derivatives of Eigenvector Components Associated with Frequency Deviation and Original versus Projected Magnitudes for $\varepsilon = 0.1 \times \gamma_0$ (10% Error)

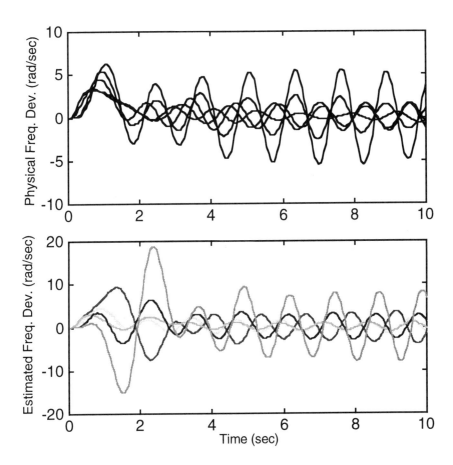

Figure 8: Frequency Deviations for $\gamma = m_8$, $\varepsilon = 20\%$, $\alpha_L = 2$, Sacrificial Machine Angle and Frequency Deviations Measured

SIMULATION RESULTS

Figure 8 shows the pulse response of the frequency deviations when m_8 is 20% larger than estimated, using an estimator to provide state feedback. The measurement set consists of the sacrificial machine angle and frequency deviations, with $\alpha_L = 2$. Control of the unstable mode is marginal. Machine 8 still participates to the largest degree, but several machines in the control group now have participations comparable to that of machine 6. The estimator shows a large initial error which decays after several seconds, followed by significant phase and amplitude errors on several machine frequencies.

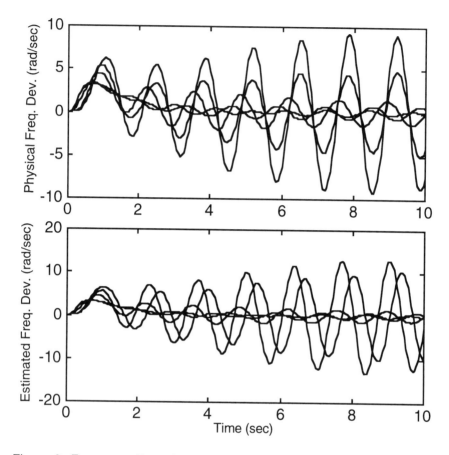

Figure 9: Frequency Deviations for $\gamma = m_8$, $\varepsilon = 20\%$, $\alpha_L = 2$, All Control Group Angles and Frequency Deviations Measured

Figure 9 shows the same test where all control machine angle and frequency deviations are supplied as measurements, again with $\alpha_L = 2$. In this example the mode is better controlled, with the frequency deviations more closely resembling those of figure 5, even as the estimate error increases to significant levels. The initial estimator response more closely tracks the physical system, indicating an overall smaller degree of excitation of estimator modes.

Figure 10 shows a typical pulse response for the frequency deviations when random parameter errors of size $\sigma = 10\%$ are applied to the inertias and time constants for machines 6 and 8. The measurement set includes all control group angle and frequency deviations. The response shows a strong resemblance to figure 5.

Participation by machine 1 is increased somewhat, but remains less than that for machines 6 and 8. Participation by machines 2 and 3 remains small. The good performance observed in this (typical) plot suggest the possibility that multiple parameter errors may tend to cancel each other in their effects on the model.

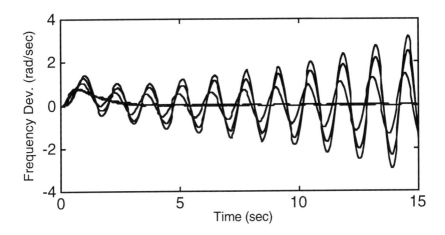

Figure 10: Frequency Deviation Response for Random Parameter Errors on Machines 6 and 8, $\sigma = 10\%$, All Control Group Angle and Frequency Deviations Measured with $\alpha_L = 6$

CONCLUSIONS

Using a model and set of assumptions more sophisticated (and presumably realistic) than that originally applied in [1], this work has demonstrated the continued possibility for a potentially subtle form of anti-competitive behavior exercised by a group of generators through their dynamic control. It has shown that the addition of governor and turbine models to the classical machine model does not impede the use of a linear feedback control to destabilize a mode and to cause participation in that mode to be significantly smaller within a control group than for machines outside the control group. It has also demonstrated the use of eigenvector derivative calculations in predicting which mode and sacrificial machine combinations are likely to yield the desired participation profile, thus making such modes easier to identify.

The ability of state estimators to reconstruct the machine states based on a limited measurement set inside the control group and to preserve the participation in the unstable mode while acting as a source of state feedback was also demonstrated. For the case studied, it was shown that if intimate knowledge of the network and control group machines is available, an estimator may be constructed with reasonable resistance to errors in the models for machines outside the control group. Furthermore, the results suggest that the estimator-driven state feedback mechanism possesses an inherent resistance to errors in governor and turbine models for these machines.

REFERENCES

[1] C. L. DeMarco, J. V. Sarlashkar, and F. L. Alvarado, "The potential for malicious control in a competitive power systems environment," *Proc. IEEE Conference on Control Applications*, Dearborn, MI, Sept. 18-20 1996.

[2] B. C. Moore, "On the Flexibility Offered by State Feedback in Multivariable Systems Beyond Closed Loop Eigenvalue Assignment", IEEE Trans. on Automatic Control, pp. 689-692, October, 1976.

[3] A. J. Wood, B. F. Wollenberg, Power Generation and Control, John Wiley & Sons, Inc., New York, 1984.

[4] J. J. D'azzo, C. H. Houpis, Linear Control System Analysis and Design, McGraw-Hill, Inc., New York, 1995.

[5] C-T. Chen, Linear System Theory and Design, Holt, Rinehart & Winston, Inc., New York, 1984.

[6] MATLAB User's Guide, The Mathworks, Inc., Natick, MA, 1993.

Contributors

Judith Cardell is in the Office of Economic Policy at the Federal Energy Regulatory Commission. She received a BS in Electrical Engineering and an AB in Government from Cornell University in 1989. She received her masters and Ph.D. in Technology and Policy in Electrical Engineering from the Massachusetts Institute of Technology in 1994 and 1997 respectively.

Stephen R. Connors is director of the Electric Utility Program and the Analysis Group for Regional Electricity Alternatives (AGREA) at the Massachusetts Institute of Technology's Energy Laboratory. Mr. Connors has a Masters in Technology and Policy from MIT, and bachelors degrees in Mechanical Engineering and Anthropology from the University of Massachusetts at Amherst. Mr. Connors' expertise is in strategic planning, renewable energy, energy conservation, environmental regulation, and analytic approaches for educating electric industry stakeholders. Since its inception in 1988, AGREA has applied its tradeoff analysis approach in New England, Switzerland, Argentina and Italy, and assisted with its application elsewhere. Prior to his work in the electric sector Mr. Connors was a Peace Corps volunteer in West Africa where he designed and tested wood conserving cookstoves as part of an appropriate technology project.

Raymond Coxe is a consulting engineer with NEES Global Transmission, Inc., the independent transmission project development company of the New England Electric System ("NEES"). He is responsible for developing independent transmission projects in the US and internationally, and for evaluating competitive energy markets. Previously, he marketed wholesale electricity and performed generation planning analyses for the NEES Companies. He received his undergraduate and doctoral engineering degrees from the Massachusetts Institute of Technology before joining the NEES Companies in 1988.

Christopher L. DeMarco received his S.B. in Electrical Engineering from MIT, in 1980, and his Ph. D. in the same subject from the University of California, Berkeley in 1985. Since 1985 he has been a member of the faculty of the Department of Electrical and Computer Engineering, University of Wisconsin-Madison.

Robert L. Earle is an associate at The Brattle Group's Washington, D.C. office. He works on issues concerning the restructuring of the electric power industry including transmission pricing and power trading. His published work includes algorithmic advances in power production costing and stochastic programming. He has a Ph.D. in Operations Research from Stanford University.

Lester H. Fink retired in 1997 as Executive Vice President of KEMA Consulting. Earlier, he conducted and managed plant and system control research at the Philadelphia Electric Company (1950-74), developed and managed the national Systems Engineering for Power program at the U.S. Energy Research and Development Administration, engaged in contract research and consulting in the private sector (1979-97), and served as adjunct professor at the University of Pennsylvania, Drexel University, and University of Maryland (passim). His publications include three U.S. patents (two on power system control) and over fifty papers. He is Life Fellow of IEEE and ISA, a member of CIGRE, and received the Meritorious Service Award of the U.S. Department of Energy.

Francisco D. Galiana was born in Alicante, Spain. He obtained his Ph.D. under Professor Fred C. Schweppe at the Massachusetts Institute of Technology in 1971. Subsequently, he worked at the Brown Boveri Research Center, Switzerland, at the Department of Electrical and Computer Engineering of the University of Michigan and, since 1977, at the Department of Electrical and Computer Engineering of McGill University, Montreal, where he is a full professor in the power engineering area. Prof. Galiana is a Fellow of the IEEE and has been active in research and teaching for over thirty years, mainly in issues related to the operation and planning of large power networks.

Thomas A. Gorski received his Bachelor of Science degree in Electrical Engineering from the University of Wisconsin-Madison in 1984, and his Master of Science degree in Electrical Engineering from the same instituition in 1995. He is currently with Ohmeda, Inc. in Madison, WI.

Frank C. Graves is a principal at The Brattle Group where he advises companies on business strategy, service design and regulatory economics. In addition, he assists senior utility management with evolving markets and industry restructuring. He helps design and implement long-range investment planning and service design for major gas and electric utilities. Mr. Graves has testified on the economics of unbundling, market entry, and new services before the FERC and state regulatory commissions. He received a B.A. in mathematics from Indiana University and an M.S. from MIT's Sloan School of Management.

Richard Green is a Senior Research Officer in the Department of Applied Economics, University of Cambridge, and a Fellow of Fitzwilliam College. He has been studying the electricity supply industry in the UK since 1989, shortly before its restructuring. He spent a year on secondment to the Office of Electricity Regulation in 1994/95.

Philip Q. Hanser is a senior consultant at The Brattle Group where he provides consulting support in the areas of economics and business analysis, and strategic planning with an emphasis on conceptual and quantitative analysis applied to transmission pricing, generation planning, tariff strategies, fuels procurement, environmental issues, forecasting, and marketing. He has published in the areas of public utilities economics, engineering-economics, and statistics. He holds an A.B. in Economics and Mathematics from Florida State University, a Phil.M. in Economics and Mathematical Statistics from Columbia University where he completed the Candidacy Requirements in Economics.

Leonard S. Hyman, a Senior Industry Advisor to Salomon Smith Barney's Global Energy and Power Group, specializes in utility and telecommunications finance and economics. He is the author of *America's Electric Utilities: Past, Present and Future* (in its sixth edition), co-author of *The New Telecommunications Industry,* (in its second edition), and editor of *The Privatization of Public Utilities.*

Marija Ilic is a Senior Research Scientist in the Department of Electrical Engineering and Computer Science at MIT, where she teaches several graduate courses in the area of electric power systems and heads research in the same area. She has twenty years of experience in teaching and doing research in this area. Prior to coming to MIT in 1997, she was an Assistant Professor at Cornell University, and tenured Associate Professor at the University of Illinois at Urbana-Champaign. Her main interest is in the systems aspects of operations, planning and economics of electric power industry.

Ralph Masiello was educated at Massachusetts Institute of Technology where he received a Ph.D. E.E. and as BSEE/MSEE. He started with ABB Systems Control in 1985 at which time he joined the company as Vice President, Energy Systems and Systems Control Divisions. He has been responsible for ABB SC's business with Electric, Water and Gas Utilities, including Energy Management and SCADA systems, applications software, software products for power systems planning and operations, power plant analysis and operations, and EPRI, DOE and other research. Before joining ABB Systems Control, Dr. Masiello was employed by Control Data Corporation where he served as Director, Projects and Planning and Manager, Systems Planning and Delivery for the Energy Management Systems Division. He currently is a vice president of the Information Systems Division responsible for business development. Within the least year and a half he played a key role in the planning and development of the California ISO.

E. Grant Read leads the Energy Modeling Research Group at Canterbury University, and acts as a Senior Advisor with Putnam Hayes and Bartlett. For the last ten years he has been very closely involved with the reform of the New Zealand electricity sector, both as a researcher and a consultant, and now plays a key advisory role in the development of the Australian national market. Dr. Read holds a B.Sc. with 1st Class Honors in Mathematics, and a Ph.D. in Operations Research, with graduate papers in Economics. He is a past president of the Operations Research Society of New Zealand.

Gerald B. Sheblé is a professor of Electrical and Computer Engineering at Iowa State University. Dr. Sheblé's research interests include artificial life techniques (expert systems, genetic algorithms, genetic programming, artificial neural networks) as applied to power systems analysis, operation and planning and other industrial domains. His research has focused on price based markets for operation and planning for the electric power and natural gas industries.

Ziad Younes is a Lebanese engineer and economist. He has graduated from the Ecole Polytechnique in Paris, and has a Master in Technology and Policy from the Massachusetts Institute of Technology. He has a D.E.A in Industrial Organization from the University of Paris IX and is currently a Ph.D. student there. Ziad's interest is mainly focused on how to deal with market imperfections in a deregulated electric power industry.

Index

Access 408-9, 419-20, 427, 431-3, 443-4
Access rights 408-9, 433
Adequacy of generation 243
Ancillary services 406, 408, 411, 422, 438
Anti-competitive 518, 545
Arc furnace 72
Area control error (ACE) 439-40
Automatic generation control (AGC) 22, 24, 26, 36, 71, 72, 99, 412, 424, 428, 432, 434, 437-8, 440-2
Available transfer capability (ATC) 19
Base load 412, 421, 424, 427-30
Bertrand 348, 350
Bertrand-Nash equilibrium (BNE) 352-3
Bilateral 16, 20, 21, 29, 30, 31, 33-37, 39, 40, 42, 78, 99
Blackout 17
California 20, 99
Call option 245, 266, 270, 272-3
Capacity payments 136-7, 141-6, 153, 163-5
Capacity planning 243-4, 253, 276
Central Electricity Generating Board 18
Charges, compatibility 98
Clean Air Act 394, 397

Closed loop price signal 459, 477-8, 482, 484, 486, 488, 490, 494, 499, 500-1
   Benefits 478, 482, 499, 500-2
   Affects of imperfect information 484, 494, 500
   Data requirements 495-7
Collusion 356
Compensation 37
Compensation, usage-based loss 15, 16, 21, 22, 27, 72, 79, 99
Competition 16-19, 21
   Ancillary services 459, 477-9, 482-3
   Energy markets 454, 459, 477-80, 482-3, 6, 501
Congestion 17, 19, 20, 78, 444, 449
Constrained zone 340
Constraints
   Generation 40, 41
   Operating 21, 39, 40, 99
   Static 22
   Stability 21
   Thermal 21
   Transmission 20, 21, 27, 98
   Voltage 21
Consumer 17, 28, 30, 31, 34, 37, 72, 78, 80

Contract 18-21, 30, 33-37, 70, 72, 79, 99
    For differences (CFDs) 34, 134, 144, 151, 162-3, 135, 147, 164
    Specification 30, 72, 73
    Firm 19, 33, 37
    Long-term 29, 31, 37, 39, 72
    Tradable 29
Control
    Area 16, 17, 22, 25, 26, 72, 414, 427, 439
    Automatic generation 16, 22
    Centers 16, 25
    Decentralized 422, 424, 427, 444
    Emergency security 16
    Excitation 22, 26
    Fringe 413-4, 421, 424, 430-1, 437, 439, 442
    Generation 405, 411, 423, 429, 434, 437, 440
    Horizon 424, 429, 431
    Manual 26
    Network 423, 428, 432-4, 443
    Preventative 16
    Primary 19, 22, 26
    Reactive power 16
    Secondary 22, 25, 26
    Schemes, decentralized 24
    Structure 405-6, 409, 423-4, 427-31, 435, 438, 440
    Tertiary 98

Coordinating councils 16, 17
Cost allocation 27
Cost minimization 17, 18, 20
Costs
    Fixed 26, 27
    O&M 26, 27, 28
    Start-up 29
    Total 22, 26, 27, 29
Cournot 348-9
Criteria 283-6, 288-9, 294, 299-301, 303-5, 311, 315, 317, 319-20, 328-31
Customers, residential 17, 19, 27, 37, 72
Decentralized control 457, 475, 479
Decision makers, financial 33
Decomposition, spatial 24
Decomposition, temporal 23,
Default regulation 412, 414, 422, 428-9, 438
Deintegration 407
Demand 24, 26, 30, 31, 99
    Actual 26, 29
    Anticipated 22, 25, 28
    Elastic 38, 99
    Fluctuations 21, 24, 26, 27, 29, 72
    Inelastic 38, 39, 40
    Predicted 21, 24, 28-31, 33, 37, 38, 71, 79, 98
    Residual 337, 339
    Scheduled 22
Demand side bidding 160
Demand-side management 388
Deregulation 405-9, 427
Deviation 15, 22, 24, 26, 37, 38, 42, 72
Deviations, expected band 72

POWER SYSTEMS RESTRUCTURING 555

Director General of Electricity Supply (the regulator) 136, 147-53, 155-8, 160, 162, 164-5
Distributed generation
    Generator modeling 463, 503
    Governor tuning 498
    Power system evolution 455, 457, 500
    Pystem modeling 463
Distribution companies (discos) 18, 19
Dynamic control 517-8, 545
Economic dispatch 22, 25, 26, 29, 30, 35, 40, 412, 414, 424-5, 430, 433, 435, 440
    Conventional 39
    Decentralized 31, 33, 36, 79
    Generalized 16, 22, 25, 26, 29-31, 33, 35-39, 42, 79
Effect, intertemporal 15, 16, 40,
Efficiency 18, 27, 33, 80
Eigenvector placement 519, 525, 538
End user 22, 28
Energy contracts 72
Energy management system 16, 25
Entities
    Generator-serving 21
    Load-serving 20, 21
    Trading 20, 21, 27
Environmental regulation 385, 388, 397-399, 401
Equipment status 22
Equivalence, theoretical 33

Estimation, state 16
Evaluation 284-5, 288-90, 294, 298, 300, 302, 304-5, 307, 310-1, 318-9, 330
Excitation system 26
Fairness 20, 98
Financing 322, 331
Flexible AC transmission system (FACTS) 21,
Flexible plants 22
Flow constraints, tie-line 27, 98
Forward market 244-6, 258, 260, 262, 276
Frequency 410, 412, 414-5, 419, 422, 424-5, 428-9, 432, 437-442, 445
    Quality 22
    Regulation 419, 437, 440
Frequency stability 459-60, 463-4, 467-70, 475, 483, 498-9
    Causes of instability 460, 468, 470, 472, 475, 515
    Sensitivity matrix 473, 475
Gaming 72, 78, 355, 409, 425
Generation dispatch 16
Generation, flexible 22
Generation, reactive power, var 19, 22
Genoa 18
Geographical position 342
Governor 22, 26, 99, 517-9, 521, 528-9, 540, 545-6
Green pricing 387, 393-4
Grid expansion strategies 366
Hedging 34

Hierarchical organization 25
Hydro storage facilities 25
Imbalances 16
Incentives 406, 409-11, 421, 426, 445
Independent power producers 17
Independent system operator (ISO) 19, 407, 412
Industry, new 29, 31, 35, 36, 73, 80, 98
Inelasticity 361
Inertia 26
Information flow 21
Integrated Resource Planning 388-90, 393
Inventories 16, 35
Investment 284-7, 289-90, 292, 294, 299, 300, 302, 304-7, 309-12, 316, 319-20, 328, 331-2
Load 19-21, 29-31, 38, 72
    Dynamics 24
    Following 414-5, 429-30, 442
    Forecasting 16
    Interruptible 19
    Self-stabilizing effect 26, 72
    Serving 409, 415, 418, 429, 431, 433
Loop flows 358
Loop, closed 24-26, 42, 71
Loop, open 24-26
Loss compensation 15, 16, 22, 412, 414-5, 419, 422, 438-9
    Usage-based 15
Loss of load probability 137, 141-2
Loss, transmission 15, 16, 19, 21, 27, 72, 79, 99
Maintenance 28
Market 15, 20, 21, 27, 30, 34, 35, 41, 99
    Clearing 31, 32, 41
    Equilibrium 15, 41
    Participants 20, 27, 29, 31, 33, 35-38, 78, 79, 80, 99
    Power 18, 27, 30, 134, 148, 150, 152-5, 163, 336, 342, 368
    Power, locational 344, 354-5
    Prediction 29
    Speculation 33
    Structure, bilateral 20
    Structure, pool 17, 20, 38
    Uncertainties 33,
    Electricity 15, 30, 34-37, 40, 41, 78, 79, 99
    Electricity, primary 15, 28, 29, 30, 72, 79
    Long-term 365
    Optimal 78
    Perfect 15
    Primary, optimal 16, 72, 79
    Real-time 15, 35
    Spot 29-31, 33-38, 42, 73, 74, 78
    Spot, clearing process 31, 32
Marketers 21, 34
Monopolistic structure 17
Nash equilibrium 356
    Inter-temporal 353

National Grid Company (NGC) 136, 138, 146-8, 150, 155-6, 159, 161, 164
National Power 136, 150, 152-3, 155, 163
Network expansion, threat 367
Network loading 415, 433
Network reinforcement 423
New England 20, 99
New York 20, 99
Nodal pricing 344
Non-dispatchable technologies 463, 478, 492, 500, 505
Objective 15-17, 22, 26-28, 30, 33, 72, 78, 79, 99
Observer 517, 531
Oligopolistic competition 347
One-part pricing 244, 246, 264, 273, 276, 278
Operation 15, 16, 20-24, 27, 40
    Efficiency 18
    Normal 19, 24
    Real-time 23, 28
    Tasks 15, 21, 22, 27
Operator parochialism 410-1, 420
Optimization 295-8, 414, 433-5
    Nonconvex 32
Ownership 28
    Multiplying 365
Performance 16, 28, 33, 36, 41, 42, 78, 79, 80
Planning 283-6, 288-95, 297-301, 304, 311, 312, 315, 317-9, 328-32
    Reserves 243-5, 248-50, 257, 276, 278
Pool 16, 17, 20, 21, 38
Poolco 16, 28, 37, 38
Portfolio standards 385, 393-4, 398
Power
    Contracted 72
    Exchange 16, 19, 20, 25, 28, 30, 36, 99
    Industry, competitive 27, 71, 78
    Plants, low cost 17
    Predicted 35
    Purchase 29, 31
    Reliable delivery of 33
    Sold 30, 33, 34
PowerGen 136, 150, 152, 154-5, 163
Price 15-20, 27, 29-35, 37-41, 72-3, 78, 80, 98-9
    Average 27
    Caps 365
    Electricity 15-6, 22, 26-7, 30, 33, 35, 37-40, 72, 78
    Volatility 15
Pricing, ex ante 37-39, 78
Pricing, ex-post 37-39, 78
Process, market-clearing, self-adjusting 31, 32
Profit 17, 33, 34, 78
Profit, actual 34
Profits, short-term 72
Project, dedicated 285, 312, 318-21, 324-5, 327, 329, 332
Project, network 285, 312, 318-21, 324-5, 327, 329, 332
Provision, reliability 16
Random deviations 22
Rates of response 24

Reaction functions 358
Reactive energy 412, 416-7
Real time 17, 26, 28, 30, 31, 35, 37, 71-73, 78, 79
Regulating units 22, 26
Regulation 19, 22, 26
    Frequency 25-27, 36, 73
    Frequency, closed-loop 24, 26
    Market for 73
    Tertiary level 22, 25, 27
Relevant market 338, 345
Reliability 244, 247-50, 257, 263, 273, 276, 277, 283-89, 292-4, 298-301, 303-6, 311, 319, 321-2, 328-32
Reliability standards 17
Renewable energy 385, 393-4
Reserves 412, 418, 422, 426, 429, 436-7
    Margin 17, 29, 419, 436, 443
    Requirement 249-50, 271
    Short-term spinning 19
    Stand-by spinning 19
Restructuring 405-9, 422, 427, 432, 435, 438, 444
Retail competiton 392-3, 398
Scheduling, generation-based 25, 38
Scheduling, tie-line flow 26
Security 410, 412, 418-20, 422-4, 426, 428-30, 431-7, 440, 443-5
    Coordinators 17
    Criterion 22
Self-adjusting 31, 42, 78, 79

    Process 32
    Effect 26, 72
Sellers 17, 19, 20
Separation 407-8
Service unbundling 19
Shareholders 18
Siting 299, 302, 327-9
Spinning reserves 418, 426
Spot price 245, 247, 253, 257-263, 266, 269, 271, 276, 353
State feedback 519, 521-35, 537-9, 543, 546
Storage 25, 99
Strategic behavior 355
Structure 16, 17, 20, 21, 24, 27, 33, 36, 37, 40, 41, 42, 78, 79, 405-7, 409-12, 421-31, 434-5, 437-8, 440-3
    Information/control, hierarchical 22
Supply function 350
    Equilibrium 349
Supply/demand mismatch 16, 26, 35, 71, 72, 78
System 15, 16, 25
    Control 406, 410, 412, 420, 422, 426, 434-435, 440, 444
    Development 283, 286, 289-90, 300, 318, 320-2, 329-30
    Expansion 283, 286, 289-90, 300, 318, 320-2, 329-30
    Marginal price 136-9
    Planning 283, 286, 300, 318

Security 410, 419-20, 423-4, 426, 430, 435, 437, 443
Structure 15, 16, 21
Disturbances 73
Systems Benefits Charges (SBCs) 393-5
Tacit collusion 351
Tariffs 17, 19
Single 18
Two-part 26, 27, 37
Tertiary level 22, 25, 27, 79
Tie lines 16, 26, 27, 98
Topology, network 343
Trading entities 21
Trading outside the pool 151-62
Transaction networks 16, 21, 22, 29
Transactions
Bilateral, physical 29, 36, 37, 42, 78
Nontradable, physical 36
Self-scheduled 30
Tradable 30, 35, 79, 99
Transco 18, 19
Transmission 132, 136-8, 146-8, 154-5, 157-8
Capacity 19
Congestion 78
Congestion contracts (TCC) 19, 370
Constraints 356
Loss compensation 15, 16,
Losses 19, 21, 27, 72, 79, 99
Network 18, 19

Two-part pricing 244-6, 273-4, 278
Unbundled 16-19
Unbundling 407-8, 411
Corporate 18, 19
Unit commitment 16, 22, 25, 26, 32, 33, 34
Simplest 33
Stochastic 34
Uplift 136-8, 142, 144, 146-8, 156, 161, 164
Utilities 16, 17
Distributed 398-400
Vertically integrated 17
Volatility 15, 72, 73, 246, 253, 257-8, 262, 271
Voltage 19, 21, 22, 26
Control 434-5
Profile 416-7, 432, 435-7
Support 412, 416-7, 429, 433, 476-7, 482, 499, 502, 514-5
Local 26